AA001042

MATERIALS RESEARCH SOCIETY
SYMPOSIUM PROCEEDINGS VOLUME 89

Diluted Magnetic (Semimagnetic) Semiconductors

December 1-6, 1986
Boston, Massachusetts, USA

Printed from e-media with permission by:

Curran Associates, Inc.
57 Morehouse Lane
Red Hook, NY 12571
www.proceedings.com

ISBN: 978-1-61738-650-3

Some format issues inherent in the e-media version may also appear in this print version.

CAMBRIDGE UNIVERSITY PRESS
Cambridge, New York, Melbourne, Madrid, Cape Town,
Singapore, São Paulo, Delhi, Tokyo, Mexico City

Cambridge University Press
32 Avenue of the Americas, New York, NY 10013-2473, USA

www.cambridge.org

Materials Research Society
506 Keystone Drive, Warrendale, PA 15086
http://www.mrs.org

©Materials Research Society 1987

This publication is in copyright. Subject to statutory exception
and to the provisions of relevant collective licensing agreements,
no reproduction of any part may take place without the written
permission of Cambridge University Press.

First published 1987

CODEN: MRSPDH

ISBN: 978-1-61738-650-3

Cambridge University Press has no responsibility for the persistence or
accuracy of URLs for external or third-part Internet Web sites referred to
in this publication and does not guarantee that any content on such Web sites
is, or will remain, accurate or appropriate.

Additional copies of this publication are available from:

Curran Associates, Inc.
57 Morehouse Lane
Red Hook, NY 12571 USA
Phone: 845-758-0400
Fax: 845-758-2634
Email: curran@proceedings.com
Web: www.proceedings.com

CURRAN ASSOCIATES INC
proceedings.com

TABLE OF CONTENTS

Synthesis of the Dilute Magnetic Semiconductor CdMnTe by Ion Implantation of Mn into CdTe 229
G.H. Braunstein, D. Heiman, S.P. Withrow, G. Dresselhaus

The Connection Between Structural Disorder and Semimagnetic Properties in Implanted CdTe 235
H.J. Jimenez-Gonzalez, A. Lusnikov, G. Dresselhaus, G.H. Braunstein

Submicron Heterostructures of Diluted Magnetic Semiconductors 241
R.L. Gunshor, L.A. Kolodziejski, N. Otsuka, S. Datta, A.V. Nurmikko

Dilute Magnetic Semiconductor Superlattices Containing $Hg_{1-x}Mn_xTe$ 247
K.A. Harris, S. Hwang, R.P. Burns, J.W. Cook Jr., J.F. Schetzina

Far Infrared Magnetoabsorption in $Hg_{1-x}Mn_xTe$/HgTe Superlattice 253
Z. Yang, M. Dobrowolska, H. Luo, J.K. Furdyna, K.A. Harris, J.W. Cook Jr., J.F. Schetzina

$Hg_{1-x-y}Mn_xCd_yTe$ Alloys for 1.3-1.8 μm Photodiode Applications 259
S.H. Shin, J.G. Pasko, D.S. Lo, W.E. Tennant, J.R. Anderson, M. Gorska, M. Fotouhi, C.R. Lu

Investigations of $Cd_{0.9}Mn_{0.1}Te$ Doped with Au, Cu, As and P Acceptors Using Optical Absorption and Photoluminescence 267
J. Misiewicz, J.M. Wrobel, P. Becla, D. Heiman

Optical Properties of Doped $Cd_{1-x}Mn_xTe$ 273
Y. Lansari, N.C. Giles, J.F. Schetzina, P. Becla, D. Kaiser

Surface and Bulk Photoconductivity of $Cd_{1-x}Mn_xTe$ 279
H. Neff, K.Y. Lay, K. Park, K.J. Bachmann

Influence of Electric Fields on Exciton Luminescence in ZnSe/(Zn,Mn) Se Superlattices 285
Q. Fu, A.V. Nurmikko, L.A. Kolodziejski, R.L. Gunshor

Optical Properties of (Pb,Eu)Te Thin Films and Superlattices 290
W.C. Goltsos, A.V. Nurmikko, D.L. Partin

Atomic Layer Epitaxy of Diluted Magnetic Semiconductors 294
M. Pessa, J. Lilja, O. Jylha, M. Ishiko, H. Asonen

Author Index

WHAT LIMITS MAGNETIC POLARON ENERGIES IN DMS?

P.A. WOLFF* AND L.R. RAM-MOHAN**
*Francis Bitter National Magnet Laboratory, MIT, Cambridge, MA 02139 USA
**Worcester Polytechnic Institute, Worcester, MA 01609 USA

I. Introduction

Magnetic polarons are ferromagnetic spin clusters created by the exchange interaction of a carrier spin (electron or hole) with localized spins imbedded in a semiconductor lattice. They were first studied in magnetic semiconductors [1]; more recently, there have been extensive investigations [2] of polaron behavior in diluted magnetic semiconductors (DMS), such as $Cd_{1-x}Mn_xTe$. DMS are favorable media for magnetic polaron studies because they have simple s-p bands and excellent optical properties. Two types of magnetic polarons have been identified in DMS - the bound magnetic polaron (BMP), whose carrier is localized by an impurity [3], and the free polaron (FP) consisting of a carrier trapped by its own, self-consistently-maintained, exchange potential [4].

For DMS with $x < 0.1$, a polaron theory [5] that assumes that unpaired Mn^{2+} spins (concentration $\bar{x} < x$) respond paramagnetically to the carrier exchange field gives good agreement with optical experiments [3,5,6]. The magnetic energies are then relatively small (10 - 20 meV), and polaron effects are only seen at low temperatures. To increase the polaron energy, and to exceed the threshold for FP formation [7], it is natural to consider crystals with larger x-values. In such materials, however, there are many nearest neighbor Mn^{2+} ions, whose spins are coupled antiferromagnetically. The result is a much more complicated magnetic behavior than that of the dilute alloys-including spin glass phases. To avoid the difficult problem of calculating the Mn^{2+} magnetization under these circumstances, we have developed a phenomenological theory of polarons that uses the measured high field magnetization of the host crystal as an input to determine the Mn^{2+} response. This approach leads to a nonlinear Schrodinger equation for the polaron wave function, that we have solved in representative cases. The solutions determine BMP energies, shapes, and moments, as well as the conditions for FP formation. The method requires measurements of the Mn^{2+} magnetization to high field, since exchange fields in acceptor-BMP exceed 50 T. Pulsed field Faraday rotation experiments [8] were used to determine the magnetization of CdMnTe and CdMnSe crystals to 45 T.

Our calculations show that for $x > 0.1$, polaron formation is severely inhibited by the antiferromagnetic, nearest neighbor Mn^{2+} - Mn^{2+} interactions. The large BMP binding energies (≈ 0.3 ev at $x = 0.3$) implied by a paramagnetic model cannot be achieved in conventional, random DMS alloys. Pure FP also do not exist in such materials, though they may be stabilized in alloy-mediated, band gap fluctuations. These conclusions are consistent with luminescence experiments [9].

To achieve large polaron energies in DMS, a host crystal with effective, free Mn^{2+} spin concentration, $\bar{x} \approx 0.2$, is required. Since next nearest neighbor Mn^{2+} - Mn^{2+} interactions are weak, this goal would be achieved if Mn^{2+} ions could be ordered on the cation sublattice, to eliminate nearest neighbor pairs. Ordering naturally occurs in stannite-type crystals [10], such as $(Cu_2MnSi)Se_4$, that are tetrahedrally-bonded, diamond-type semiconductors with Mn^{2+} spacing $\sqrt{2}$ times that of nearest neighbor Mn^{2+} pairs in CdMnTe. In effect, stannites are DMS with $x = 0.25$, and no nearest neighbor Mn^{2+} ions. Magnetic ion ordering might also be achieved by MBE

Mat. Res. Soc. Symp. Proc. Vol. 89. © 1987 Materials Research Society

growth of $(Cd_3Mn)Te_4$. Our calculations indicate that polaron energies in these stannite structures could be five times that of a random $(Cd_{0.75}Mn_{0.25})Te$ alloy.

II. The Polaron Wave Equation

Magnetic polaron wave functions are usually approximated [5] by products of orbital and spin wave functions:

$$\psi(r,s;S_j) \simeq \phi(r)\,\chi\,(s;S_j), \tag{1}$$

where (r,s) are the coordinate and spin of the localized carrier, and S_j the spins of Mn^{2+} ions at sites R_j. The Schrodinger equation of the simplest magnetic polaron, the donor - BMP, then takes the form:

$$\frac{-\hbar^2\nabla^2\phi}{2m^*} - \frac{e^2\phi}{\varepsilon r} - J\sum_j[\delta(r-R_j)\langle s.S_j\rangle]\phi = E\phi, \tag{2}$$

where J is the electron - Mn^{2+} exchange constant. The thermal average over spin states, $\langle s.S_j\rangle$, is evaluated from the spin Hamiltonian:

$$H_{spin} = -J\sum_j[|\phi(R_j)|^2(s.S_j)] \equiv -\sum_j[K_j(s.S_j], \tag{3}$$

and is a functional of ϕ.

The correlation function, $\langle s.S_j\rangle$, has been calculated [5] for the case of noninteracting Mn^{2+} spins (small x). At low temperatures the result is:

$$\langle s.S_j\rangle \approx 5/4\,B_{5/2}\left(\beta K_{j/2}\right), \tag{4}$$

where $B_{5/2}$ is the Brillouin function. In the continuum limit, the corresponding nonlinear Schrodinger equation takes the form:

$$\frac{-h^2\nabla^2\phi}{2m^*} - \frac{e^2\phi}{\varepsilon r} - 5/4\ x\ (JN_0)B_{5/2}[\beta J|\phi(r)|^2/2]\phi = E\phi. \tag{5}$$

This formula is equivalent, for small values of the Brillouin function argument, to one derived by SPALEK [11]. In addition, it correctly describes Mn^{2+} spin saturation - that is crucial in FP formation - as $\beta J|\phi(r)|^2$ becomes large.

Equation (4) determines the alignment of the Mn^{2+} spin at site R_j in the local exchange field, $\mu g B_{exch}(R_j) = J|\phi(R_j)|^2/2$. The corresponding factor $5/2(xN_0)B_{5/2}$ in (5) is $(\mu g)^{-1}$ times the susceptibility of the noninteracting Mn^{2+} spin system. These observations suggest a generalization of (5) to the case of interacting Mn^{2+} spins. The proper recipe is to make the replacement:

$$5/2\,xN_0B_{5/2}\,[\beta\mu g B_{exch}(r)] \rightarrow (\mu g)^{-1}M[B_{exch}(r)] \tag{6}$$

in (5), where M(B) is the measured magnetization of the crystal. This result can be derived, with reasonable approximations, from the Hamiltonian of the interacting carrier and Mn^{2+} spin systems.

To test the theory, we have numerically integrated (5) to determine the energy and form of $\phi(r)$ in the noninteracting Mn^{2+} spin case. The nonlinearity of the equation complicates the analysis since, for an arbitrary starting value, $\phi(0)$, the resulting wave function is usually not properly normalized. Thus, in effect, one must search a two dimensional parameter space $[E,\phi(0)]$ to find a normalized eigenfunction of (5). In practice, this search was performed by an iterative procedure that converges rapidly. Figure 1 compares polaron wave functions at temperatures T = 1 K, 10 K, and 100 K (negligible polaron effect) for the interesting case x = 0.1, $m^* = 0.4\ m_0$, $(JN_0) = -0.88$ ev that provides a hydrogenic approximation to the acceptor - BMP in CdMnTe. The calculated energy, $E_{BMP} \approx 80$ meV, is larger than the measured [6] value, $E_{BMP} \approx 40$ meV, because Mn^{2+} spin pairing reduces the free spin concentration in the crystal to $\bar{x} = .035$.

We have also used (5), without the Coulomb term, to calculate the properties of FP in p - CdMnTe (again for x = 0.1). The FP is weakly bound (E_{FP} = 21 meV) at 1.5 K, and becomes unstable above 2.5 K. These results imply that the threshold for FP formation in p - CdMnTe is x > 0.1. Since x < 0.08 in alloys with random Mn^{2+} distribution, pure FP are probably not stable in most DMS [9].

III. Phenomenological Calculations of BMP Properties

To calculate the properties of BMP in the large - x regime, we replace the free spin magnetization in (5) by the measured values for CdMnTe and CdMnSe crystals [8]. Typical magnetization data are illustrated in Figure 2. M(B) can be accurately decomposed into a Brillouin-like term, that saturates for B > 15 T, plus a linear term. The former results from alignment of small spin clusters whose number decreases rapidly above the percolation threshold (x = 0.17); the latter is reminiscent of the magnetization of an antiferromagnet.

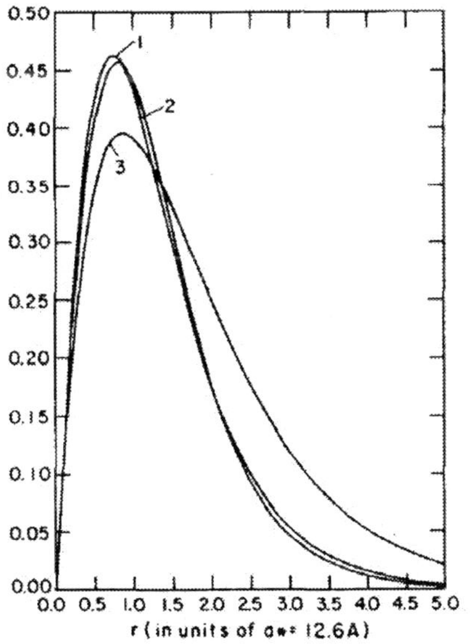

Figure 1 BMP wave functions vs. radius (r/a_0^*) in p-$Cd_{0.9}Mn_{0.1}Te$.
1)T = 1 K, 2)T = 10 K, 3)T = 100 K

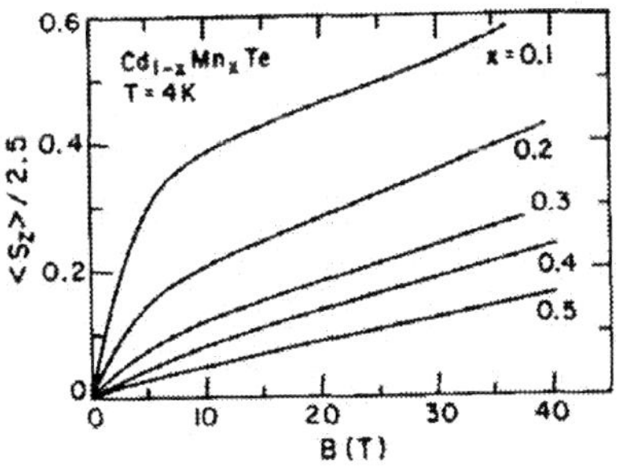

Figure 2 Mn^{2+} spin alignment field (in tesla) for $Cd_{1-x}Mn_xTe$ at 4 K

Note that in the x = 0.5 sample, $\langle S_z \rangle$ extrapolates to its maximum value at B = 240 T. The small slope of the M vs B curve drastically reduces BMP energies. Table 1 lists the calculated radii and magnetic energies of acceptor - BMP in CdMnTe. Beyond x = 0.2, the polaron energy decreases with increasing x, whereas in the absence of Mn^{2+} - Mn^{2+} interactions a linear increase is predicted. For large x, most of the Mn^{2+} spins are "frozen out" of the polaron problem because the exchange field is not strong enough to break up their antiferromagnetic alignment. These conclusions are consistent with recent luminescence experiments of ZAYHOWSKI[9].

Table I

Calculated Acceptor - BMP Properties in $Cd_{1-x}Mn_xTe$

x	$E_{magnetic}$[meV]	Radius [A]
0.1	31	10.1
0.2	38	9.5
0.3	37	9.2
0.4	29	9.6
0.5	22	9.9

IV Ordered DMS

The calculations outlined above imply that Mn^{2+} - Mn^{2+} interactions severely inhibit magnetic polaron formation in DMS alloys with x > 0.2. Calculated and measured polaron energies are a factor of five smaller than those anticipated in a material of comparable x - value without Mn^{2+} - Mn^{2+} interactions. This disappointing conclusion suggests a need for DMS having a sizable concentration of essentially noninteracting Mn^{2+} spins. Since next nearest neighbor Mn^{2+} - Mn^{2+} interactions are quite small (in CdMnTe, $J_{nnn} \approx 0.1\ J_{nn}$), such a material could be realized by ordering of the Mn^{2+} ions on the cation sublattice, to eliminate nearest neighbor Mn^{2+} pairs. In such a crystal the large, hole exchange field (B_{exch} = 30 - 100 T) of the acceptor - BMP will overcome weak, next nearest neighbor Mn^{2+} - Mn^{2+} interactions to cause fairly complete polaron formation.

There is a large class of tetrahedrally bonded, diamond-type semiconductors - the stannites - that satisfies these conditions [10]. Examples include Cu_2FeSnS_4, Cu_2MnGeS_4, Cu_2MnSnS_4, $Cu_2MnSiSe_4$, etc. These crystals are effectively DMS with x = 0.25 and no nearest neighbor transition metal pairs (the nearest Mn^{2+} - Mn^{2+} distance in Cu_2MnGeS_4 is approximately equal to the next nearest neighbor spacing in CdMnTe). The stannites have small Neel temperatures ($T_N \approx 10$ K), as expected of antiferromagnets with weak exchange interactions.

An ordered Mn^{2+} sublattice, similar to that found in stannites, might also be fabricated by MBE - growth of suitably alternated CdTe and MnTe layers. Such a structure would be produced in superlattices grown in the (120) direction, with three CdTe layers alternating with a single MnTe layer (x = 0.25 for the overall structure). As indicated in Figure 3, which shows two layers of the CdMnTe cation lattice, the (120) superlattice has no Mn^{2+} ions in every other (010) plane, and square arrays of Mn^{2+} ions, at the next nearest neighbor distance, in the intervening ones. The structure has no nearest neighbor Mn^{2+} ions.

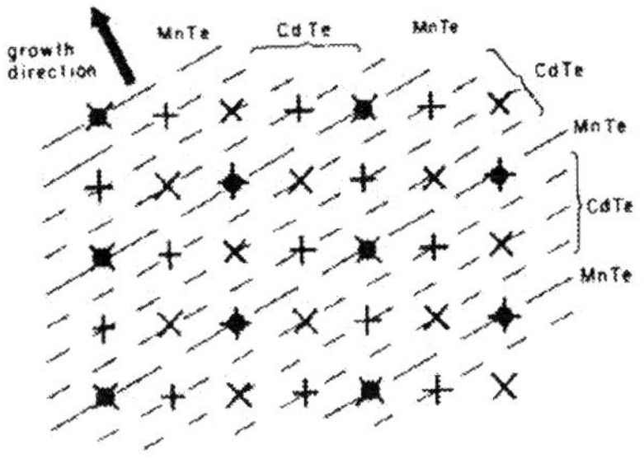

Figure 3 Cation layers in ordered $Cd_{0.75}Mn_{0.25}Te$ grown by (120) MBE. (+ = top layer; x = next layer)

To estimate the acceptor - BMP energy of the stannites, we have solved the nonlinear Schrodinger equation assuming a constant susceptibility below saturation; the susceptibility was determined from the Neel temperature ($T_N \approx 10$ K). Other parameters in the calculation were chosen equal to those of p - CdMnTe, namely $m^* = 0.4\ m_0$ and $(JN_0) = -0.88$ ev. At T = 0, the polaron has total energy $E_{BMP} = -224$ meV and radius $r_0 = 7.9$ A. The binding is more than five times stronger than that of an acceptor - BMP in $Cd_{0.75}Mn_{0.25}Te$.

This research was supported by NSF Grant DMR-8504366.

Bibliography

[1]See E. Nagaev, Physics of Magnetic Semiconductors (MIR Publishers, Moscow, 1983).
[2]P. Wolff and J. Warnock, J. Appl. Phys. 55, 2300 (1984).
[3]A. Golnik, J. Gaj, M. Nawrocki, R. Planel, and C. Benoit a la Guillaume, Proc. XV Intl. Conf. Phys. Semiconductors, Kyoto, 1980 (J. Phys. Soc. Japan, Suppl. A49, pg. 819); M. Nawrocki, R. Planel, G. Fishman, and R. Galazka, ibid, pg. 823.
[4]A. Golnik, J. Ginter, and J. Gaj, J. Phys. C16, 6073 (1983).
[5]T. Dietl and J. Spalek, Phys. Rev. Letters 48, 355 (1982); T. Dietl and J. Spalek, Phys. Rev. B28, 1548 (1983); D. Heiman, P. Wolff, and J. Warnock, Phys. Rev. B27, 4848 (1983).
[6]T. Nhung and R. Planel, Proc. XVI Intl. Conf. Phys. of Semiconductors, Montpellier, Physica 117B - 118B, 488 (1980).
[7]T. Kasuya, A. Yanase, and T. Takeda, Sol. State Comm. 8, 1543 (1970).
[8]D. Heiman, E.D. Isaacs, P. Becla, and S. Foner (to be published); D. Heiman, E.D. Isaacs, S. Foner, A. Wold, K. Dwight, and D. Ridgely (to be published).
[9]J. Zayhowski, Ph.D. Thesis, MIT, 1985.
[10]W. Schafer and R. Nitsche, Mat. Res. Bull. 9, 645 (1974).
[11]J. Spalek, Phys. Rev. B30, 5345 (1984).

ANTIFERROMAGNETIC EXCHANGE CONSTANTS BETWEEN Mn^{2+} IONS IN II-VI SEMIMAGNETIC SEMICONDUCTORS

R.L. AGGARWAL*
Francis Bitter National Magnet Laboratory and Department of Physics
Massachusetts Institute of Technology, Cambridge, MA 02139

ABSTRACT

Mn^{2+} ion spins in II-VI compound semimagnetic semiconductors (SMS) such as $Cd_{1-x}Mn_xTe$ interact with one another through an antiferromagnetic exchange which is responsible for many interesting properties of Mn-alloyed II-VI SMS. In particular, this interaction leads to unique magnetization (M) behavior of SMS as a function of applied magnetic field (B). The nearest-neighbor (NN) Mn^{2+}-Mn^{2+} interaction has been shown to yield a series of five steps in M vs. B curves above about 10 T, providing a direct determination of the NN exchange constant J_{NN}. A detailed analysis of the magnetization steps has yielded values for the next NN exchange constant J_{NNN}. The magnetization steps also yield information on the distribution of Mn^{2+} ions. Inelastic neutron scattering and Raman scattering provide alternative methods for the direct determination of the exchange constants.

INTRODUCTION

Mn^{2+} ions in II-VI compound semimagnetic semiconductors (SMS) such as $Cd_{1-x}Mn_xTe$, $Cd_{1-x}Mn_xSe$, etc. have orbital angular momentum L=0 and spin angular momentum S=5/2, which arises from the half-filled 3d shell. These Mn^{2+} spins interact with one another through an antiferromagnetic (AF) exchange interaction which is responsible for many unique properties of these materials, such as the magnetic phase diagram [1-4] and the occurrence of bound magnetic polarons [5-7].

Until about three years ago, values for the nearest-neighbor (NN) exchange constant J_{NN}, as deduced from indirect measurements varied from ~1 K to 10 K for the same material. This order-of-magnitude discrepancy prompted our group at the M.I.T. Francis Bitter National Magnet Laboratory to undertake direct measurement of these exchange constants. Parts of this work have been carried out in collaboration with several groups from other institutions.

In the absence of an external magnetic field, the spin Hamiltonian for an isolated pair of Mn^{2+} ions at lattice sites i and j is given by

$$H_p = -\sum_{i \neq j} J_{ij} \vec{S}_i \cdot \vec{S}_j \quad . \tag{1}$$

For $S_i = S_j = 5/2$, the corresponding energy levels are given by

$$E_p = -J_{ij}\left[S_p(S_p+1) - \tfrac{35}{2}\right] \tag{2}$$

where S_p = 0, 1, 2, 3, 4 or 5 is the spin of the pair. Since J_{ij} is negative for AF interaction, the ground state of the pair corresponds to S_p=0 with five excited states at energies $2|J_{ij}|$, $6|J_{ij}|$, $12|J_{ij}|$, $20|J_{ij}|$, and $30|J_{ij}|$ lying above it. The resulting energy level diagram is shown in Fig. 1. Clearly, a measurement of the energy difference between any two levels of the pair will provide a direct determination of $|J_{ij}|$. For

Mat. Res. Soc. Symp. Proc. Vol. 89. © 1987 Materials Research Society

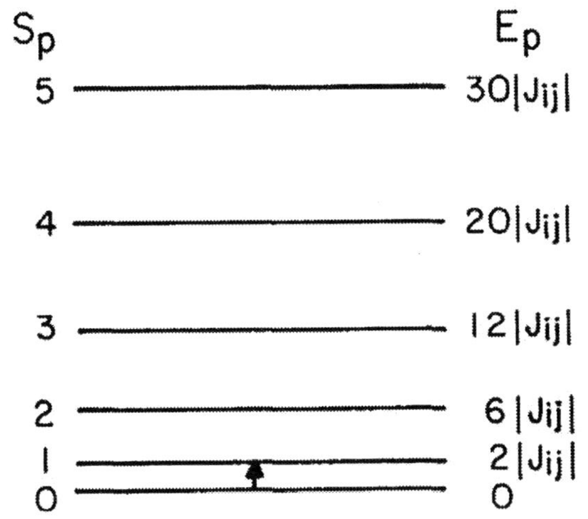

Figure 1. Energy level diagram for an isolated pair of Mn^{2+} ion spins.

example, a transition from the ground state to the first excited state may be observed at low temperatures via 1) inelastic neutron scattering, or 2) Raman scattering.

In the presence of an external magnetic field, the spin Hamiltonian for an isolated pair is given by

$$H_p(B) = H_p(0) + g_{Mn}\mu_B\vec{B}\cdot\sum_i \vec{S}_i \;, \tag{3}$$

where g_{Mn} is the g-factor for the Mn^{2+} ion. The corresponding energy levels of the pair in a magnetic field are given by

$$E_p(B) = E_p(0) + g_{Mn}\mu_B B M_p, \tag{4}$$

where M_p is the component of the pair spin along $\vec{B}$; M_p has $2S_p+1$ values for a given S_p. Since the ground state corresponds to $S_p=M_p=0$, its energy is independent of the applied magnetic field. The first excited state corresponds to $S_p=1$ so that it splits into three Zeeman components with $M_p=1,0,-1$. The energy of the component with $M_p = -1$ will decrease with increasing B and will cross the ground state at a field B_1 given by [8,9]

$$g_{Mn}\mu_B B_1 = 2|J_{ij}| \;. \tag{5}$$

Thus a measurement of the crossing point will provide a <u>direct</u> determination of $|J_{ij}|$.

MAGNETIZATION STEPS

At the crossing point B_1, the ground state switches from a nonmagnetic state with $M_p=0$ to a magnetic state with $M_p = -1$ and, therefore, gives rise to a magnetization step at $B=B_1$. Continuing this argument further, we would expect five steps in magnetization at magnetic fields

$$B_n = nB_1 \tag{6}$$

where n = 1, 2, 3, 4, 5, as shown schematically in Fig. 2. At the last step, the pair achieves complete ferromagnetic alignment.

While the position of the steps provides a direct determination of $|J_{11}|$, the height of the step provides a measure of the concentration of the pairs which, in turn, depends upon the distribution of the Mn^{2+} ions in the lattice. The experimental results as discussed later are consistent with the predictions of the random distribution model. Prior to our work there was considerable uncertainty on this issue. Furthermore, the position of the steps for a given pair is sensitive to its local environment; therefore, the shape of the steps can also provide information on the microscopic distribution of Mn^{2+} ions.

Magnetization steps have been observed using a) high-field magnetization measurements [8,9], and b) optical measurements via Zeeman splittings of the exciton [8] or Faraday rotation [10] at high magnetic fields.

The exciton in II-VI SMS with zinc-blende structure splits into four Zeeman componets, as shown in Fig. 3 for $Zn_{0.95}Mn_{0.05}Te$. The components a and b are observed in the σ_+ polarization, and the components c and d are observed in the σ_- polarization. The large splitting $\Delta E_{3/2} \equiv E_d - E_a$ between the a and d components is due to the exchange interaction between the Mn^{2+} ions and the band electrons, and can be written as [11,12]

$$\Delta E_{3/2} \approx -x\langle S_z\rangle N_0(\alpha-\beta), \tag{7}$$

where x is the molar fraction of Mn^{2+} ions, $N_0\alpha$ and $N_0\beta$ are the exchange constants between Mn^{2+} ions and the conduction and valence band electrons, respectively, and $\langle S_z\rangle$ is the average z-component of the Mn^{2+} spin and can be written as

$$\langle S_z\rangle = -\frac{5}{2}\left(\frac{\bar{x}}{x}\right)B_{5/2}\left(\frac{5g_{Mn}\mu_B B}{2k(T+T_0)}\right) - \frac{1}{2}P_p\sum_n\left\{1+\exp\left[\frac{g_{Mn}\mu_B}{kT}(B_n-B)\right]\right\}^{-1}, \tag{8}$$

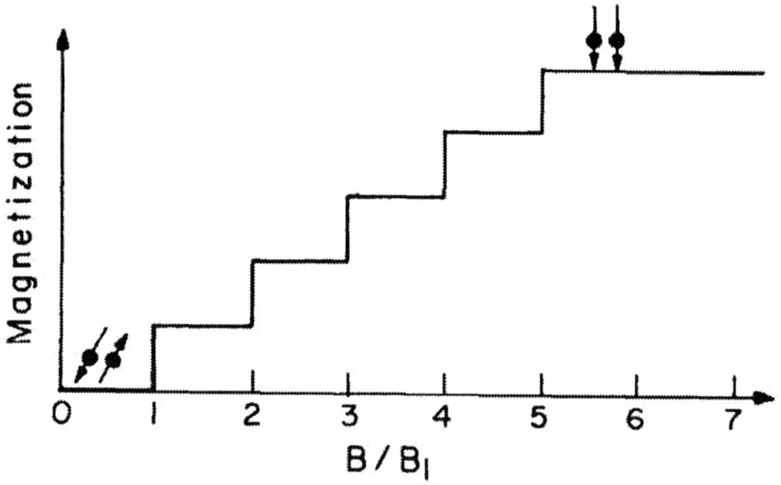

Figure 2. Schematic illustration of magnetization steps due to an isolated Mn^{2+} ion spin pair at T=0 K.

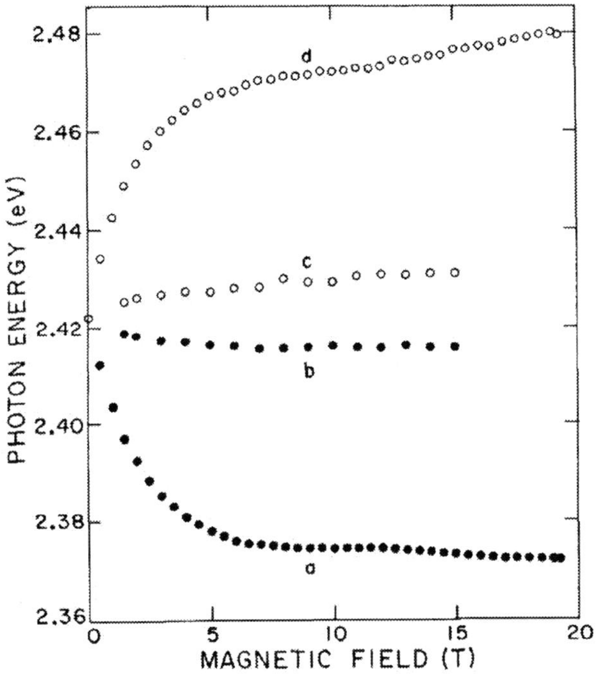

Figure 3. Magnetic field dependence of the energies for the Zeeman split component of the 1s exciton in $Zn_{0.95}Mn_{0.05}Te$ at 1.4 K, observed in magnetoreflectance in the Faraday configuration (after Aggarwal et al. [13]).

The first term in Eq. (8) represents the paramagnetic contribution of the isolated Mn^{2+} ions and the Mn^{2+}-ion clusters excluding pairs. The second term in Eq. (8) represents the steps arising from the AF-coupled NN pairs, with P_p being the probability for pairing.

The magnetic field dependence of $\Delta E_{3/2}$ for $Zn_{0.95}Mn_{0.05}Te$ is shown in Fig. 4. The solid circles represent data points and the solid curve represents the least-squares fit to Eq. (7). The dashed curve is a continuation of the best fit between 0 and 12 T, using only the Brillouin function contribution for $\langle S_z \rangle$. Relative to this reference level, the first step in

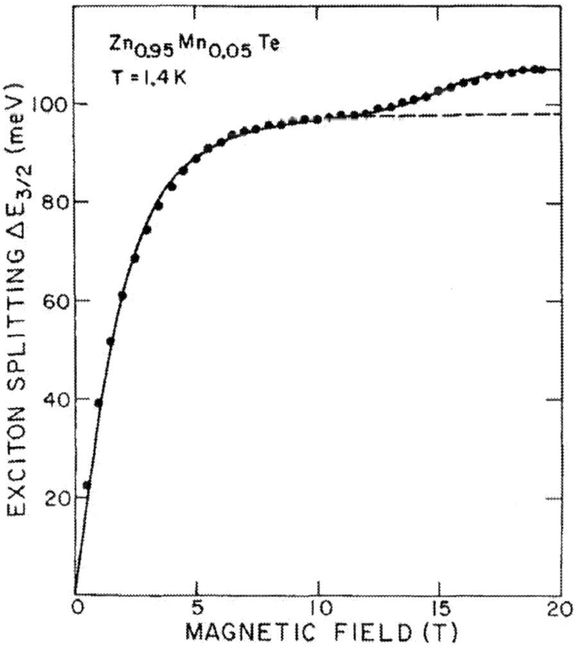

Figure 4. Magnetic field dependence of the Zeeman splitting $\Delta E_{3/2}$ for the exciton in $Zn_{0.95}Mn_{0.05}Te$ at T = 1.4 K (after Aggarwal et al. [13]).

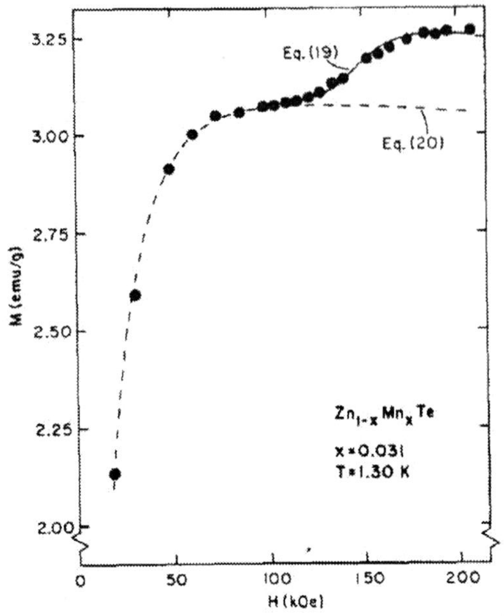

Figure 5. High-field magnetization curve of $Zn_{0.96}Mn_{0.04}Te$ at 1.48 K (after Shapira et al. [16]).

$\Delta E_{3/2}$ is clearly evident between 12 and 19 T and is centered at $B_1=15.0\pm0.6$ T, which yields $J_{NN} = -10.1\pm0.4$ K, using Eq. (5) which ignores the internal fields due to neighboring Mn^{2+} ions on the pairs. Taking the internal field effects into account, Barilero et al. [14] have deduced a more accurate value $J_{NN} = -9.25\pm0.3$ K. For a random distribution of Mn^{2+} ions, P_p is given by [15]

$$P_p = 12x(1-x)^{18}. \tag{9}$$

The observed step height of 9.0 meV corresponds to $P_p=0.27$ which is somewhat higher than the value of 0.24 obtained from Eq. (9) for x = 0.05.

Figure 5 shows the first magnetization step as observed in the high-field magnetization curve of $Zn_{0.96}Mn_{0.04}Te$ [16] centered at $B_1=14.7\pm0.7$ T, in excellent agreement with the value obtained from the Zeeman splittings of the exciton. Similar good agreement between high-field magnetization and optical measurements has been obtained in $Cd_{1-x}Mn_xTe$, $Cd_{1-x}Mn_xSe$ and $Cd_{1-x}Mn_xS$.

Two magnetization steps have been observed in several materials. Optical results for $Cd_{0.95}Mn_{0.05}Te$ are given in Fig. 6 which shows the pair contribution to exciton splitting vs. magnetic field; the first and second steps are centered at $B_1=11.0\pm0.5$ T, and $B_2=19.5\pm1.0$ T. Our simple model for isolated pairs predicts $B_2=B_1$. The observation of $B_2 < 2B_1$ can be understood by considering the effect of other Mn^{2+} ions on the pairs in terms of the internal exchange fields exerted by the neighbors on members of the pair [17]. This has the effect of pushing all the steps to higher fields by the same amount. Therefore, J_{NN} can be determined from the difference of the two steps using the following equation

$$g_{Mn}\mu_B(B_2-B_1) = 2|J_{NN}|. \tag{10}$$

Again assuming random distribution, this model [17] yields a better fit to the data, as illustrated in Fig. 7 for $Cd_{0.95}Mn_{0.05}Te$, and yields $J_{NN} = -6.3\pm0.3$ K, $J_{NNN} = -1.9\pm1.1$ K and $J_{NNNN} = -0.4\pm0.3$ K. Similar analysis has been carried out for $Cd_{0.95}Mn_{0.05}Se$ [17].

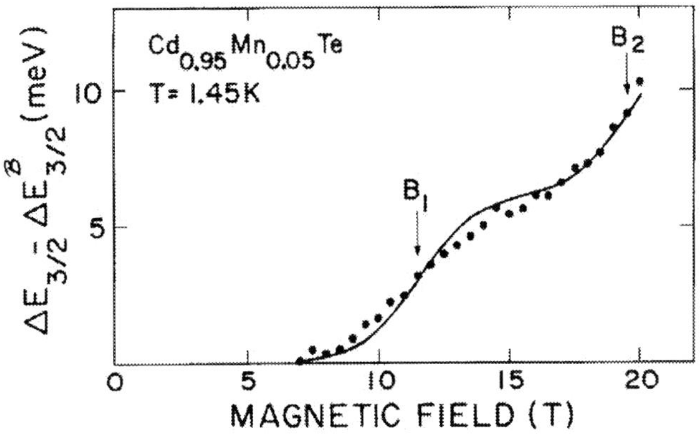

Figure 6. Pair contribution to the exciton splitting in $Cd_{0.95}Mn_{0.05}Te$ at 1.45 K showing two magnetization steps. Solid circles (•) denote the data points. The solid curve is the best fit for B_1=11.5 T, B_2=19.5 T and P_p=0.21, assuming only NN interaction (after Aggarwal et al. [12]).

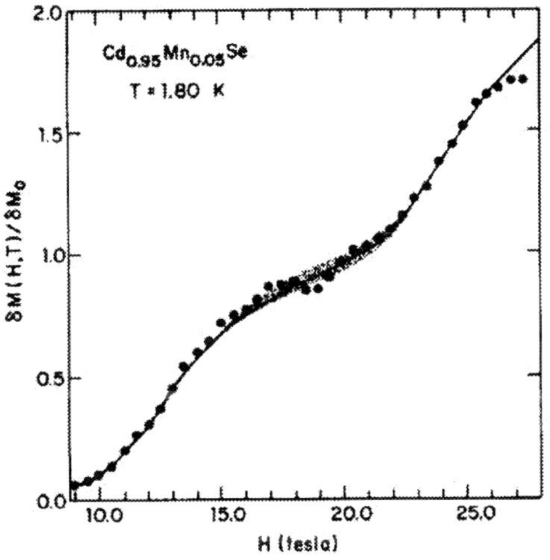

Figure 7. Pair magnetization δM scaled to $\delta M_0 = g\mu_B P_p$ for $Cd_{0.95}Mn_{0.05}Te$ 1t 1.45 K. Solid circles denote the data points. The solid curve is obtained from the model incorporating the internal magnetic fields due to the neighbors on the members of the pair (after Larson et al. [17]).

INELASTIC NEUTRON SCATTERING

Neutron scattering is a powerful tool for probing the magnetic excitations of the SMS. In particular, inelastic neutron scattering has been used to observe the transition from the ground state to the first excited state of the pair [18-19]. This is illustrated in Fig. 8 for $Zn_{0.97}Mn_{0.03}Te$ [18]. This spectrum yields J_{NN} = -8.79±0.14 K; this value is well outside the range J_{NN} = -10.1±0.4 K and J_{NN} = -10.0±0.7 K obtained from magnetization steps in exciton splitting and high-field magnetization curves, respectively.

RAMAN SCATTERING

More recently, the transitions among the energy levels of pairs have been observed in Raman scattering spectra of $Cd_{1-x}Mn_xSe$ and $Cd_{1-x}Mn_xS$ [20]. Figure 9 shows the Raman spectra of $Cd_{0.875}Mn_{0.125}S$ at 5 K and B=20 kG,

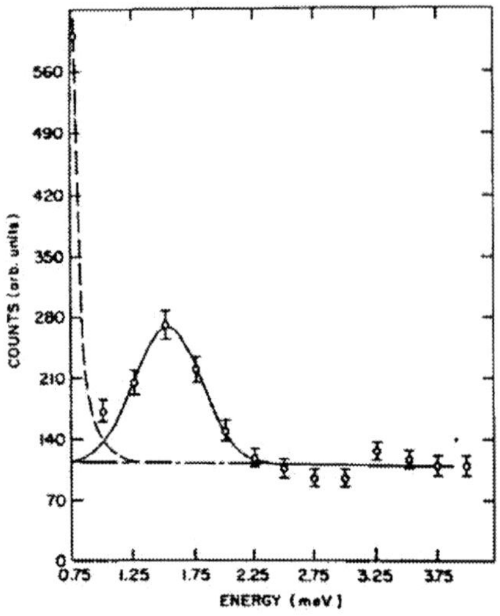

Figure 8. Spectrum of inelastic neutron scattering for $Zn_{0.97}Mn_{0.03}Te$ at 4.2 K, obtained at Brookhaven National Laboratory (after Corliss et al. [18]).

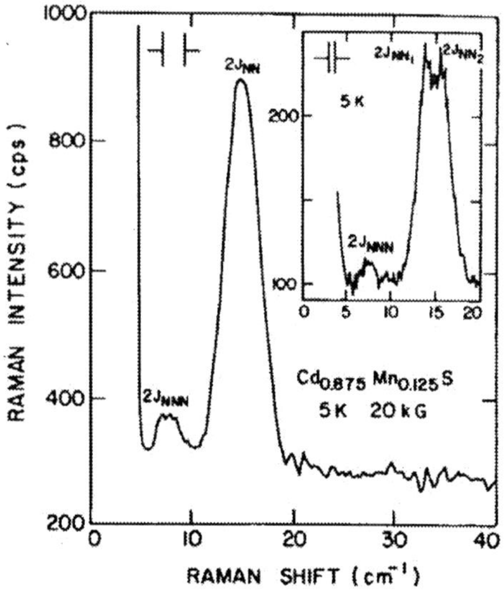

Figure 9. Raman spectra for $Cd_{0.875}Mn_{0.125}S$ at 5 K showing the transitions from the ground state of the pairs (after Bartholomew et al. [20]).

using an argon-ion laser line at 496.5 nm. The peak labeled $2J_{NN}$ is due to the transition from the ground state to the first excited NN pair and yields $J_{NN} = -10.6\pm0.2$ K, in excellent agreement with the value $J_{NN} = -10.5\pm0.8$ K deduced from the first magnetization step in $Cd_{0.977}Mn_{0.023}S$.

At higher temperatures, say 40 K, transitions between the first and second excited states are also observed at twice the energy for the transition between the ground state and the first excited state.

CONCLUSIONS

Values of the nearest-neighbor exchange constant J_{NN} for the Mn^{2+} ions in wide-gap II-VI semimagnetic semiconductors have been established with an accuracy of better than 10% for a number of materials. Values for the next nearest-neighbor exchange constant have been deduced for $Cd_{0.95}Mn_{0.05}Te$ and $Cd_{0.95}Mn_{0.05}Se$ with an accuracy of 60-90%. Recent theoretical calculations of the electronic structure and magnetic interactions [21-22] show that the exchange interaction is largely due to superexchange and that it is antiferromagnetic at least up to the third nearest neighbor. Predictions of these theoretical analyses for J_{NN} are in most cases in good agreement with experiment.

ACKNOWLEDGEMENTS

I wish to thank my collaborators Professor S.N. Jasperson, Dr. Y. Shapira, Professor A. Wold, Dr. K. Dwight, Professor N. Miura, Dr. K.C. Hass, B.E. Larson, Professor A.K. Ramdas, and Professor J.K. Furdyna for participating in parts of the work reported in this paper. Part of the work at MIT is supported by National Science Foundation grant no. DMR-85-04366. The Francis Bitter National Magnet Laboratory is supported by the National

Science Foundation through its Division of Materials Research under Cooperative Agreement No. 85-11789.

REFERENCES

*Also, MIT Lincoln Laboratory, Lexington, MA 02173.

1. R.R. Galazka, S. Nagata, and P.H. Keesom, Phys. Rev. B22, 3344 (1980).
2. S. Oseroff, Phys. Rev. B25, 6484 (1982).
3. M.A. Novak, O.G. Symko, D.J. Zheng, and S. Oseroff, Phys. Rev. B29, 1310 (1984).
4. C.J.M. Denissen, H. Nishihara, J.C. van Gool, and W.J.M. de Jonge, Phys. Rev. B33, 7637 (1986).
5. A. Golnik, J. Ginter, and J.A. Gaj, J. Phys. C: Solid State Phys. 16, 6073 (1983).
6. D. Heiman, P.A. Wolff, and J. Warnock, Phys. Rev. B27, 4848 (1983).
7. P.A. Wolff and L.R. Ram-Mohan, this Proceedings.
8. R.L. Aggarwal, S.N. Jasperson, Y. Shapira, S. Foner, T. Sakakibara, T. Goto, N. Miura, K. Dwight, and A. Wold, in Proceedings of the 17th International Conference on the Physics of Semiconductors, San Francisco, 1984, edited by J.D. Chadi and W.A. Harrison, (Springer, New York, 1985), p. 1419.
9. Y. Shapira, S. Foner, D.H. Ridgley, K. Dwight, and A. Wold, Phys. Rev. B30, 4021 (1984).
10. D. Heiman, E.D. Isaacs, P. Becla, and S. Foner, this Proceedings.
11. J.A. Gaj, R. Planel, and G. Fishman, Solid State Commun. 29, 435 (1979).
12. R.L. Aggarwal, S.N. Jasperson, P. Becla, and R.R. Galazka, Phys. Rev. B32, 5132 (1985).
13. R.L. Aggarwal, S.N. Japserson, P. Becla, and R.R. Galazka, Phys. Rev. B34, 1789 (1986).
14. G. Barilero, C. Rigaux, N.H. Hau, J.C. Picoche, and W. Giriat, this Proceedings.
15. R.E. Behringer, J. Chem. Phys. 29, 537 (1958).
16. Y. Shapira, S. Foner, P. Becla, D. Domingues, M.J. Naughton, and J.S. Brooks, Phys. Rev. B33, 356 (1986).
17. B.E. Larson, K.C. Hass, and R.L. Aggarwal, Phys. Rev. B33, 1789 (1986).
18. L.M. Corliss, J.M. Hastings, S.M. Shapiro, Y. Shapira, and P. Becla, Phys. Rev. B33, 608 (1986).
19. T.M. Giebultowicz, J. Rhyne, and J. K. Furdyna, to be published.
20. D.U. Bartholomew, E.-K. Suh, S. Rodriguez, and A.K. Ramdas, Solid State Commun. 62, 235 (1987).
21. B.E. Larson, K.C. Hass, H. Ehrenreich, and A.E. Carlsson, Solid State Commun. 56, 347 (1985).
22. H. Ehrenreich, K.C. Hass, B.E. Larson, and N.F. Johnson, this Proceedings.

HIGH FIELD MAGNETIZATION STEP IN $Zn_{1-x}Mn_xTe$

G.BARILERO*, C.RIGAUX*, NGUYEN HY HAU*, J.C.PICOCHE**, W.GIRIAT***
*Groupe de Physique des Solides de l'Ecole Normale Supérieure, 24 rue Lhomond, 75231 Paris Cedex 05, France
** Service National des Champs Intenses, CNRS, 38042 Grenoble Cedex, France
***Centro de Fisica, Instituto Venezolano de Investigaciones Cientificas Caracas, Apto, 1827, Venezuela

ABSTRACT

High field magnetization experiments are reported for dilute $Zn_{1-x}Mn_xTe$ alloys ($x \simeq 0.03$) at T=1.3K. The first magnetization step is analyzed including effects of internal field due to the interactions of Mn-Mn pairs with the remaining spins. An accurate estimate of the nearest neighbor (NN) exchange constant is obtained which agrees quite well recent neutron scattering experiments.

I. INTRODUCTION

In dilute magnetic semiconductors (DMS), such as $Zn_{1-x}Mn_xTe$, each Mn^{2+} ion has a total spin angular momentum S=5/2, originating from the half filled 3d shell. The Mn^{2+} spins are coupled by a short range antiferromagnetic (AF) interaction [1]. This interaction has been shown [2-5] to yield a series of five steps in the low temperature magnetization curves M(H) of DMS above $H \simeq 10$ Tesla. This new phenomenon results from the progressive alignment of the NN Mn pair spin component along the applied magnetic field. Including only NN AF interactions, equidistant steps are expected to occur at fields $H_n = 2|J|/g\mu_B$, where μ_B is the Bohr magneton and g=2 is the Landé factor of the Mn^{2+} ion. The first observation of the magnetization step in $Zn_{1-x}Mn_xTe$ was reported by Shapira et al.[5] and the NN AF exchange constant J=-10K was determined from the simple NN cluster model [3][4]. In $Cd_{1-x}Mn_xTe$ the second step observed in magneto-optical experiments [3] occurs at a field smaller than twice the field of the first, implying that magnetic interactions of pairs with more distant neighbors have to be accounted for in the quantitative analysis of magnetization data. Larson et al.[6] developed a generalized cluster model considering a distribution of local fields due to different spin configurations in the environment of the pair.

In this contribution we analyze the first magnetization step observed at $T \simeq 1.3K$ in dilute $Zn_{1-x}Mn_xTe$ alloys ($x \simeq 0.03$). We treat interactions of the pairs with the remaining spins in the crystal, using the molecular field approximation (MFA). We show that the internal field due to the interactions of pairs with more distant spins than NN can be expressed in terms of the parameter T_0 deduced from the analysis of the low field magnetization ($H < 2|J|/g\mu_B$). A comparison is given between the molecular field obtained from our simplified approach with the mean effective field calculated from the explicit model of Larson et al.[6]. The inclusion of effects of internal fields on the pair magnetization yields a more accurate determination of the NN exchange constant than previously reported [5]. An excellent agreement is obtained with recent neutron scattering experiments [7].

Mat. Res. Soc. Symp. Proc. Vol. 89. © 1987 Materials Research Society

II. EXPERIMENTS

Magnetization experiments were carried out at 1.31K for an alloy of composition x=2.9% and at 1.28K for x=3.4%, in magnetic fields up to 19T, using an extraction method [8]. Fig.1 shows the magnetization as a function of magnetic field for x=0.034 at T=1.28K. Two different regions in the magnetization curve can be distinguished : below 11T, the magnetization tends to an apparent saturation and exhibits a step centered at H≃15T as shown in Fig.2.

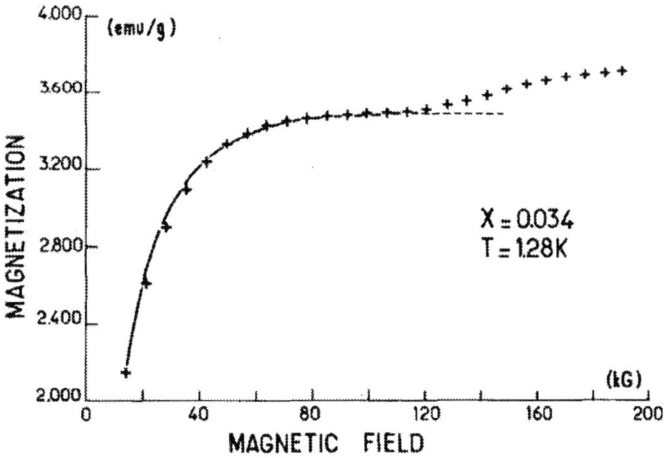

Fig.1 - Magnetization curve
Crosses : Experiments
Solid line : Theoretical fit using Eq.(19) with M_0=3.38 emu/g, T_0=1.80K

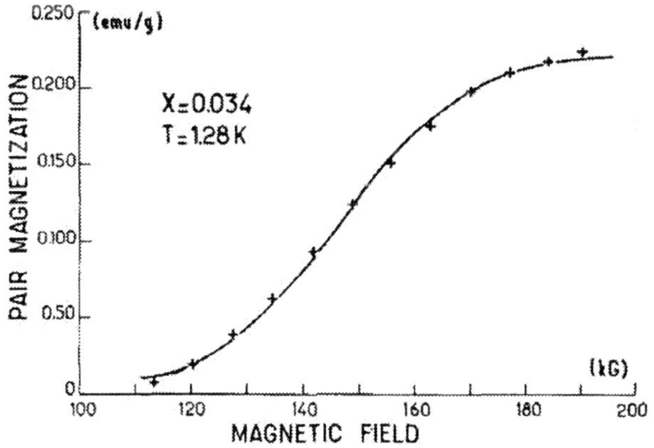

Fig.2 - Pair magnetization vs H
Crosses : Experiments
Solid line : Theoretical fit using Eq.(20) with ∂M=0.217 emu/g ; $H^{(1)}$=146.8KG ; T_{eff}=1.55K (M_0=3.38 emu/g), T_0=1.80K)

III. LOW TEMPERATURE MAGNETIZATION

The existence of short range AF interactions between Mn^{2+} leads to the formation of NN clusters. The exchange interactions between more distant spins are expected to be much smaller than NN exchange interaction [9]. It was shown previously that the magnetization of very dilute alloys is well approximated by the sum of the contributions of four types of clusters : singles, pairs, closed and open triangles [4]. The magnetization per unit mass is :

$$M = -g\mu_B x N_A \langle\langle S_z \rangle\rangle / m(x) \tag{1}$$

where $m(x)=(1-x)m_{ZnTe}+xm_{MnTe}$ is the molar mass of the compound $Zn_{1-x}Mn_xTe$ and N_A is the Avogadro number. $\langle\langle S_z\rangle\rangle$ denotes the thermal average as well as the spatial average of the spin component along the magnetic field. The spatial average over the random distribution of localized spins $\langle\langle S_z\rangle\rangle$ is obtained by weighting the thermal average of each cluster $\langle S_z\rangle_i$ by the probability $P_i(x)$ that a Mn spin belongs to the cluster i.

$$\langle\langle S_z\rangle\rangle = P_1\langle S_z\rangle_1 + P_2\langle S_z\rangle_2 + P_3\langle S_z\rangle_3 + P_4\langle S_z\rangle_4 \tag{2}$$

For a random distribution, $P_1=(1-x)^{12}$ (singles), $P_2=12x(1-x)^{18}$ (pairs), $P_3=18x^2(1-x)^{23}(7-5x)$ (open triangles), $P_4=24x^2(1-x)^{22}$ (closed triangles).

1. Low field magnetization ($H < 11T$)

The energy level scheme for various clusters was discussed by Nagata et al.[10]. The energy levels of a pair are :

$$E = -J[S_T(S_T+1)-35/2] + g\mu_B mH$$

where S_T is the total spin of the pair ($0\leqslant S_T\leqslant 5$), and $-S_T\leqslant m\leqslant S_T$. In magnetic field $H<2|J|/g\mu_B$, the pairs are frozen in their non-magnetic ground state ($S_T=0$). As $|J|/k\sim 10K$, the pair magnetization is negligible at $T\sim 1.3K$ in fields $H<11T$.

As shown in Ref.[4], CT and OT remain in the low field ground states in fields up to $3|J|/g\mu_B$ and $7|J|/g\mu_B$ respectively. In the low field region, the contribution per spin to the magnetization is $\langle S_z\rangle_3=\langle S_z\rangle_1/3$ for OT and $\langle S_z\rangle_4=\langle S_z\rangle_1/15$ for CT. Thus, the average Mn spin component is :

$$\langle\langle S_z\rangle\rangle = (P_1+P_3/3+P_4/15)\ \langle S_z\rangle_1 \tag{3}$$

The Hamiltonian for a single spin $\vec{S}_1$ is :

$$\mathcal{H}_{single} = g\mu_B S_{1z}H - 2\sum_{j\geqslant 3} J_{1,j}\ \vec{S}_1.\vec{S}_j \tag{4}$$

where $J_{1,j}$ is the exchange constant interaction between the spin $\vec{S}_1$ and unfrozen spins $\vec{S}_j$ (at sites j) more distant from $\vec{S}_1$ than nearest neighbors. For random Mn distribution, the summation over all occupied sites is replaced by its spatial average [11] :

$$\sum_{j\geqslant 3} J_{1,j} \approx x\sum_{\text{all } j\geqslant 3} J_{1,j} \simeq x(6J_2+24J_3) \tag{5}$$

when second and third neighbours are only considered. Treating the interactions between $\vec{S}_1$ and $\vec{S}_j$ in the MFA,

$$\langle S_z \rangle_1 = - SB_s[Sg\mu_B(H-h_m)/kT] \quad (6)$$

where $B_S(x)$ is a normalized Brillouin function for a spin S=5/2 and

$$g\mu_B h_m = 2 \langle\langle S_z \rangle\rangle \sum_{j \geqslant 3} J_{1,j} \quad (7)$$

Using Eqs.(3) and (6),

$$\langle\langle S_z \rangle\rangle = - S(P_1+P_3/3+P_4/15)\ B_s[Sg\mu_B(H-h_m)/kT] \quad (8)$$

In practice, the magnetization is usually described by a modified Brillouin function :

$$\langle\langle S_z \rangle\rangle = - S(P_1+P_3/3+P_4/15)\ B_s[Sg\mu_B H/k(T+T_0)] \quad (9)$$

and the parameter T_0 which account for the long range AF interactions should be identified to

$$kT_o = -[S(S+1)/3]\ (P_1+P_3/3+P_4/15)\ 2x\ (6J_2+24J_3) \quad (10)$$

by comparing the low magnetic field limits of Eqs.(8) and (9). The low field magnetization per unit mass is

$$M_{LF} = g\mu_B x N_A(P_1+P_3/3+P_4/15)[m(x)]^{-1} SB_s[Sg\mu_B H/k(T+T_0)] \quad (11)$$

2. <u>High field magnetization</u>

The Hamiltonian describing the correlated motion of a pair $(\vec{S}_1,\vec{S}_2)$ in the magnetic field H//z is :

$$\mathcal{H}_{pair} = -2J\ \vec{S}_1.\vec{S}_2 + g\mu_B(S_{1z}+S_{2z})H - 2\sum_{j \geqslant 3} J_{1,j}\vec{S}_1.\vec{S}_j - 2\sum_{j \geqslant 3} J_{2,j}\vec{S}_2.\vec{S}_j \quad (12)$$

where J is the NN exchange interaction. $J_{1,j}$ and $J_{2,j}$ are the exchange constants between each spin of the pair and the remaining spins $\vec{S}_j$ more distant from S_1 and $\vec{S}_2$ than NN. Treating these interactions within the MFA,

$$\mathcal{H}_{pair} = -2J\ \vec{S}_1.\vec{S}_2+g\mu_B S_{1z}(H-h_1)+g\mu_B S_{2z}(H-h_2) \quad (13)$$

where $g\mu_B h_1=2\langle\langle S_z \rangle\rangle \sum_j J_{1,j}$, and similarly for h_2.

For random Mn site occupation,

$$g\mu_B h_1 = g\mu_B h_2 = g\mu_B h'_m \simeq 2x\langle\langle S_z \rangle\rangle \sum_{all\ j} J_{1,j} \quad (14)$$

where j involves all lattice sites more distant from the pair than NN. The energy states of the pair are

$$E(S_T,m) = -J[S_T(S_T+1)-35/2] + g\mu_B m(H-h'_m) \quad (15)$$

Since the fraction of pairs is small compared to the number of contributing spins, we neglect the contribution of pairs in the molecular field expression (14). At field $H \simeq 11T$, singles and triangles (in their low field state) are nearly field-aligned so that :

$$g\mu_B h'_m \simeq -2xS(P_1+P_3/3+P_4/15)(4J_2+20J_3) \qquad (16)$$

Comparing Eqs.(10) and (16), the molecular field h'_m can be directly related to the parameter T_0 :

$$g\mu_B h'_m = 2kT_o(1+J_3/J_2+4J_3)/S+1 \qquad (17)$$

At T=1.3K, only the first two levels (1,-1) and (0,0) are expected to be populated and the pair magnetization is :

$$\partial M_{pair} = [g\mu_B xN_A/m(x)](P_2/2)\ \{1+\exp -[2J+g\mu_B(H-h'_m)]/kT\}^{-1} \qquad (18)$$

The high field magnetization is then $M_{HF}=M_{LF}+\partial M_{pair}$ where M_{LF} is given by Eq.(11).

It is interesting to compare the results of our simplified approach with those obtained from the model of Larson et al.[6]. Since the internal fields h_r due to singles in the environment of the pair are only considered in this model, the effective probability for Mn site occupation $x_{eff}=xP_1(x)$ for a single (instead of x) should be introduced in the probabilities P_r of local environment. Making this correction, the mean effective field $\bar{h}$ calculated from Ref.[6] is in excellent agreement with the molecular field $h'_m = -5xP_1(x)(4J_2+20J_3)/g\mu_B$ obtained from our treatment when only interactions of pairs with singles are considered. For instance, at x = 0.034, one gets $g\mu_B h'_m=0.45J_2+2.2J_3$ from our analysis, whereas the mean effective field calculated from the method of Larson et al. is $g\mu_B\bar{h}=0.43J_2+2.18J_3$.

IV. COMPARISON BETWEEN THEORY AND EXPERIMENTS

1) In the low field range ($H<11T$), the magnetization data are analyzed neglecting the contribution of pairs. The magnetization was fitted to the experimental data using the following expression :

$$M_{LF} = M_0 B_{5/2}[5g\mu_B H/2k(T+T_0)] \qquad (19)$$

where M_0 and T_0 are fitting parameters. The comparison between the calculated and experimental magnetization is shown in Fig.1. Numerical values of M_0 and T_0 obtained from the best fits are :

$M_0 = 3.05\pm0.03$ emu/g, $T_0 = 1.65\pm0.06$ K for x = 0.029

$M_0 = 3.38\pm0.03$ emu/g, $T_0 = 1.80\pm0.06$ K for x = 0.034

The saturation values M_0 obtained from the best fits are in good agreement with the theoretical saturation value given by Eq.(11) for a random distribution of Mn ions : ($M_{th}=3.02\pm0.15$ emu/g and 3.36 ± 0.15 emu/g for x=0.029 and 0.034 respectively).

2) The high field magnetization ($H>11T$) was fitted to the experimental data using the following expression :

$$M_{HF} = M_0+\partial M\{1+\exp[g\mu_B(H^{(1)}-H)/kT_{eff}]\}^{-1} \qquad (20)$$

M_0 is the saturation value obtained from the best fits of the low field magnetization data. ∂M and $H^{(1)}$ are ajustable parameters.

As the width of the step is not well reproduced by the thermal broadening, we introduce an effective temperature to account phenomenologically for other possible broadening mechanisms. The Fig.2 shows the comparison between $M_{HF}-M_0$ for x=0.034. The fitting parameters ∂M and $H^{(1)}$ obtained for the best fits are :

x=0.029 : $\partial M=0.174\pm0.007$ emu/g, $H^{(1)}=146.6\pm2.6$ KG, $T_{eff}=1.55$ K

x=0.034 : $\partial M=0.217\pm0.007$ emu/g, $H^{(1)}=146.8\pm2.5$ KG, $T_{eff}=1.55$ K

The height of the step ∂M obtained from the fits agree quite well with the theoretical value $P_2 g\mu_B x N_A/2m(x)=0.171\pm0.012$ emu/g for x=0.029 and 0.214 ± 0.013 emu/g for x=0.034, calculated for a random distribution. As previously shown by Shapira et al.[5] for dilute $Zn_{1-x}Mn_xTe$, this result together with the saturation value M_0 provides evidence of the random distribution of Mn ions over the fcc sublattice.

From the numerical values of $H^{(1)}$ and T_0 obtained from the theoretical fits in the high and low field regions respectively, we determine the NN exchange constant J using the relation :

$$g\mu_B H^{(1)} \simeq -2J+2kT_0/S+1$$

the last term in Eq.(17) was neglected. For $J_2<J_3<0$, this approximation leads to an uncertainty on J smaller than 0.1K. The exchange constant $J/k=-9.25\pm0.3$ K determined from the present study is in excellent agreement with the value $J/k=-9.5\pm0.2$ K obtained from recent neutron scattering experiments [7].

REFERENCES

1. J.Spalek, A.Lewicki, Z.Tarnawski, J.K.Furdyna, R.R.Galazka, Z.Obuszko, Phys. Rev. B 33, 3407 (1986).
2. R.L.Aggarwal (this proceeding).
3. R.L.Aggarwal, S.N.Jasperson, P.Becla, R.R.Galazka, Phys. Rev. B 32, 5132 (1985).
4. Y.Shapira, S.Foner, D.H.Ridgley, K.Dwight, A.Wold, Phys. Rev. B 30, 4021 (1984).
5. Y.Shapira, S.Foner, P.Becla, D.N.Domingues, M.J.Naughton, J.J.Brooks, Phys. Rev. B 33, 356 (1986).
6. B.E.Larson, K.C.Hass, R.L.Aggarwal, Phys. Rev. B 33, 1789 (1986).
7. T.M.Giebultowicz, J.J.Rhyne, J.K.Furdyna (this proceeding).
8. G.Barilero, C.Rigaux, M.Menant, Nguyen Hy Hau, W.Giriat, Phys. Rev. B 32, 5144 (1985).
9. M.Escorne, A.Mauger, R.Triboulet, J.L.Tholence, Physica 107B, 309 (1981).
10. S.Nagata, R.R.Galazka, D.P.Mullin, H.Akbarzadeh, G.D.Khattack, J.K.Furdyna, P.H.Keesom, Phys. Rev. B 22, 3331 (1980).
11. G.Bastard, C.Lewiner, J. Phys. C 13, 1469 (1980).

MAGNETIZATION OF RANDOM AND NONRANDOM (Cd,Mn)Te ALLOYS

D. HEIMAN, E.D. ISAACS,* P. BECLA AND S. FONER*
Francis Bitter National Magnet Laboratory
Massachusetts Institute of Technology, Cambridge, MA 02139

ABSTRACT

Pulsed-field magnetization measurements up to 40 T by Faraday rotation were made for random alloy $Cd_{1-x}Mn_xTe$ with $0.1 \leq x \leq 0.5$. At liquid helium temperature the magnetization can be separated into the sum of two components: a paramagnetic Brillouin-like part that saturates by B=15 T, and a part linear in B. For large x, the linear part dominates and becomes independent of x. Mean-field calculations suggest that an enhanced magnetization can be obtained with new ordered-alloy crystal structures which have reduced antiferromagnetic cancellations.

I. INTRODUCTION

$Cd_{1-x}Mn_xTe$ is a diluted-antiferromagnetic system which exhibits a wide range of magnetic behavior: paramagnet-like response at low x; frustrated spin-glass-like behavior; and short-range antiferromagnetism [1-3]. These magnetic properties ultimately determine the s-d exchange-enhanced Zeeman splittings of the semiconductor bands [1], and the properties of magnetic polarons [4]. At low temperatures the large band splittings imply an exchange-enhanced magnetic field 1000 times larger than the applied field. In addition to these splittings, the magnetic response at high-fields governs the properties of magnetic polarons, which can have self-induced internal exchange fields as large as 100 tesla.

The magnetization M(B) of $Cd_{1-x}Mn_xTe$ for $x \lesssim 0.3$ was measured at liquid helium temperatures by Gaj, Planel and Fishman in magnetic fields to B=15 T [5]. They fit the average z-component of manganese spin by the empirical equation

$$\langle S_z \rangle = \bar{S}B_{5/2}[5\mu_B B/k(T+T_o)], \qquad (1)$$

where $B_{5/2}$ is the Brillouin function for spin S=5/2, B is the applied field, μ_B the Bohr magneton, k the Boltzmann constant, and T the temperature. $\bar{S}$ and T_o are treated as fitting parameters which vary with x. $\bar{S}(x)$ is less than S due to strong nearest-neighbor antiferromagnetic coupling between Mn^{++} ions, and $T_o(x,T) > 0$ arises from weaker more-distant neighbor coupling. Subsequent measurements of M(B) for x > 0.2 revealed significant deviations from this modified Brillouin function [6], and the present work at fields above 15 T shows even larger deviations [7].

We show that M(B) can be described by the sum of two contributions: (i) a paramagnetic modified-Brillouin function that saturates at low fields; and (ii) a high-field susceptibility that is linear in B. The relative contributions of these two parts vary with manganese concentration: at low-x the saturating part is largest, whereas at high-x the linear contribution dominates. This trend is related to both the decreasing fraction of isolated clusters with increasing x, and to the random network of coupled spins. This network limits the susceptibility above x=0.3. We also discuss new ordered-alloy crystal structures which are expected to show increased susceptibility due to reduced antiferromagnetic cancellations.

Mat. Res. Soc. Symp. Proc. Vol. 89. © 1987 Materials Research Society

II. EXPERIMENTAL RESULTS

Magnetic fields, furnished at the Francis Bitter National Magnet Laboratory pulsed-field facility, were produced by multilayer, steel-reinforced, copper solenoids operating at liquid nitrogen temperature. A field of 45 tesla with a pulse length of 10 msec is furnished by this facility. The sample temperature was influenced slightly by the rapidly changing field, but was within ±1 K of the bath temperature for the data presented, as determined by fits to Eq. 1. Faraday rotation measurements were made using a fiber-optical arrangement to transfer light to the dewar, as described previously [8]. The sample holder, immersed in liquid helium, held a "sandwich" consisting of a mirror, sample (~1mm thick), and plastic linear polarizer. With the large rotations obtained using light somewhat resonant with the bandgap (within a few hundred meV), the output light signal was oscillatory with a period for every 180 deg of rotation. Typically, 20 to 40 rotations were observed by 40 T. We found that M and the Faraday rotation were approximately proportional (within ±10%) between B = 0 and 20 T. Assuming a linear relation at all fields, the relative M(B) was derived from the rotation angle as a function of field. M was scaled using dc magnetization measurements at B=5 T. The connection between $\langle S_z \rangle$ and M, in units of emu/g, is $M = (2\mu_B xA/W)\langle S_z \rangle$, where A is Avogadro's number and W is the molecular weight of $Cd_{1-x}Mn_xTe$.

Figure 1 shows M(B) for samples with x=0.1 and 0.4 at T=4 K. Although M does not change markedly with x, $\langle S_z \rangle$ is a strong function of the manganese concentration, as seen in Fig. 2. There is a substantial decrease in the average spin for increasing x, resulting from the increased antiferromagnetic interactions. The most remarkable feature of these data is the exceedingly linear behavior (within a few percent) above 15 T. Linear-like behavior at high fields has been predicted by numerical Monte Carlo calculations [9].

Over the present range of fields, up to 40 T, M(B) can be decomposed into a Brillouin-like part and a linear part, described by

$$M(B) = M_s B_{5/2}[5\mu_B B/k(T_{eff})] + \chi_{HF} B. \qquad (2)$$

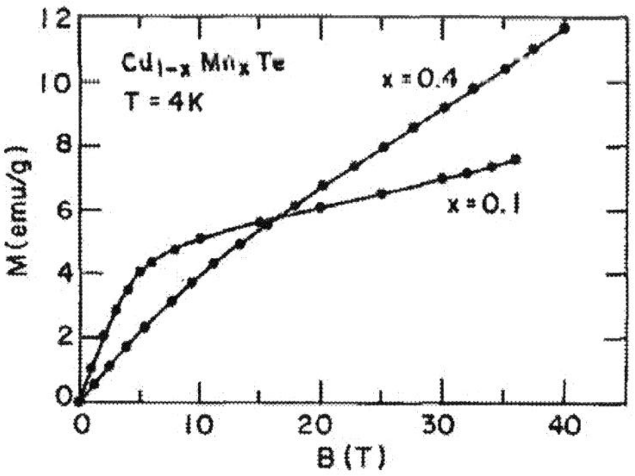

Fig. 1 Magnetization M versus applied magnetic field B for $Cd_{1-x}Mn_xTe$, x=0.11, and 0.38. From ref. 7.

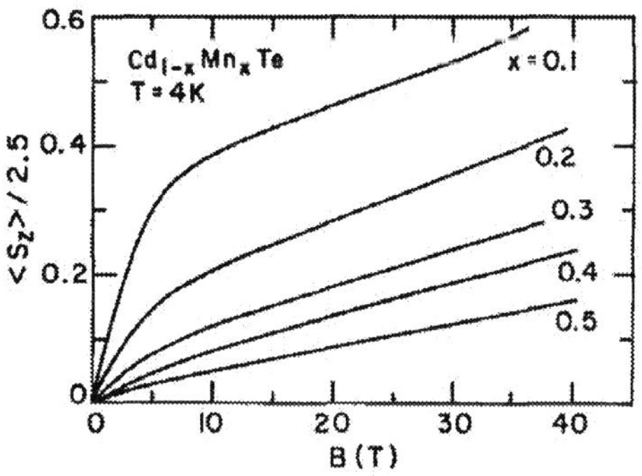

Fig. 2 Average z-component of Mn^{++} ion spin $\langle S_z \rangle$ versus applied magnetic field B, for $Cd_{1-x}Mn_xTe$, x=0.11, 0.20, 0.30, 0.38, and 0.49.

Table I. Parameters fit to Eq. (2). Uncertainties are estimated at ±5% for x and T_{eff} and ±20% for M_s and χ_{HF}.

x	T_{eff} (K)	M_s (emu/g)	χ_{HF} (emu/g-T)
0.11	6.5	4.5	0.08
0.20	7.3	3.5	0.17
0.30	9.2	2.4	0.19
0.38	12.7	1.8	0.25
0.49	6.2	1.0	0.24

Good agreement between experiment and Eq. (2) was obtained for $0.1 \leq x \leq 0.5$. The best-fit values for $\chi_{HF}(x)$, $M_s(x)$ and $T_{eff}(x)$ are compiled in Table I. We see that as x increases M_s decreases, while χ_{HF} increases up to x=0.3, where it saturates at χ_{HF}=0.24. These features are discussed below.

III. DISCUSSION OF RESULTS

At low concentrations (x < 0.1), the magnetization at low temperatures is well understood [10]. M_s is less than M_o (obtained at full saturation) because the antiferromagnetic coupling energy of nearest-neighbors is greater than both kT and $5\mu_B B$. When the Mn^{++} ions are distributed randomly, M_s is determined by: (a) the probability of finding ions in singlets, pairs, triplets and larger clusters; and (b) the total spin of the ground state for each of these clusters. For x=0.1, agreement with the data is achieved using clusters of up to three spins. However, for higher x, M_s cannot be determined by this simple model because the majority of spins are in clusters larger than triplets.

Although there is no simple theory of M(B) for x > 0.1, numerical calculations have been made for a microscopic model of a diluted isotropic Heisenberg antiferromagnet [11]. This model shows a linear susceptibility in the large-x limit and a substantial nonlinear component at smaller x. This work concludes that small isolated clusters cannot account for the majority of the <u>nonlinear</u> M(B) response. Additional contributions arise from the gradual saturation of local ferrimagnetic variations, where each Mn^{++} spin feels an exchange field of a different magnitude and direction.

The <u>linear</u> contribution to M(B) for x > 0.2 arises mainly from changes in the internal magnetization of large clusters and the infinite chain. The exceedingly linear behavior of our data is qualitatively similar to the perpendicular component in antiferromagnets [12] and other dilute antiferromagnets.[13] The derived values for χ_{HF} can be compared to the isotropic susceptibility, χ_{AF}, in the mean-field approximation for an antiferromagnet at T=0. Assuming nearest-neighbor interactions only, a face-centered cubic array of Mn^{++} ions has four noninteracting sublattices, giving $\chi_{AF} = -A(g\mu_B)^2/12WyJ$, where g=2, y=4 is the number of sublattices and J the nearest-neighbor exchange constant [14]. This relation also holds for the diluted case x<1. Using J/k = -6.0 K [15], and x=0.5, then χ_{AF} = 0.25 emu/g-T. The good agreement with χ_{HF} = 0.24 is fortuitous, since we have neglected higher-neighbor interactions.

IV. ORDERED-ALLOY SUPERLATTICES

As we have seen from the experiments, the magnetic response is inhibited by the antiferromagnetic coupling of Mn^{++} ions on nearest-neighbor (NN) sites. In these random alloys most of the ion spins are antiferromagnetically locked-out on NN lattice sites at low temperature due to the large number of first-NN sites of the fcc lattice, 12. On the other hand, a nonrandom diluted alloy with an appropriately ordered structure of Mn^{++} ions has the advantage of enhancing the magnetic response, both by reducing the number of NN pairs and by increasing the NN distance. Structures with lower dimen- sionality usually have less NN pairs. Indeed, measurements on ultra-thin 2D layers of MnSe suggest an enhancement of the magnetic properties as a result of the reduced dimensionality of the layers [16]. In addition, the Mn^{++}-Mn^{++} exchange energy J is a strong function of distance; for next-NN J_{NNN}~0.1 J_{NN}. In view of this, we have pointed out that the magnetic response will be significantly increased when the Mn^{++} ions are ordered on higher-NN fcc sites [17]. We examine the details for increasing the magnetic susceptibility by both reducing the number of NN pairs and increasing the pair separation.

Ordered-alloy superlattices may be created by selectively depositing atomic monolayers along specific crystal directions. First consider the simple cases of layering ordered $Cd_{1-x}Mn_xTe$ along [100], [110], and [111] directions. Figure 3 shows the in-plane structure of Mn^{++} ions for each of these growth directions. Here, every n-th cation layer contains all Mn^{++} ions and is chosen to eliminate first-NN interactions between layers. The straight lines connect the central ion to its NNs. The coordination number Z is the number of NN for a given Mn^{++} ion, and is 4, 2, and 6 for the respective planes. Table II lists the in-plane symmetry of Mn^{++} ions, the closest distance between Mn^{++} ions $d = a/\sqrt{2}$ (a is the lattice constant), and Z. Since the magnetic susceptibility is proportional to 1/Z, the linear chains of [110]:n=3 are the most favorable.

By layering along the [210] direction in pairs, a chalcopyrite structure is obtained — Fig. 4a shows $CdMnTe_2$ made from a CdCdMnMn cation sequence. Using another sequence, a stannite-like structure of Cd_3MnTe_4 containing Mn atoms at every fourth layer has been proposed [17] and is shown in Fig. 4b. The stannite structure shows only weak antiferromagnetism since the closest Mn^{++}-Mn^{++} distance corresponds to second-NN sites in $Cd_{1-x}Mn_xTe$.

The [311] polar surface is also attractive because it is easily produced (in GaAs) without faceting and is used for growing high-quality epitaxial layers [21]. The [311]:n=3 structure has linear chains similar to the [110]:n=3 structure.

Table II. Parameters for the nonrandom diluted alloy $Cd_{1-x}Mn_xTe$. Refer to text for explanation of symbols. Structures in parentheses are planar.

Growth Direction	n	x	Z'	d/a	Crystal Structure	J (K)	χ/χ_0
	1	1.00	12	0.71	zincblende	6	1.0
[100]	2	0.50	4	0.71	tetragonal	6	2.6
	3	0.33	4	0.71	(fcc)	6	2.5
[110]	2	0.50	4	0.71	-	6	2.6
	3	0.33	2	0.71	(linear)	6	5.0
[111]	2	0.50	6	0.71	(hexagonal)	6	1.7
	2	0.50			-		
[210]	2,2	0.50	4	0.71	chalcopyrite	6	2.6
	4	0.25	4	1.00	stannite-type	0.5	28
[311]	3	0.33	2	0.71	(linear)	6	5.0

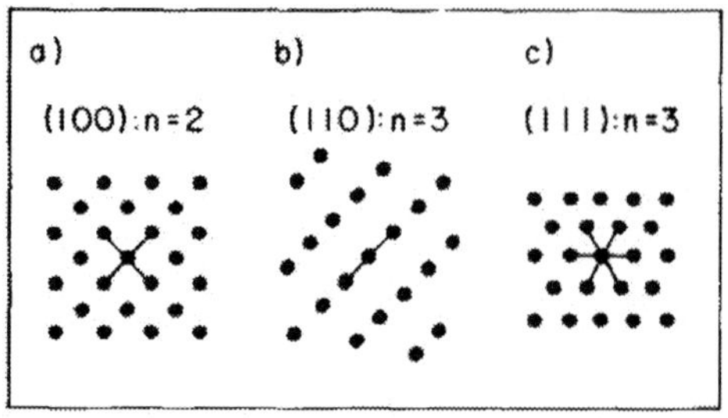

Fig. 3 In-plane lattice sites of the face-centered cubic cation sublattice of the zinceblende structure for [100], [110] and [111] planes The lines connect the central atom to its first nearest neighbors.

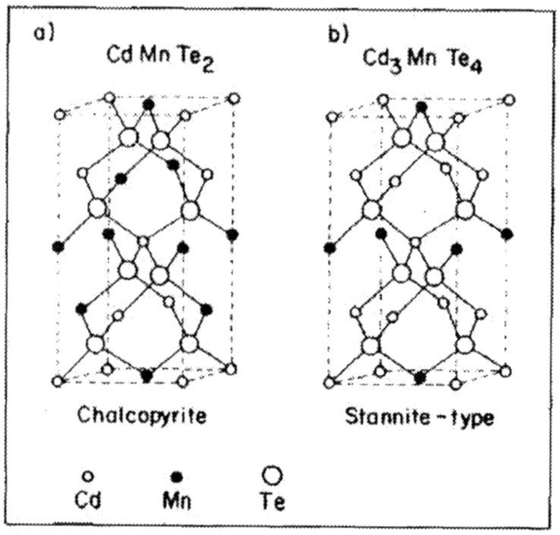

Fig. 4. Schematic diagram of a unit cell of the chalcopyrite and stannite-type structure.

A relative figure-of-merit for these DMS ordered structures is determined by calculating the magnetic susceptibility from a mean-field model. In the over-simplified case of a two-sublattice diluted antiferromagnet at T=0 K [18], the isotropic susceptibility is

$$\chi(\mathrm{emu/g\text{-}T}) = A(g\mu_B)^2/3Z'WJ \quad , \qquad (3)$$

where Z' is the number of closest neighbors. The Mn^{++}-Mn^{++} exchange energy J varies dramatically with distance d, given empirically by [19] $J = 70\exp(-4.89d^2/a^2)$, in K units for up to third-NN. This strong variation makes it more advantageous to maximize χ by increasing d than by reducing Z'. Table II displays Z', J, and the ratio χ/χ_o for various ordered structures, where χ_o ~ 0.4 emu/g-T is the value for x=1. The magnetic enhancements for the structures [110]:n=3 and [210]:n=4 are 5, 28, respectively. Enhancements larger than about 10-100 are probably not realistic for temperatures above a few degrees K. The high temperature susceptibility is enhanced by $(T+\theta)/T$, where $\theta=2xZJ_{NN}S(S+1)/3$ K is the Curie temperature. For x=0.25 and z=12 the enhancement is 1.3 at room temperature and 2.4 at T=77 K.

In summary, measurements of random-alloyed $Cd_{1-x}Mn_xTe$ show that the high-field susceptibility becomes small and independent of x for x > 0.3 due to antiferromagnetic cancellations. On the other hand, an ordered structure of a nonrandom alloy is expected to show a significant magnetic enhancement. This translates into sizeable bandgap tuning and Faraday rotations at low fields. Additional advantages of ordered structures results from:
(a) their lack of alloy potential fluctuations, which will considerably sharpen exciton resonances; (b) new crystal structures (Fig. 3) with different band structures due to zone folding; and (c) enhanced structural stability relative to the random configuration [20].

ACKNOWLEDGEMENTS

We thank Y. Shapira, P.A. Wolff and J.J. Zayhowski for stimulating and helpful discussions, and J. Conlon for technical assistance. This work was supported by the National Science Foundation DMR-8504366, the Office of Naval Research N00014-83-K-0454, and the U.S. Army ARDEC DAAL03-86-D-0001. The Francis Bitter National Magnet Laboratory is supported by the National Science Foundation through its Division of Materials Research. E.D. Isaacs is supported by an AT&T Bell Laboratories Ph.D. Scholarship.

REFERENCES

*Also, Department of Physics.

1. N.B. Brandt and V.V. Moshchalkov, Adv. Phys. 33, 193 (1984); J.K. Furdyna, J. Appl. Phys. 53, 7637 (1982); and references cited therein.
2. S. Oseroff and P.H. Keesom, in "Semiconductors and Semimetals," ed. by R.K. Willardson and A.C. Beer, (Academic Press, Inc., New York, 1986).
3. R.R. Galazka, S. Nagata and P.H. Keesom, Phys. Rev. B22, 3344 (1980).
4. P.A. Wolff, in "Semiconductors and Semimetals," ed. by R.K. Willardson and A.C. Beer, (Academic Press, Inc., New York, 1986).
5. J.A. Gaj, R. Planel and G. Fishman, Solid State Commun. 29, 435 (1979).
6. D. Heiman, Y. Shapira, S. Foner, B. Khazai, R. Kershaw, K. Dwight, and A. Wold, Phys. Rev. B29, 5634 (1983).
7. D. Heiman, E.D. Isaacs, P. Becla, and S. Foner (Phys. Rev. B).
8. D. Heiman, Rev. Sci. Instrum. 56, 684 (1985).
9. J.J. Zayhowski, Ph.D. Thesis, Massachusetts Institute of Technology (1986).
10. Y. Shapira, S. Foner, D. Ridgley, K. Dwight, and A. Wold, Phys. Rev. B30, 4021 (1984); and Y. Shapira, S. Foner, P. Becla, D.N. Domingues, M.J. Naughton, and J.S. Brooks, Phys. Rev. B33, 356 (1986).
11. A. Brooks-Harris and S. Kirkpatrick, Phys. Rev. B16, 542 (1977).
12. J.S. Smart, Effective Field Theories of Magnetism, (W.B. Saunders, Philadelphia, 1966).
13. D.J. Breed, K. Gilijamse, J.W.E. Sterkenburg, and A.K. Miedema, Physica 68, 303 (1973).
14. P.W. Anderson, Phys. Rev. 79, 705 (1950).
15. R.L. Aggarwal, S.N. Jasperson, P. Becla, and R.R. Galazka, Phys. Rev. B32, 5132 (1985).
16. L.A. Kolodziejski, R.L. Gunshor, N. Otsuka, B.P. Gu, Y. Hefetz, and A.V. Nurmikko, Appl. Phys. Lett. 48, 1482 (1986); A.V. Nurmikko, D. Lee, Y. Hefetz, L.A. Kolodziejski, and R.L. Gunshor, Proc. 18th Int. Conf. Phys. Semicond., Stockholm, 1986, (to be published).
17. P.A. Wolff, D. Heiman, E.D. Isaacs, P. Becla, S. Foner, L.R. Ram-Mohan, D.H. Ridgley, K. Dwight, and A. Wold, Proceedings of the International Conference on the Application of High Magnetic Fields in Semiconductor Physics, Wurzburg, August 1986 (to be published).
18. Y. Shapira and S. Foner, Phys. Rev. B1, 3083 (1970).
19. H. Ehrenreich, K.C. Hass, N.F. Johnson, B.E. Larson, and R.J. Lempert, Proceedings of the 18th Int. Conf. Phys. Semicond., Stockholm, 1986, (to be published).
20. G.P. Srivastava, J.L. Martins, and A. Zunger, Phys. Rev. B31, 2561 (1985); T.S. Kuan, T.F. Kuech, W.I. Wang, and E.L. Wilkie, Phys. Rev. Lett. 54, 201 (1985).
21. C.B. Duke, C. Mailhiot, A. Paton, A. Kahn, and K. Stiles, J. Vac. Sci. Tech. A4, 947 (1986).

UNIVERSALITY OF THE SPIN-GLASS TRANSITION IN THE $Cd_{1-x}Mn_xTe$ SYSTEM

T. DATTA*, J. AMIRZADEH**, A. BARRIENTOS*, E.R. JONES*, AND J.F. SCHETZINA***
* University of South Carolina, Dept. of Physics, Columbia, SC 29208, U.S.A.
** Morris College, Sumter, SC 29150, U.S.A.
*** North Carolina State University, Dept. of Physics, Raleigh, NC 27695

ABSTRACT

Universality of the spin-glass (SG) transition in bulk $Cd_{1-x}Mn_xTe$ was investigated by determining the scaling behavior of the spin-glass order parameter q as a function of the reduced temperature $t = (T_g-T)/T_g$, where T_g is the transition temperature. q(T) was determined from the SQUID magnetometric data both above and below the transition. It can be shown that, $q(T) = 1 + T[|\theta|-C/\chi(T)]^{-1}$. C and θ, the Curie and Curie-Weiss parameters, were obtained from a non-linear regression of the dc low field susceptibility $\chi(T)$ above T_g. q(T) thus obtained exhibits the cannonical order parameter criteria, viz. $q(T>T_g)=0$, $q(T\to T_g)\to 0$, and $q(T\to 0)\to 1$. In range of Mn concentration studied $(0.3 \leq x \leq 0.55)$, T_g ranged between 13 K and 23.5 K. Thus q as a function of absolute temperature behaves differently for different x. However, universality is clearly evidenced when the dependence on t is determined. q(t) exhibits a universal scaling law. We observe $q \sim t^{\beta}$ with $\beta \simeq 0.95$. Overall good agreement was noted with the Sherrington-Kirkpatrick (SK) infinite range SG model. Observed value of β is in excellent agreement with the model prediction β -1. But in the $Cd_{1-x}Mn_xTe$ system we find $\theta<0$, indicating a net antiferromagnetic interaction and $|\theta|/T_g > 1$ as opposed to the SK model. We believe this to be additional evidence that the spin interactions are not distributed as a Gaussian function.

INTRODUCTION

The ternary alloy, $Cd_{1-x}Mn_xTe$ is an example of a dilute magnetic semiconductor (DMS). [1-9]. Crystals of this alloy with $x \lesssim 0.7$ have the zinc-blend structure. The Mn atoms randomly occupy sites in the fcc Cd sublattice with the probability x. There is some controversy as to the exact nature of the low-temperature magnetic ordering [2,4,5] in this system. Perhaps, some of the peculiarities arise from the fact that unlike the long range RKKY spin-spin interaction in the better known metallic alloys [10-12], the interaction is short ranged for the semiconductors [2,5]. It has been predicted that the spin-disorder and the lattice induced frustration present in this structure would create a spin-glass (SG) phase at a low enough temperature [14,15]. Evidence for such a SG phase has been reported [2,4,9,10]. We have found additional evidence for such ordering in $Cd_{1-x}Mn_xTe$, for four different Mn concentrations: x = 0.30, 0.40, 0.45 and 0.55.

EXPERIMENTAL

Measurements were made with a fully automated computer-controlled superconducting quantum interference device (SQUID), variable-temperature susceptometer (VTS) [16]. The specimens were randomly oriented small crys-

Mat. Res. Soc. Symp. Proc. Vol. 89. © 1987 Materials Research Society

talites with combined mass not exceeding several tenths of grams. The stoichiometry was checked by both spot and area measurements with an energy dispersive x-ray micro-analysis (EDAX). Each set of experiments were queued in ascending order of the magnetic field B. The experiments were performed over a field range of $0.1 \lesssim B \lesssim 100$ mT and temperature range of $1.5 \lesssim T \lesssim 400$ K.

With the specimen initially at room temperature, three experimental protocols were followed: (i) zero-field-cooled warm up (ZFCW) [11,12]; in this case the sample was introduced into the VTS cryostat and was allowed to thermally equilibrate to ~ 4K, then the dc field was turned on. The data were collected by holding B constant and raising the temperature in a programed sequence. (ii) In the field-cooled warm up (FCW), the sample entered the VTS with a 10mT field already on. After a waiting period, the field was lowered to the value desired for measurement and data were taken as the temperature was raised. (iii) At the end of a warm up (ZFCW or FCW) additional field-cooled cooldown (FCC) measurements were performed in the same field as the temperature was cycled down.

ANALYSIS AND DISCUSSION

We observed the well-known magneto-thermo-hysteresis between FCW and FCC data [11,13] associated with the SG ordering. FCW displays the critical behavior in the low-temperature ordered phase. For the FCC, criticality is observed only above the freezing, when the system is in equilibrium [17-18]. Figure 1 shows some typical $\chi(T)$ data for both FCW and FCC for $x = 0.30$. Notice the single cusp particularly prominent in FCW. This cusp appears to

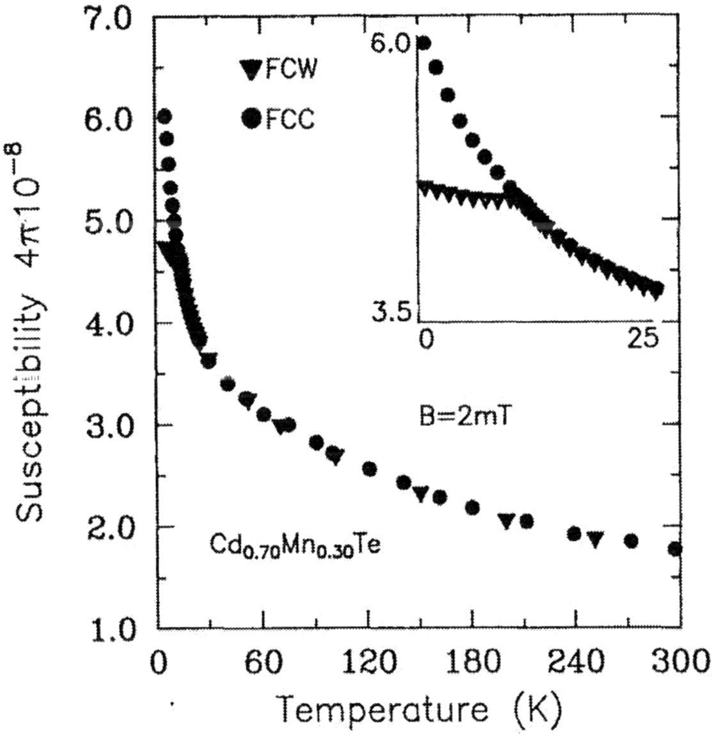

Fig. 1. Longitudinal mass susceptibility vs temperature for $Cd_{0.7}Mn_{0.3}Te$ in external field of 2 mT. Circles represent FCC and triangles represent FCW. Inset show an enlarged view of the cusp region.

be broader and progressively shifted to higher temperature as the Mn concentration, x, is increased [9]. The fractional upward shift $\Delta T_p/T_p$ in this peak temperature T_p, decreased with higher Mn concentration [2,4,9]. Another point of interest is the overlap of the onset of strong irreversibility between zero field and field-cooled data with the T_p [12]. Both the cusp and the irreversibility exhibited sensitivity to B.

These behaviors are generic SG signatures and the results confirm earlier reports. However, we do not observe a low temperature $\sim|\ln T|^n$ type $\chi(T)$ dependence predicted for incommensurate spin-glasses [5]. Neither did we observe the reported Curie law behavior of $\chi(T)$ at low (T < 4 K) temperatures [2,5]. In the present case we searched as low as 1.5 K. From this we conclude, that the present specimens are of higher purity than those reported earlier [2,5] which contained $\sim 10^{18}$ cm^{-3} paramagnetic impurities.

The high temperature ($T_p \lesssim T < 400$ K) data were numerically analyzed using a non-linear regression package [10]. The diamagnetic susceptibility of pure CdTe is less than 1% of the measured susceptibility in our samples. Corrections for this diamagnetism was not necessary for the low-temperature analysis and was not done.

In the temperature region above the cusp the longitudinal dc susceptibility may be expressed as follows:

$$\chi(T)_x = \frac{S_x}{B} B_{5/2} \left[\frac{5}{2} \cdot \frac{g\mu_B B}{k(T+\theta)} \right]. \qquad (1)$$

In Eq. (1) S_x is an x dependent parameter representing the effective moment per unit mass; $B_{5/2}$ is the appropriate Brillouin function; g is the g factor, μ_B is the Bohr magneton and k is the Boltzmann constant. For small fields Eq. (1) reduces to the Curie-Weiss limit

$$\chi(T)_x = \frac{C_x}{T + \theta} , \qquad (2)$$

where the effective Curie constant C_x is given by

$$C_x = \frac{C}{M_{Te} + M_{Cd} + x(M_{Mn} - M_{Cd})} \qquad (3)$$

The parameter θ represents the effective spin-spin interaction. In the mean field approximation C and θ are characteristic material constants independent of both T and B. In a real system, expecially near the critical points, T and B dependence in these parameters may be expected. In first order, θ is much more sensitive to temperature than is C. We observe that above about 50 K, C and θ are insensitive to T. Spalek et al. [19] have measured the magnetic susceptibility of $Cd_{1-x}Mn_xTe$ for temperatures between 77 K and 300 K, in this region for all samples studies they found C and θ to be constant. At lower temperatures we observe a strong temperature dependence of θ. The value of θ decreases as T is reduced below 50K and the temperature dependence becomes stronger as the cusp is approached. Similar behavior has been observed in manganese doped ZnTe [20].

In order to discuss the behavior below the transition we need the values of C and θ near the temperature of the transition. These were obtained from analysis of the data for T between T_p and 50 K. The results for applied field B = 2 mT are: (i) x = 0.30, $C_x = 8.46 \times 10^{-3}$ K m^3 kg^{-1} and $\theta = -47$K; (ii) x = 0.40, $C_x = 9.39 \times 10^{-3}$ K m^3 kg^{-1} and $\theta = -48$ K; (iii) x = 0.45,

$C_x = 1.15 \times 10^{-2}$ K m³ kg^{-1} and $\theta = -48$K;(iv)x = 0.55,$C_x = 4.45 \times 10^{-2}$ K m³kg^{-1} and $\theta = -269$ K. Clearly the replacement of Cd atoms by Mn increases the effective magnetic moment. This is indicative of collective effects such as clustering in the Mn distribution [2]. The negative values of θ indicate a dominant antiferromagnetic interaction. This antiferromagnetism becomes stronger with increasing x as indicated by larger |θ|. In the sample with highest concentration (x = 0.55), θ is considerably higher than it is for the three lower x samples. Such anomalous behavior for high (x = 0.5) Mn concentration samples had been previously observed in the C (T) data [2]. In the zinc-blend structure, the next nearest neighbor exchange integrals J_{NNN} are much smaller than the nearest neighbor J_{NN} integrals. Hence the observed antiferromagnetism is primarily attributable to the, on the average, antiparallel nearest neighbor spin alignements between the Mn^{2+} ions.

It is convenient to employ a Fischer type expression for χ(T) [21,22] to quantitatively discuss the critical region and the low temperature phase. Generalized for net anti-ferromagnetic interaction [10,12] χ(T) is given by

$$\chi(T) = \frac{C\,\lambda(T)}{T + |\theta|\lambda(T)} \quad . \qquad (4)$$

In terms of the Edwards-Anderson [23] spin-glass order parameter, q (T), λ can be expressed as

$$\lambda(T) = 1 - q(T), \qquad (5)$$

where, q=0 above the transition. From the measured low temperature χ(T) (ZFCW) and the values of C and θ discussed earlier, the SG order parameter may be determined from Eqs. (4) and (5). That is [13],

$$q(T) = 1 + T[|\theta| - C/\chi(T)]^{-1}, \qquad (6)$$

The SG freezing temperature T_f, such that $q(T \geq T_f) = 0$, was self consistently determined by the numerical analysis. In all cases T_f was in agreement with T_{ir} the temperature where FCW and FCC irreversibility could be experimentally first detected.

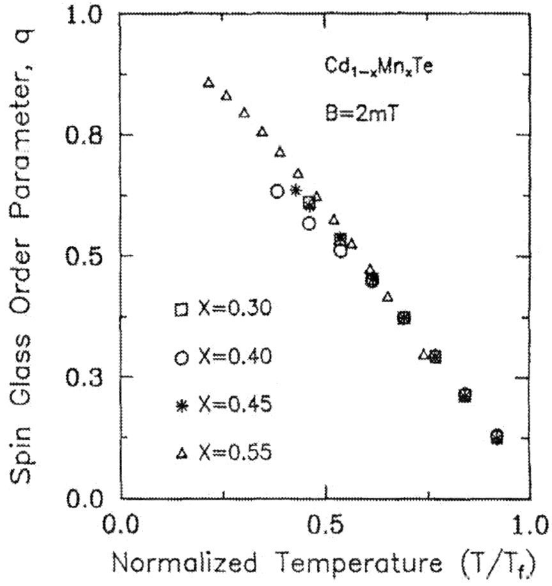

Fig. 2. Spin glass order parameter of $Cd_{1-x}Mn_xTe$ as a function of normalized temperature for four values of x. All data follow the same curve indicating universality when displayed as functions of $t = T/T_f$ irrespective of T_f.

Universality of the SG ordering in this system becomes apparent when the order parameter values for all the four Mn concentrations are expressed as a function of the normalized temperature $t = T/T_f$ as shown in Fig. 2. The critical scaling property of q as a function of the reduced temperature (1-t) i.e. $q \sim (1-t)^{\beta}$, is clearly exhibited in Fig. 3. The observed value of the scaling exponent $\beta = 0.9$ is in good agreement with the mean-field prediction $\beta = 1$.

For all of these specimens we determined $|\theta|/T_f > 1$. This is at variance with the SK model [10,12,24] and is a long standing difficulty. Since $J_{NNN} \sim 0.1\ J_{NN}$, the next nearest neighbor interactions cannot be ignored. This may render the two parameter (normal) distribution approximation of the models [23,25] unrealistic.

Fig. 3. Spin glass order parameter of $Cd_{1-x}Mn_xTe$ as a function of reduced temperature for four values of x.

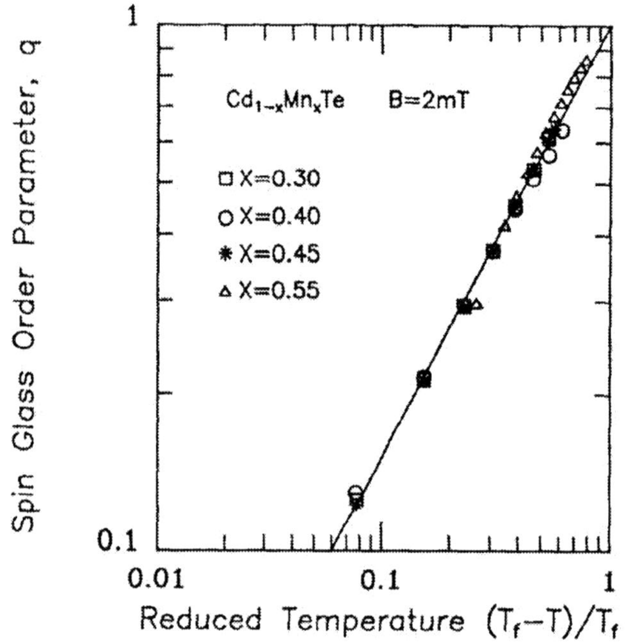

In these systems classical scaling behaviors [24] of M(T,B) and C (T,B) are not observed [4]. In contrast, the order parameter appears to follow mean-field SK model scaling quite satisfactorily. This may indicate that compared with the length scales of the system, viz., spin-spin correlation length, cluster size, etc., the average interaction distance (in the condensed phase) is effectively infinite. On the other hand this may also imply that some of the predictions of the model [23,24] are general and are not restricted to infinite range interactions.

CONCLUSIONS

We believe that the coincidence of the onset of thermomagnetic hysteresis with the single cusp in $\chi(T)$ indicates that the hysteresis and the cusp are controlled by the same mechanism. The lack of additional structure in the susceptibility and the well-defined behavior of the onset of irreversibility with magnetic field [11,26] is evidence of a de Almeida-Thouless transition [25]. Microscopically, the de Almeida-Thouless transition represents the transition from paramagnetic to replica-symmetry-broken spin-glass ordered phase.

In summary, we observe generic, SG-like static magnetic behavior and critical scaling of q, in $Cd_{1-x}Mn_xTe$ for $0.3 \leq x \leq 0.55$. This is the first observation of such universal scaling of the EA, SG order parameter in DMS system. It is argued that the observed behavior is associated with a de Almeida-Thouless type transition.

ACKNOWLEDGEMENTS

We wish to thank Professor C. P. Poole, Jr. for many useful suggestions. The work was partially supported by NSF-ISP-80-4451 and USC-RPSC#13070E127, at USC; NSF-DMR 82-1306 and DARPA/ARO-DAAG29-83-K-0102 at N.C. State.

REFERENCES

1) For a very readable review please see: J.K. Furdyna, J. Appl. Phys. 53, 7637 (1982).
2) R.R. Galazka, S. Nagata and P.H. Keesom, Phys. Rev. B22 3344 (1980).
3) S. Nagata, R. R. Galazka, G.D. Khattak, C.D. Amarasekara, J.K. Furdyna and P.H. Keesom, Physica 107B, 311 (1981); M. Escorne, A. Manger, R. Triboulet and J.L. Tholence, ibid., 309 (1981); M.A. Novak, S. Oseroff and O.G. Symko, ibid., 313 (1981).
4) S. Oseroff and F. Acker, Solid State Commun. 37, 19 (1980).
5) M. Escorne and A. Manger, Phys. Rev. B25, 4674 (1982).
6) G. D. Khattak, A. Twardowski and R.R. Galazka, Phys. Status Solidi A87, K87 (1985).
7) A. Barrientos, C. Almasan, T. Datta, E.R. Jones Jr., R.N. Bicknell and J.F. Schetzina, Bull. Am. Phys. Soc. 31, 252 (1986).
8) J.R. Anderson, M. Gorska, L.J. Azevedo and E.L. Venturini, Bull. Am. Phys. Soc. 31 253 (1986).
9) T. Datta, A. Barrientos, J. Amirzadeh, E.R. Jones, Jr. and J.F. Schetzina, submitted for publication.
10) T. Datta, S.D. Levine, D. Thornberry and E.R. Jones, Jr., Phys. Status Solidi B121, K125 (1984).
11) T. Datta, D. Thornberry, E.R. Jones, Jr., and H.M. Ledbetter, Solid State Commun. 52, 515 (1984).
12) T. Datta, D. Thornberry, C. Almasan and E.R. Jones, Jr., Solid State Commun. 56, 523 (1985).
13) A.P. Malozemoff, S.E. Barnes, and B. Barbara, Phys. Rev. Lett. 51, 1704 (1983).
14) J. Villain, Z. Phys. B33, 41 (1979).
15) C.P. Poole and H.A. Farach, Z. Phys. B47, 55 (1982).
16) Model # VTS-805, BTI Corp. San Diego, CA, USA.
17) S. Fishman and A. Aharony, J. Phys. C12, L729 (1979).
18) A.R. King, J.A. Mydosh and V. Jaccarino, Phys. Rev. Lett. 56, 2525 (1986).
19) J. Spalek, A. Lewicki, Z. Tarnawski, J.K. Furdyna, R.R. Galazka, Z. Obuszko, Phys. Rev. B 33, 3407 (1986).
20) Y. Shapira, S. Foner, P. Becla, D.N. Domingues, M.J. Naughton, J.S. Brooks, Phys. Rev. B 33, 356 (1986).
21) K.H. Fischer, Phys. Rev. Lett. 34 1438 (1975).
22) J.A. Hertz, Phys. Rev. Lett. 52, 1880 (1983).
23) D. Sherrington and S. Kirkpatrick, Phys. Rev. Lett. 35, 1792 (1975).
24) J. Soultie and R. Tournier, Jr. Low Temp. Phys. 1, 95 (1969).
25) J.R.L. de Almeida, K.J. Thouless, J. Phys. A11, 983 (1978).
26) I.A. Campbell, D. Arvanitis, A. Fert, Phys. Rev. Lett. 51, 57 (1983).

Magnetic Specific Heat of $Cd_{1-x}Mn_xTe$ at Low Temperatures and High Magnetic Fields

W. Y. CHING* AND D. L. HUBER**
*Department of Physics, University of Missouri-Kansas City, Kansas City, MO 64110
**Department of Physics, University of Wisconsin-Madison, Madison, WI 53706

ABSTRACT

Theoretical predictions for the magnetic specific heat of $Cd_{1-x}Mn_xTe$:x=0.20, 0.35, 0.50, and 0.65 are reported for B=0, 10T, and 100T. The analysis applies to the low-temperature regime, $T \lesssim 10K$, where the fundamental excitations are harmonic magnons. The calculations use values for the exchange interactions which were inferred from fits to the dynamic structure factor describing inelastic neutron scattering at low temperatures. For x=0.35, 0.50, and 0.65 the specific heat is only weakly affected by applied fields up to 10T. At the lowest concentration the application of a field of 10T leads to a significant reduction in the specific heat. Results are also presented for the distribution of magnon modes at various concentrations and fields.

I. INTRODUCTION

Recently, reports have been published on the effects of an applied magnetic field on the low-temperature specific heat of the insulating spin glass $Eu_xSr_{1-x}S$ [1-3]. Complementary theoretical calculations have shown that the harmonic magnon or, equivalently, the independent boson approximation can account quantitatively for the magnetic component of the specific heat when the calculations are carried out with empirically determined values for the exchange integrals [3-6].

The purpose of this paper is to extend the theoretical studies of Refs. 3-6 to another class of insulating spin glasses - the dilute magnetic semiconductors as represented by the system $Cd_{1-x}Mn_xTe$. Previously, we have shown that the harmonic magnon model reproduces the magnetic specific heat of $Cd_{0.5}Mn_{0.5}Te$ in zero field [7]. In this paper, along with x=0.50, we consider the concentrations x=0.20, 0.35, and 0.65. Results are reported not only for B=0 but also for B=10T and 100T.

Like the $Eu_xSr_{1-x}S$ compounds, the exchange couplings in $Cd_{1-x}Mn_xTe$ are dominated by the interaction between nearest-neighbors on a fcc magnetic lattice. Unlike $Eu_xSr_{1-x}S$, however, the nearest-neighbor interaction is antiferromagnetic in sign leading to the appearance of short-range Type III antiferromagnetic order at low temperatures. Correspondingly, the magnetic field has a much smaller effect on the specific heat in $Cd_{1-x}MnTe$ than it does in $Eu_xSr_{1-x}S$ which is ferromagnetic at T=0 for x>0.70.

II. ANALYSIS

Apart from the values of the exchange integrals, our calculations are similar to the ones we reported in Ref. 6 for $Eu_xSr_{1-x}S$. Equilibrium spin configurations are obtained by minimizing the energy of a classical array of spins with both exchange and Zeeman interactions. The energies of the magnons are inferred from the eigenvalues of a dynamical matrix

Mat. Res. Soc. Symp. Proc. Vol. 89. © 1987 Materials Research Society

characterizing the linearized equations of motion of the spin operators. In this approach the magnetic specific heat is given by

$$C_H = (kT^2)^{-1} \int_0^\infty dE\ N(E)E^2 e^{E/kT}(e^{E/kT}-1)^{-2}, \quad (1)$$

where N(E) is the magnon density of states.

Although there is still some controversy over the exact value of the exchange interactions in $Cd_{1-x}Mn_xTe$, it is generally agreed that the interaction between nearest-neighbors is on the order of 1 meV. In our calculations we have used the values $J_{nn} = 10J_{nnn} = 12.5K$, which have been inferred from theoretical fits to the dynamic structure factor characterizing the inelastic neutron scattering [8]. The calculations were carried out for finite arrays of spins, N=325-480, distributed at random on fcc supercells, with averages taken over 2-3 configurations. The analysis was limited to $T \leq 10K$, which we believe is within the range of applicability of the theory. It should be noted, however, that we have not taken anisotropy into account. Although $Cd_{1-x}Mn_xTe$ is weakly anisotropic, the anisotropic terms in the Hamiltonian will affect the low-frequency magnon modes. As a consequence, our results are not particularly accurate at very low temperatures where the specific heat is strongly influenced by the anisotropy. From our previous analysis of $Cd_{0.5}Mn_{0.5}Te$, it appears that this happens when T<1-2K [7].

III. RESULTS

Our results for the magnetic specific heat per spin (in dimensionless units since energy is measured in units of temperature) are shown in Figs. 1-4 for x=0.20, 0.35, 0.50, and 0.65, respectively, with field values B=0, 10T, and 100T. We have also calculated the specific heat for B=1T. The results, however, are graphically indistinguishable from the zero-field data.

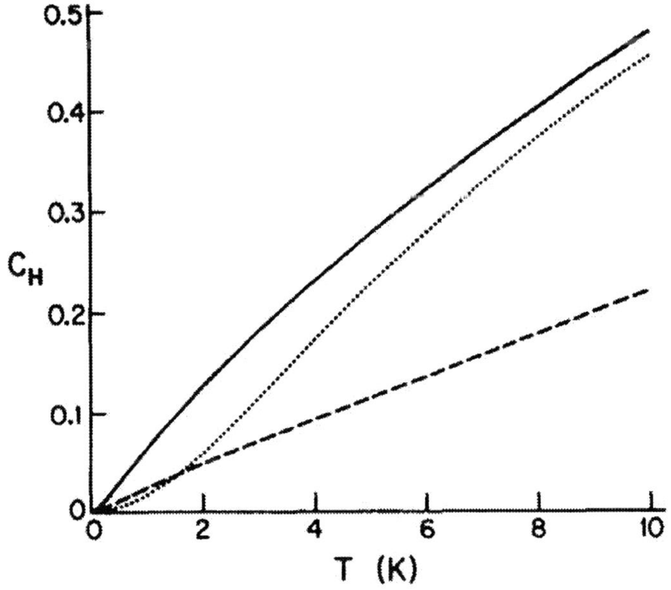

Fig. 1. Magnetic specific heat per spin vs temperature; x=0.20. ___ B=0; ... B=10T; _ _ _ B=100T.

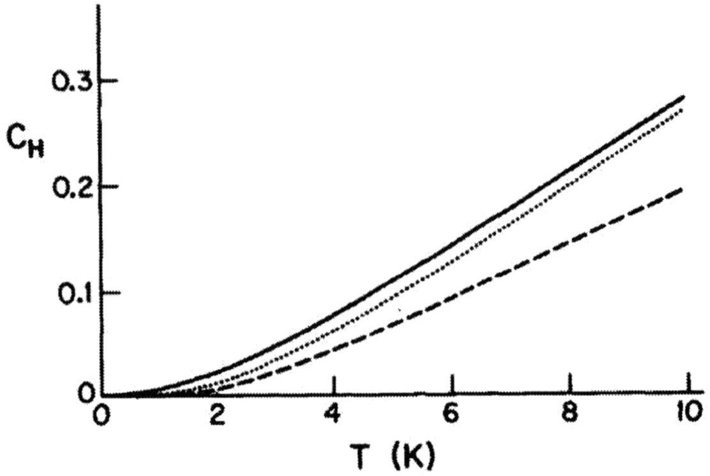

Fig. 2. Magnetic specific heat per spin vs temperature; x=0.35 . ___ B=0; ... B=10T; _ _ _ B=100T.

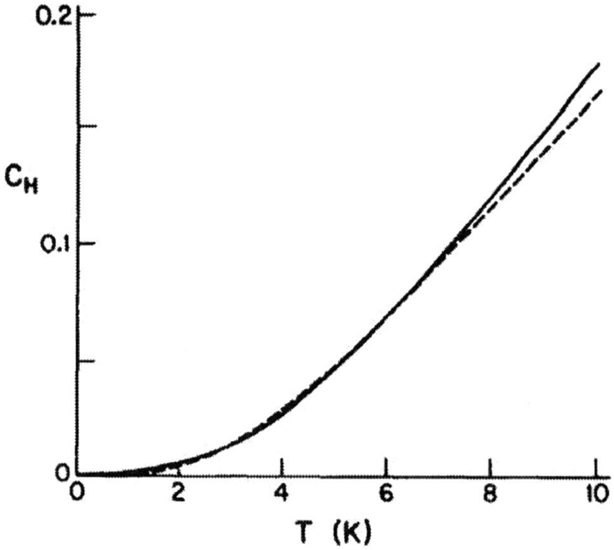

Fig. 3. Magnetic specific heat per spin vs temperature; x=0.50. ___ B=0; _ _ _ B=100T. Results for B=10T are graphically indistinguishable from the curve for B=0.

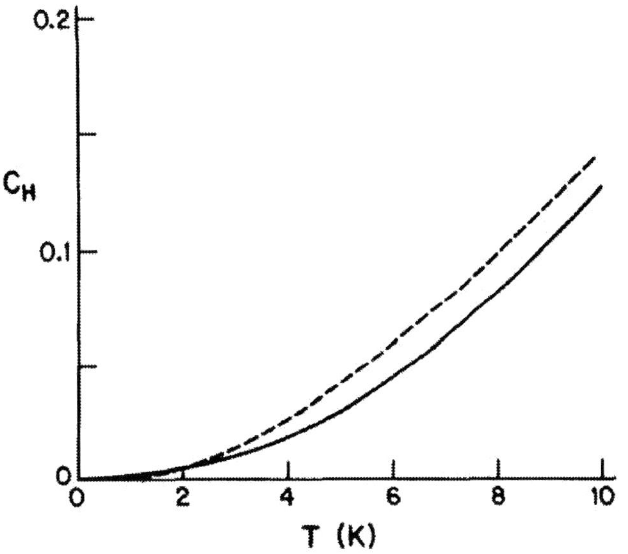

Fig. 4. Magnetic specific heat spin vs temperature; x=0.65. ___ B=0; _ _ _ B=100T. Results for B=10T are graphically indistinguishable from the curve for B=0.

IV. DISCUSSION

Perhaps the most surprising feature of Figs. 1-4 is the relative insensitivity of the specific heat to applied fields. Only in the most dilute system are there significant differences between B=0 and B=10T. Generally speaking, this behavior can be traced to the "rigidity" of the magnon distribution in the presence of applied fields which are weak in comparison with the mean nearest-neighbor exchange field. In the case of Type III order the latter takes on the values 19T(x=0.20), 33T(x=0.35), 47T(x=0.50), and 60T(x=0.65). For $x \geq 0.35$ N(E) is globally insensitive to applied field for $0 \geq B \geq 10$T. Only in the most dilute system (x=0.20) are there qualitative changes in N(E) over this interval (cf Figs.5-7).

The behavior displayed in Figs. 1-4 stand in contrast to what has been found $Eu_xSr_{1-x}S$ [1-6]. As mentioned, the difference can be traced to a difference in the sign of the exchange interaction between nearest-neighbors. In $Eu_{0.25}Sr_{0.75}S$ and $Eu_{0.54}Sr_{0.46}S$ there is a gap in the density of states for $B \geq 1$-2T and complete alignment of the spins for B=4T [6]. Further increases in the field only produce a uniform shift in the eigenvalue distribution.

Since we have used a proven theory with experimental values for the exchange interactions, we expect our results to be qualitatively if not quantitatively accurate. Indications that this is the case come from comparisons with the results of Galazka et al.[9] who reported measurements of the specific heat for x = 0.2, 0.3, 0.5, and 0.7 with $0 \leq B \leq 2.8$T. In agreement with Figs. 3 and 4, they find no field dependence at x = 0.5 and 0.7 whereas the change in C_H at lower concentrations is qualitatively similar to that shown in Figs. 1 and 2.

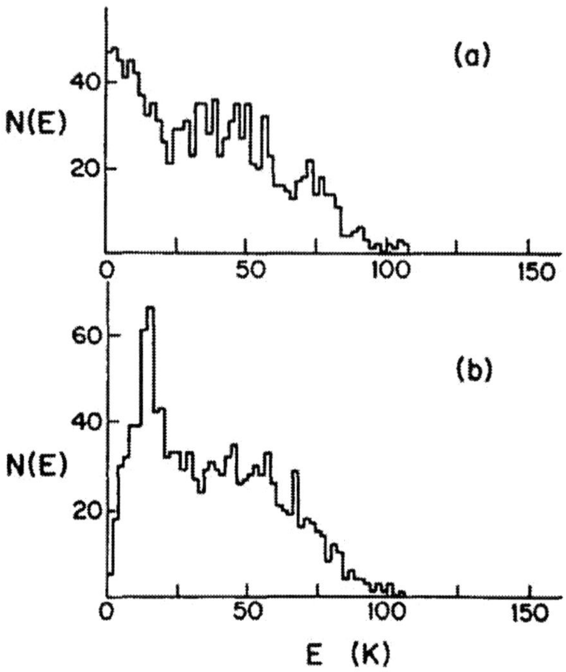

Fig. 5. Distribution of magnon modes; x=0.20 (a) B=0 ; (b) B=10T. Zeeman interaction (B=10T) = 13.4K.

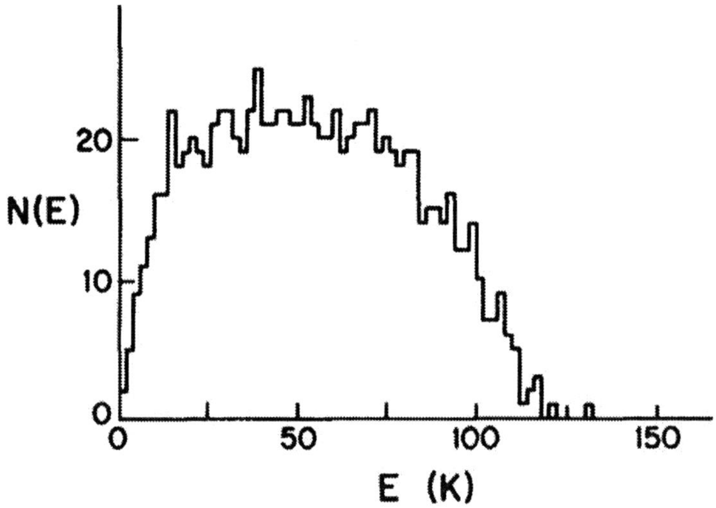

Fig. 6. Distribution of magnon modes; x=0.35; B=0. Virtually the same distribution is obtained with B=10T.

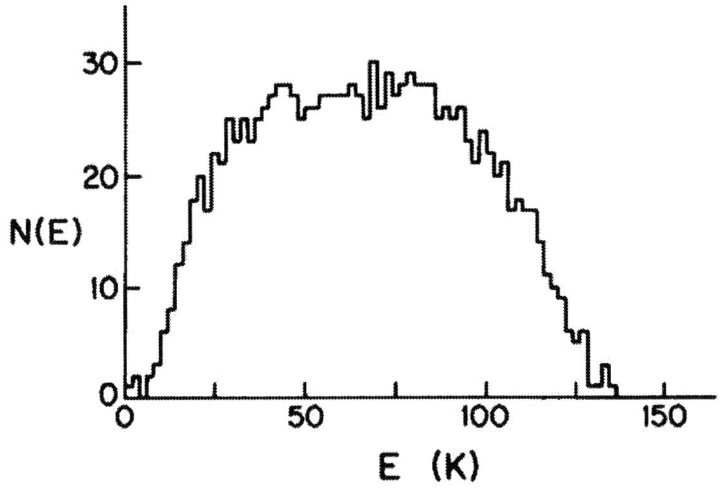

Fig. 7. Distribution of magnon modes; x=0.50; B=0. Virtually the same distribution is obtained with B=10T.

ACKNOWLEDGMENTS

Research partially supported by the NSF under the grant DMR-8203704. Computer time on the Cray X-MP was provided by the Office of Basic Energy Research, DOE.

REFERENCES

1. H. v. Lohneysen, R. van der Berg, G. V. Lecomte, and W. Zinn, Phys. Rev. B31, 2920 (1985).

2. H. v. Lohneysen, R. van der Berg, J. Wosnitza, G. V. Lecomte, and W. Zinn, J. Magn. Magn. Mater. 54-57, 189 (1986).

3. J. Wosnitza, H. v. Lohneysen, W. Zinn, and U. Krey, Phys. Rev. B33, 3436 (1986).

4. W. Y. Ching, D. L. Huber, and K. M. Leung, Phys. Rev. B21, 3708 (1980).

5. U. Krey, Z. Phys. B38, 243 (1980); 42, 231 (1981); J. Magn. Magn. Mater. 28, 231 (1982); J. Phys. (Paris) Lett. 46, L845 (1985).

6. W. Y. Ching and D. L. Huber, Phys. Rev. B34, 1960 (1986).

7. W. Y. Ching and D. L. Huber, Phys. Rev. B30, 179 (1984).

8. T. M. Giebultowicz, J. J. Rhyne, W. Y. Ching, D. L. Huber, and R. R. Galazka, J. Magn. Magn. Mater. 54-57, 1149 (1986).

9. R. R. Galazka, S. Nagata and P. H. Keesom, Phys. Rev. B22, 3344 (1980).

FROM CONCENTRATED TO DILUTE MAGNETIC SEMICONDUCTOR: OLD CONCEPTS AND NEW IDEAS

S. von MOLNAR*
*IBM Thomas J. Watson Research Center, Yorktown Heights, NY 10598

INTRODUCTION

Magnetic semiconductors[1,2] have enjoyed a renaissance with the discovery by Komarov et al. in 1977[3] of very large enhancements of magneto-optical properties in Mn doped CdTe. Although these magnetic semiconductors are dilute, that is to say they are solid solutions of II-VI or IV-VI semiconductors in which the cations have been replaced (randomly) by divalent transition metal or rare earth ions, their physical properties may be understood, in general, with concepts developed for the concentrated systems, for example the europium chalogenides. It is the purpose of this paper to review some of these ideas and point out new developments which have broadened our understanding, particularly as regards magnetic polarons, the highly correlated ferromagnetic region surrounding trapped carriers in an otherwise paramagnetic (P.M.) or antiferromagnetic (A.F.) semiconducting hosts.

Towards this end, the body of the paper begins with a review of direct ion-ion magnetic exchange in both concentrated and dilute systems. It is shown that the interactions between ionic moments and conduction carriers is vital in explaining direct exchange in Eu compounds and that it introduces a range of electronic phenomena such as giant magnetoresistance, strong magneto-optical effects, metal-non metal transitions and magnetic polaron formation, all of which are absent in conventional semiconductors. The discussion ends with a simple description of magnetic polarons. Evidence for these particles in inelastic light scattering, luminescence and transport will also be briefly reviewed. Much of the new physics in magnetic semiconductors is evolving from these latter studies, in particular the dynamical optical and magnetic studies.

EXCHANGE IN MAGNETIC SEMICONDUCTORS:

1) Ion-Ion Exchange

Magnetic properties of the canonical magnetic semiconductors, the europium chalcogenides (EuX, where X=0, S, Se, Te), have been examined in detail by many methods including Faraday balance,[4] neutron diffraction,[5] and most recently spin resonance.[6] EuX has the rock-salt structure. The seven 4f electrons of the Eu^{2+} ion are very tightly bound to the core with radial probability peaking at approximately 1 Å[1], and the Hundt's rule coupled ground state is a spherically symmetrical spin only $^8S_{7/2}$. Thus crystalline field anisotropies are negligible. The exchange Hamiltonian may be expressed in the Heisenberg form, that is

$$\mathcal{H}_{ex} = -\sum_{i,j} J_{ij}\vec{S}_i \cdot \vec{S}_j, \qquad (1)$$

Mat. Res. Soc. Symp. Proc. Vol. 89. © 1987 Materials Research Society

where J_{ij} are the exchange constants coupling the two body spin-spin interactions $\bar{S}_i \cdot \bar{S}_j$ with $|S| = 7/2$. The analysis of neutron data by Passell, et al.[5] for nearest (nn) and next nearest neighbor (nnn) interactions, J_1 and J_2, are shown in Fig. 1 as a function of anion-cation separation. This analysis does not consider longer range exchanges.[6] J_1 is seen to be associated with a ferromagnetic direct exchange between Eu ions which decreases dramatically with increasing Eu-Eu distance, whereas J_2 is an indirect exchange through the anions, which is less distance sensitive, and is negative for all the compounds except EuO.

The microscopic theory describing the variation of J_1 and J_2 was given by Kasuya[7] and includes, among other things, a prediction for positive J_2 in EuO (before the sign had been established experimentally). For J_1 the theory involves intrasite transfer, M, of a 4f electron to an empty 5d state of the cation. In the excited state the electron experiences d-f exchange, I_{df}, with neighboring Eu cations and then relaxes back to the ground $4f^7$ configuration. Such tortuous exchange is necessary because direct f-f overlap leading to ferromagnetic coupling is impossible with a 1 Å radial extent. Thus

$$J_1 = \frac{2I_{df}|M|^2}{SU^2}, \tag{2}$$

where U is the energy difference between the f and d states. The dependence of J_1 on lattice constant clearly depends both on M through the overlap integral and on U through crystal field splitting of the 5d states. Kasuya[7] estimated a dependence proportional to $\exp(-8R/R_0)$, and, with R_0 the nearest neighbor Eu distance for EuO, he was able to fit the data for J_1 in Fig. 1. The theory for J_2 is more complex, involving both the ordinary Kramers - Anderson superexchange via the occupied p orbitals of the chalcogen ligands and higher order terms. These include virtual excitations of the p states to nearest

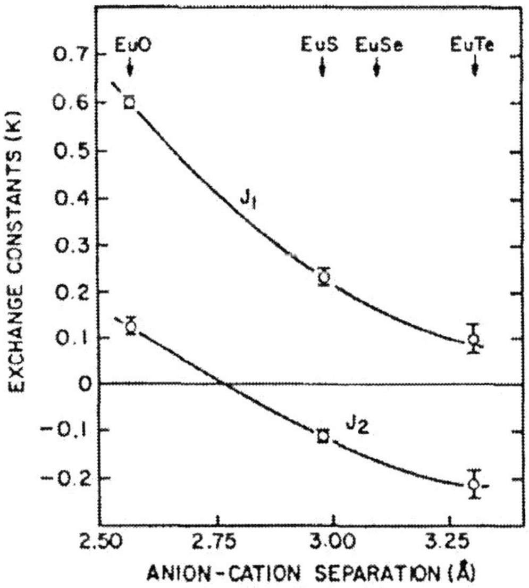

Fig. 1. Exchange constants in Eu monochalcogenides versus anion-cation separation. The solid lines have no theoretical significance (from ref. 5).

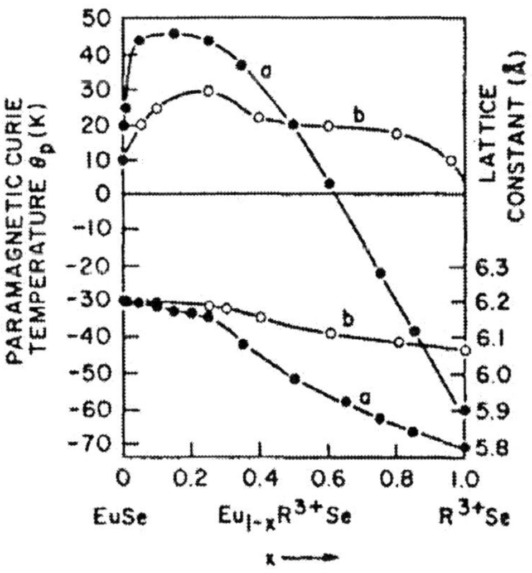

Fig. 2. Paramagnetic Curie temperature and lattice constant for: (a) $Eu_{1-x}Gd_xSe$ and (b) $Eu_{1-x}La_xSe$ versus x (from ref. 15).

neighbor Eu s or d states and the subsequent spin alignment through the strong $I_{s,d-f}$ exchange. A cross term between these two mechanisms is also considered. It turns out that the second and third terms are the most important for fcc EuX, so that both J_1 and J_2 depend strongly on $I_{s,d-f}$. The importance of $I_{s,d-f}$ and its analogue in the $II_{1-x}Mn_xVI$ dilute magnetic semiconductors (DMS), I_{sp-d}, cannot be overemphasized. It distinguishes the magnetic from ordinary semiconductors.

The magnetic phase diagram of $Cd_{1-x}Mn_xTe$ is complex. In contrast to the case for Eu compounds both nn and nnn exchange, J_1 and J_2, are antiferromagnetic. The II-VI materials become spin glasses at concentrations much smaller than percolation arguments would suggest, indicating that longer range then nn play a role in the freezing process.[8] No evidence for long range AF order is observed, even near $x \lesssim 0.7$,[9] above which Mn concentration the material undergoes a structural transformation from zinc-blende to nickel-arsenide. Magnetic measurements, in particular the original work by Oseroff,[10] constituted the principal early experimental evidence for spin glass behavior. Shapira, et al.[11] and Aggarwal, et al.[12] derived values for J_1 and J_2 from steps in the field dependence of magnetization and magnetoreflectance. Lewicki, et al.[13] were able to conclude from their analysis of the dependence of paramagnetic Curie temperature, θ, as a function of cation-cation distance, that the dominant contribution to J_1 in $II_{1-x}Mn_xVI$ alloys is superexchange, comparable to J_2 in EuX. A summary of experimental magnetic parameters for both concentrated and dilute systems in given in Table I.

The first evidence, however, for the dominance of superexchange for J_1 and J_2 in $II_{1-x}Mn_xVI$ alloys were theoretical arguments presented by Larson, et al.[14] who found that valence band hybridization with the Mn d states makes the most important contribution. They predict that

$$J_1 = J_{d-d} \propto [I^h_{sp-d}]^2. \tag{3}$$

Here the superscript "h" refers to holes in the valence band and points towards one of the clear differences between EuX and DMS, which is that valence band interactions are stronger and have much greater influence on the physical properties in DMS. For the Eu chalcogenides electron exchange, $I_{s,d-f}$, is most important (see Table I).

TABLE I

Experimental values for various exchange constants in magnetic semiconductors.

	J_1(nn) (K)	J_2(nnn) (K)	$I_{s,d-f}$ (eV)	$I^e_{s,d-f}$ (eV)	$I^h_{s,p-d}$ (eV)
EuX	0.2 to 0.6*	-0.2 to (+0.12)*	0.05 to 0.1+		
$Cd_{1-x}Mn_xTe$	−6.3†	−1.9†		0.22Δ	−0.88Δ

* from ref. 5   + from ref. 18   Δ from ref. 27
† B. E. Larson, K. C. Haas and R. L. Aggarwal, Phys. Rev. B 33, 1789 (1986).

2) Ion-Carrier Exchange

Since, according to Eqs. (2) and (3) even direct exchange depends on $I_{s,d-f}$ or I^h_{sp-d}, it is expected that the presence of conduction electrons (holes) can modify the average magnetic properties or, alternatively, that the magnetic state of the system will affect the carriers. Holtzberg, et al.[15] were the first to explore this idea by studying the EuX – R.E.$^{3+}$X alloy system. The trivalent rare earth, R.E.$^{3+}$, acts as a donor in EuX and the initial rise in θ (see Fig. 2) verifies the powerful effect of the electrons. Experiments based on similar concepts have recently been performed in PbSnMnTe[16] in which the influence of p-type charge carriers in excess of $3 \times 10^{20}\,cm^{-3}$ converts a dilute paramagnetic alloy containing only 3% Mn into an ordered ferromagnet.

There are numerous examples of the effects of magnetism on carrier properties in EuX and the present selection is necessarily limited and subjective. Among transport effects, some of the most dramatic are in defect doped EuO. The first observation of a sharp resistance decrease by a factor of 10^9 in cooling EuO through the FM ordered state by Oliver, et al.[17] Penney et al.[18] made a direct correlation between the activation energy for transport and magnetization in the ordered state and determined $I_{s,d-f}$ directly from their data. Shapira, et al.[19] were able to extend the analysis to the paramagnetic region by including short range order effects.

In it simplest form, the interaction Hamiltonian from the perspective of the carrier spin, $\vec{s}$, is given by

$$\mathcal{H}_I = -I\sum_j [(\vec{s} \cdot \vec{S}_j)\delta(\vec{r} - R_j)], \tag{4}$$

which, in the molecular field approximation, reduces to an energy splitting of the conduction band $\Delta E = I < S_z >$ for magnetically ordered EuX. Thus the activation energy for transport was interpreted as the difference between localized donor states and the conduction band bottom, which varies linearly with $< S_z >$ and thus the magnetization. Another dramatic, and perhaps the first direct observation of band splitting effects came from the temperature dependence of the optical absorption in EuX[20,21] (see Fig. 3) in which the red shift below the magnetic ordering temperature, T_c, is a direct measure of $I = \Delta E/ < S_s > \sim 0.1\,eV$. Although there remains some controversy as to the exact nature of the optical transition, whether band or excitonic,[22] the result of the magnetic interaction is incontrovertible. Other direct evidence for Eq. (4) comes from Fowler-Nordheim tunneling.[23] This technique exploits the fact that electrons from one electrode of a very thin capacitor device, appropriately biased, may tunnel into the empty conduction band states of the insulator (in this case EuS or EuSe). The I-V characteristics thus produced give a direct measure of the energy difference between the electrode Fermi level and the conduction band bottom, ϕ. As magnetic order occurs, either with decreasing temperature or increasing magnetic field, B, (see also the B dependence of the optical absorption in Fig. 3) the energy difference ϕ decreases, which is reflected in the I-V characteristics, specifically the applied voltage at constant current (see Fig. 4). It is also worth noting that there should exist bias conditions for which the current is carried by electrons of one spin polarization only.

All of the examples described for EuX have an analogue in DMS. The materials are generally paramagnets or spin glasses, however, and an applied magnetic field is necessary to develop a net magnetic moment, $< S_z >$. This is not necessarily true when one considers only the local environment and further comments concerning local fluctuations will be deferred until a discussion of magnetic polarons in DMS. In a magnetic field a

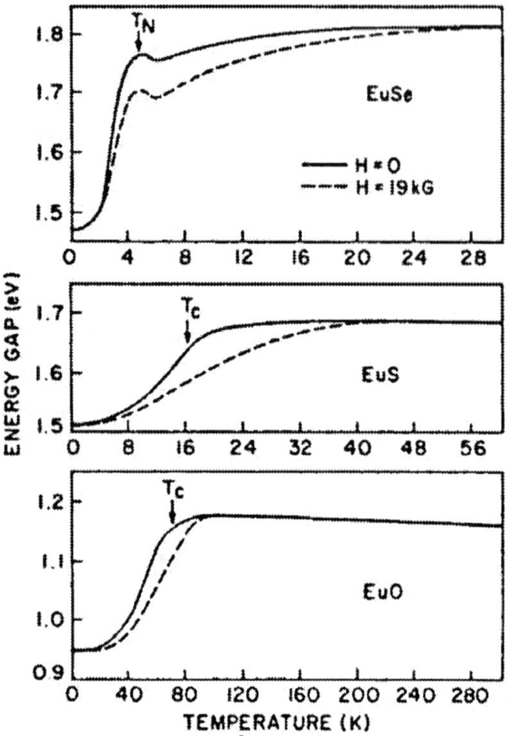

Fig. 3. The red shift of the optical absorption with magnetization (from ref. 21).

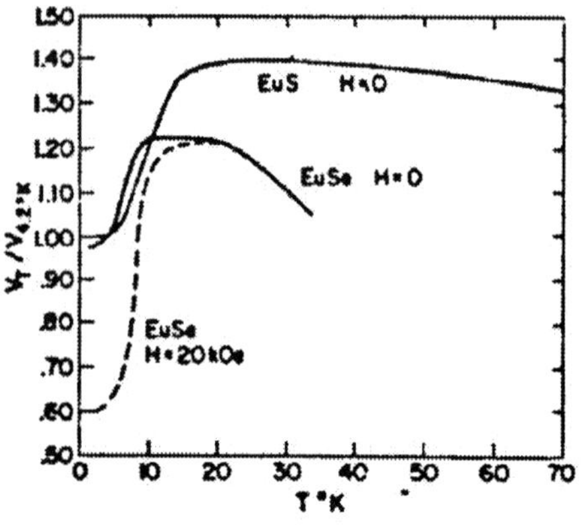

Fig. 4. $V_T/V_{4.2K}$, the voltage at a given current level (normalized to its value at 4.2K), versus temperature (from ref. 23).

term $-\mu_B g^*(\bar{s}\cdot\bar{B})$ has to be added to Eq. (4) where μ_B =Bohr magneton and g^* is the spectroscopic splitting factor for the carriers. Once again utilizing the mean field approximation (ignoring fluctuations) the energy splitting may be expressed as[24]

$$\Delta E \simeq 2\mu_B[g^* + \frac{I}{kT_{eff}} \times \frac{35}{12}\bar{x}g_{Mn}]\bar{s}\cdot\bar{B} \equiv \mu_B\tilde{g}B \tag{5}$$

where $< S_z >$ has been evaluated in the small field limit. Here T_{eff} and $\bar{x}$ are parameters which depend on the true transition metal (Mn^{2+}) concentration, x. Although g^* varies between 0.5 and 2, the second term in the bracket can be as high as several hundred at low temperature, which is the principal reason that these materials display physical properties very similar to the concentrated EuX. Magnetotransport properties of $Cd_{1-x}Mn_xSe$[25,26] initially studied by Shapira et al.,[25] can be explained qualitatively in terms of a band splitting model.[19] An elegant analysis by Gaj et al.[27] extracted values for $I^e_{s,p-d}$ and $I^h_{s,p-d}$ from the magneto-optical properties of free excitons in $Cd_{1-x}Mn_xTe$. In effect they ignored the first very small term in Eq. (5) and incorporated $< S_z >$ explicitly, with the result that $\Delta E_c = I^e_{s,p-d}x < S_z >$. Simultaneous measurements of magnetoreflectivity and magnetic moment consequently not only confirmed the exchange interaction model but also provided precise values for the exchange constants (see Fig. 1 and Table I). Fowler-Nordheim tunneling has also been observed in $Cd_{1-x}Mn_xTe$.[28] The total splitting is, however, much smaller than in EuX because $x < S_z >$ is reduced by at least a factor of 10.

MAGNET POLARONS

As was alluded to earlier, local fluctuations and magnetic correlations in the neighborhood of charged centers may play an important role in magnetic semiconductors. Magnetic polarons were first suggested to explain transport effects in EuX by von Molnár and Methfessel[29] and were studied theoretically in a particularly simple case by Kasuya.[30] It was imagined that an electron is injected into an otherwise perfect magnetic insulator. With this hypothesis it could be shown that in ferromagnets, the polaron can exist only very close to T_c, where the susceptibility diverges. In real lattices, however, both coulombic potentials and magnetic interactions coexist and a more appropriate description must include coulombic forces. The earliest description of these "magnetic impurity states" was given by Kasuya and Yanase.[31]

In antiferromagnetic lattices magnetic polarons, i.e., essentially ferromagnetic clusters surrounding charge carriers, are expected to be stable below the Néel temperature, T_N, and therefore amenable to study.[32] A field induced metal-insulator phase transition observed in AF $Gd_{3-x}v_xS_4$ [33,34] was shown to be due to such polaronic effects and Penney et al.[34] determined the polaron size by comparison of transport and magnetic properties. Ultra-low temperature studies exploited this field induced transition to clarify the three dimensional localization transition.[35] Two examples in EuTe ($T_N \sim 10K$) stand out as other vivid demonstrations of the existence of polarons. The first was the observation by Busch et al.[36] that the photoluminescence of EuTe in the A.F. state decreases in intensity with both increasing temperature and field. They interpreted these results as being due to polaron unbinding, the recombination probability decreasing as the optically excited electron became free to move farther away from the (localized) 4f hole. Similar effects are evident in transport studies of lightly doped EuTe by Shapira et al.[37] (see Fig. 5).

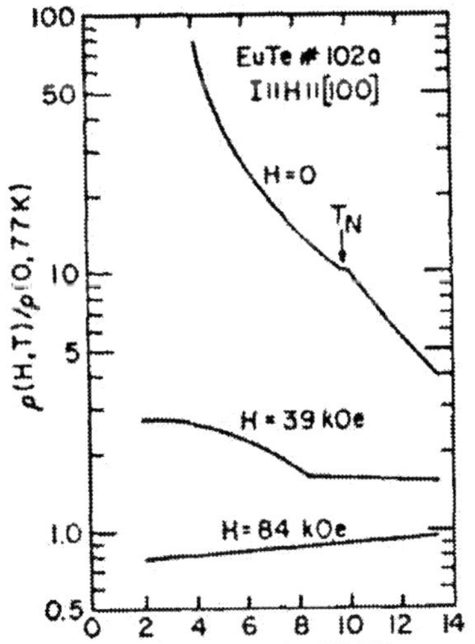

Fig. 5. Temperature variation of the resistivity for EuTe containing 8×10^{18} carriers/cm^3 (from ref. 37).

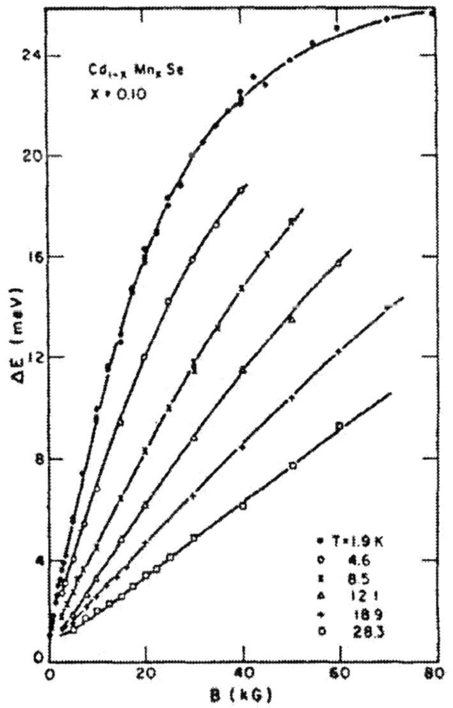

Fig. 6. Spin-flip Stokes energy ΔE vs applied field B from Raman scattering $Cd_{0.9}Mn_{0.1}Se$ (from ref. 24).

Although there exist recent magnet-transport studies in Mn doped II-VI semiconductors which are interpretable in terms of magnetic polaron formation,[38,39] the most direct evidence has come from optical studies. Both Dietl et al.[40] and Heiman et al.[24] independently developed theories involving magnetic polarons to explain spin-flip Raman data,[24,41] an example of which is shown in Fig. 6. It is to be noted, in particular, that the spin-flip energy is finite at zero field. This result proves the existence of local magnetization fluctuations and bound magnetic polarons, since spin-flip processes of the carriers require finite energy only in the presence of a field. The work of Heiman et al.,[24] furthermore, successfully describes both the average energy of the local field as well as the shape of the distribution, i.e., the linewidths of the Raman intensity. Another important contribution to an understanding of polaron formation has come from the detailed studies of the polarization of luminescence on $II_{1-x}Mn_xVI$ crystals on excitation with circularly polarized light. Warnock et al.,[42] formulated a realistic model for the statistics of polaron formation in a random alloy with locally fluctuating magnetic fields. Both chemical and thermal magnetic effects contribute to the binding energy, polaron size and net polarization, leading to predicted maximum polarization values of up to 50% and field dependencies which agree well with experiment.

Arguably the most novel developments in the physics of magnetic semiconductors have come with the advent of molecular beam epitaxy (MBE) and time resolved techniques. Among the former, the photoluminenscence studies of Zhang et al.[43] and Warnock et al.[44] established the importance of interfaces on polaron formation in MBE superlattices. Extensive reviews appear in this conference. Dynamical optical processes in transmission[45] and luminescence[46] have been studied at picosecond time scales and have been rèlated to polaron formation by monitoring the time evolution of the polaron binding energy. Direct observation of the spin dynamics responsible for polarons, however, have only recently been made possible by techniques developed by Awschalom et al..[47,48] The first of these, time resolved Faraday rotation,[47] measured the time evolution of magnetization induced optically by excitation near band edge in the presence of an applied field. An example of the data, Fig. 7, shows the polaron formation time of order 200 to 400 psec, followed by a decay of about 1 n sec which is related to the exciton recombination.

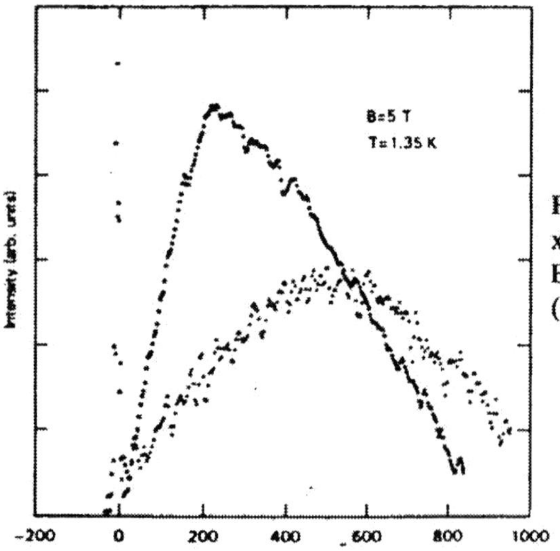

Fig. 7 Time-resolved Faraday rotation for x=0.18 (circles) and x=0.115 (triangles). E_{pump} = 1.839 and 1.754 eV, respectively (from ref. 47).

The second[48] employs a planar dc SQUID as a detector to measure the static and dynamic induced magnetization as a function of impinging radiant energy.[49] This magnetic spectroscopy promises to be an important new tool in magnetic research in general, and magnetic semiconductors in particular. Previous static measurements of optically induced magnetization at fixed energy have already been successfully performed with rf SQUIDS.[49]

It is clear that DMS research has borrowed heavily from earlier work in concentrated systems. Above all it has been shown that carrier-ion exchange dominates the physics of magnetic semiconductors, including magnetic polaron formation. Examples of new experimental techniques and improved theoretical descriptions applied to DMS demonstrate that magnetic semiconductors continue to be an important part of magnetic research.

I am grateful to my colleagues D. D. Awschalom, T. Penney and J. Warnock for their many instructive comments and discussions.

REFERENCES

1. For reviews, see, e.g.: S. Methfessel and D. C. Mattis, Handbuch der Physik, Vol. 18, ed. S. F. Flugge (Springer Verlag, Berlin, 1968); F. Holtzberg, S. von Molnár and J. M. D. Coey, in Handbook on Semiconductors, Vol. 3, ed. T. S. Moss (North-Holland, Amsterdam, 1980), pp. 803-856.
2. J. K. Furdyna, J. Appl. Phys. **53**, 7637 (1982); see also N. B. Brandt and V. V. Moshalkov, Adv. Physics **33**, 193 (1984).
3. A. V. Komarov, S. M. Ryabschenko, O. V. Terletskii, I. I. Zheru and R. D. Ivanchuck, Zh. éksp. teor. Fiz. **73**, 608 (1977).
4. T. R. McGuire, B. E. Argyle, M. W. Shafer and J. S. Smart, J. Appl. Phys. **34**, 1345 (1963).
5. L. O. Passell, O. W. Dietrich and J. O. Als-Nielsen, Phys. Rev. B **14**, 4897 (1976); see also J. O. Als-Nielsen, O. W. Dietrich and L. O. Passell, ibid., 4908, and O. W. Dietrich, J. O. Als-Nielsen and L. O. Passell, ibid., 4923.
6. H. G. Bohn, W. Zinn, B. Dorner and A. Kollmar, Phys. Rev. B **22**, 5447 (1980).
7. T. Kasuya, IBM J. Res. Develop. **14**, 214 (1970).
8. W. J. M. de Jonge, A. Twardowski, C. J. M. Denissen and J. H. M. Swagten, paper Q3.12, this conference; see also C. J. M. Denissen and W. J. M. de Jonge, Solid State Commun. **59**, 503 (1986).
9. T. M. Giebultowicz and T. M. Holden, paper Q1.8, this conference.
10. S. B. Oseroff, Phys. Rev. B **25**, 6584 (1982).
11. Y. Shapira, S. Foner, D. H. Ridgley, K. Dwights and A. Wold, Phys. Rev. B **30**, 4021 (1984); R. L. Aggarwal, S. N. Jasperson, P. Becla and R. R. Galazka, Phys. Rev. B **32**, 5132 (1985).
12. R. L. Aggarwal, S. N. Jasperson, J. Stankiewicz, Y. Shapira, S. Foner, B. Khazai and A. Wold, Phys. Rev. B **28**, 6907 (1983).
13. A. Lewicki, J. Spalek, J. K. Furdyna and R. R. Galazka, J. Mag. Mag. Mat. **54-57**, 1221 (1986).
14. B. E. Larson, K. C. Haas, H. Ehrenreich and A. E. Carlsson, Solid State Commun. **56**, 347 (1985); J. Masek and B. Velicky, private communication.

15. F. Holtzberg, T. R. McGuire, S. Methfessel and J. C. Suits, Phys. Rev. Lett. **13**, 18 (1964).

16. T. Story, R. R. Galazka, R. B. Frankel and P. A. Wolff, Phys. Rev. Lett. **56**, 777 (1986).

17. M. R. Oliver, J. O. Dimmock, A. L. McWorter and T. B. Reed, Phys. Rev.B **5**, 1078 (1972).

18. T. Penney, M. W. Shafer and J. B. Torrance, Phys. Rev. B **5**, 3669 (1972).

19. Y. Shapira, S. Foner, R. L. Aggarwal and T. B. Reed, Phys. Rev. B **8**, 2316 (1973).

20. S. Methfessel, Z. angew. Phys. **18**, 414 (1965); S. Methfessel, F. Holtzberg and T. R. McGuire, IEEE Trans. Mag. **MAG-2**, 305 (1966); M. J. Freiser, S. Methfessel and F. Holtzberg, J. Appl. Phys. **39**, 900 (1968).

21. G. Busch and P. Wachter, Phys. Kondens. Mat. **5**, 232 (1966), Z. angew. Phys. **26**, 1 (1968).

22. T. Kasuya, C. R. C. Critical Reviews in Solid State Science **3**, 131 (1972).

23. L. Esaki, P. J. Stiles and S. von Molnár, Phys. Rev. Lett. **19**, 852 (1967).

24. D. Heiman, P. A. Wolff and J. Warnock, Phys. Rev. B **27**, 4848 (1983).

25. Y. Shapira, D. H. Ridgley, K. Dwight, A. Wold, K. P. Martin and J. S. Brooks, J. Appl. Phys. **57**, 3210 (1985).

26. J. Stankiewicz, S. von Molnár and W. Girat, Phys. Rev. B **33**, 3573 (1986).

27. J. A. Gaj, R. Planel and G. Fishman, Solid State Commun. **29**, 435 (1979).

28. T. Siegrist, S. von Molnár and F. Holtzberg, Appl. Phys. Lett. **47**, 1087 (1985).

29. S. von Molnár and S. Methfessel, J. Appl. Phys. **38**, 959 (1967).

30. T. Kasuya, A. Yanase and T. Takeda, Solid State Commun. **8**, 1543 (1970).

31. T. Kasuya and A. Yanase, Rev. Mod. Phys. **40**, 684 (1968).

32. T. Kasuya, Solid State Commun. **8**, 1635 (1970).

33. S. von Molnár, F. Holtzberg, T. R. McGuire and T. J. A. Popma, A.I.P. Conf. Proc. **5**, 869 (1972); S. von Molnár and F. Holtzberg, ibid, **10**, 1259 (1973).

34. T. Penney, F. Holtzberg, L. J. Tao and S. von Molnár, A.I.P. Conf. Proc. **18**, 908 (1974).

35. S. von Molnár and T. Penney, in Localization and Metal - Insulator Transitions, ed. H. Fritzsche and D. Adler (Plenum Publ. Corp., New York, 1985), p.183.

36. G. Busch, P. Streit and P. Wachter, Solid State Commun. **8**, 1759 (1970).

37. Y. Shapira, S. Foner, N. F. Olivera, Jr. and T. B. Reed, Phys. Rev. B **5**, 2647 (1972).

38. M. Sawicki, T. Dietl, J. Kossut, J. Igalson, T. Wojtowicz and W. Plesiewicz, Phys. Rev. Lett. **56**, 508 (1986).

39. T. Wojtowicz, T. Dietl, M. Sawicki, W. Plesiewicz and J. Jaroszynski, Phys. Rev. Lett. **56**, 2419 (1986).

40. T. Dietl and J. Spalek, Phys. Rev. Lett. **48**, 355 (1982).

41. M. Nawrocki, R. Planel, G. Fishman and R. R. Galazka, Phys. Rev. Lett. **46**, 735 (1981).

42. J. Warnock, R. N. Kershaw, D. Ridgely, K. Dwight, A. Wold and R. R. Galazka, J. Luminescense **34**, 25 (1985).

43. X. C. Zhang, S. K. Chang, A. V. Nurmikko, L. A. Kolodziejski, R. L. Gunshor and S. Datta, Phys. Rev. B **31**, 4056 (1985).

44. J. Warnock, A. Petrou, R. N. Bicknell, N. C. Giles-Taylor, D. K. Blanks and J. F. Schetzina, Phys. Rev. B **32**, 8116 (1985).

45. J. H. Harris and A. V. Nurmikko, Phys. Rev. Lett. **51**, 1472 (1983).

46. J. J. Zayhowski, C. Jagannath, R. N. Kershaw, D. Ridgley, K. Dwight and A. Wold, Solid State Commun. **55**, 941 (1985).

47. D. D. Awschalom, J. M. Halbout, S. von Molnár, T. Siegrist and F. Holtzberg, Phys. Rev. Lett. **55**, 1128 (1985).

48. D. D. Awschalom, J. Warnock and S. von Molnár, to be published; D. D. Awschalom and J. Warnock, contribution Q2.4, this conference.

49. H. Krenn, W. Zawadzki and G. Bauer, Phys. Rev.Lett. **55**, 1510 (1985).

RAMAN SCATTERING BY MAGNETIC EXCITATIONS IN DILUTED MAGNETIC SEMICONDUCTORS

A.K. RAMDAS AND S. RODRIGUEZ
Department of Physics, Purdue University, West Lafayette, IN 47907 U.S.A.

ABSTRACT

Raman scattering provides significant information on the nature of the magnetic excitations of diluted magnetic semiconductors (DMS). Transitions involving an exchange of a quantum of angular momentum between the system and the radiation field result in Raman lines with shifts equal to the energy of the magnetic excitations in which the total spin of the crystal changes by $\hbar$. Such transitions have a one-to-one correspondence with those seen in magnetic resonance but observed in the optical rather than in the microwave region of the electromagnetic spectrum. In the Mn-based II-VI DMS a variety of magnetic excitations have been observed in their Raman spectra. Microscopic models underlying these phenomena provide powerful insights into the nature of the magnetism of DMS.

INTRODUCTION

The incomplete d-shell of the magnetic atoms in a diluted magnetic semiconductor (DMS) gives rise to a variety of properties in which their localized magnetic moments play important roles, either individually or collectively through their mutual interactions. For sufficiently low concentrations, x, of the magnetic ions, or at temperatures above a critical temperature T_N, the material is in a paramagnetic phase. Temperature T_N is, of course, a function of x. A transition to an ordered phase occurs below T_N. In these materials, the interaction between neighboring magnetic ions is antiferromagnetic, giving rise to a spin glass or to an antiferromagnetic phase [1].

The theory of Raman scattering by magnetic excitations is similar to that by phonons [2]. However, because of the axial nature of the magnetic field and of the magnetization $\vec{M}$, the selection rules for Raman scattering differ from those associated with symmetric polarizability tensors.

In a magnetic system, the electric susceptibility ($\underset{\sim}{\chi}$) is a functional of the magnetization ($\vec{M}$) as well as of other variables describing the internal motion. The macroscopic electric polarization ($\vec{P}$) is then given by $\underset{\sim}{\chi}(\vec{M})\cdot\vec{E}_L$, where $\vec{E}_L = \vec{E}_0 \exp[-i\omega_L t]$ is the electric field of the incident radiation.

Of the several microscopic mechanisms for $\underset{\sim}{\chi}(\vec{M})$, exchange interactions with itinerant or localized electrons, having energies comparable to electrostatic interactions, are expected to be most important. This suggests that

Mat. Res. Soc. Symp. Proc. Vol. 89. ©1987 Materials Research Society

the Raman features associated with magnetic excitations in DMS should exhibit strong resonance enhancement when the energy of the incident radiation is near the energy of an electronic transition e.g., the direct energy gap.

The phenomenological theory of the modulation of $\vec{P}$ by magnetic excitations yields the term linear in $\vec{M}$

$$\vec{P}^{(1)} = iG\,\vec{M} \times \vec{E}_L \quad , \tag{1}$$

where G is a constant and a scattering cross-section (σ) for the first order Raman process of the form

$$\sigma = C\left|(\hat{\epsilon}_S \times \hat{\epsilon}_L)\cdot\vec{M}\right|^2 \tag{2}$$

where C is an appropriate function of ω_L and ω_S, the incident and scattered frequencies, respectively, and $\hat{\epsilon}_L$ and $\hat{\epsilon}_S$, the corresponding directions of polarization. Thus Raman scattering does not occur when the polarizations of the incident and scattered radiation are parallel. In the presence of a magnetic field $\vec{H}$, $\vec{M}$ varies according to the Bloch equation $d\vec{M}/dt = \gamma\vec{M} \times \vec{H}$, where γ = gyromagnetic ratio; $\gamma = -ge/2mc$ is negative for electrons. The solutions for the Bloch equation, considering only inelastic scattering and $\vec{H} \| \hat{z}$, are such that $M_x \pm iM_y$ vary as $\exp[\mp i\gamma Ht]$, respectively. Let $\Omega = -\gamma H$ be the Larmor frequency and θ, the angle between $\vec{M}$ and $\vec{H}$; consider circularly polarized radiation $\hat{\sigma}_+$(+ helicity) and $\hat{\sigma}_-$(- helicity) incident along $\hat{z} \| \vec{H}$ i.e., with $\vec{E}_L = (\hat{x} \pm i\hat{y})E_0 \exp[-i\omega_L t]$. One can then show

$$\vec{P}^{(1)} = \mp \hat{z}\; G\; M\; E_0 \sin\theta \exp[-i(\omega_L \mp \Omega)t] \quad . \tag{3}$$

Thus in this geometry a Stokes line appears in $(\hat{\sigma}_+,\hat{z})$ and an anti-Stokes in $(\hat{\sigma}_-,\hat{z})$. In a similar way, for incident light linearly polarized along $\hat{z}$ propagating normal to $\vec{H}$ i.e. for $\vec{E}_L = E_0\,\hat{z}\exp[-i\omega_L t]$,

$$\vec{P} = \frac{1}{2} G\; M\; E_0 \sin\theta\, [(\hat{x} - i\hat{y}) \exp[-i(\omega_L - \Omega)t] - (\hat{x} + i\hat{y}) \exp[-i(\omega_L + \Omega)t]] \quad . \tag{4}$$

Thus the Stokes and the anti-Stokes lines occur in the geometries $(\hat{z},\hat{\sigma}_-)$ and $(\hat{z},\hat{\sigma}_+)$, respectively. The two cases described by Eqs. (3) and (4) are related to one another by time reversal symmetry.

In order to describe the magnetic excitations observed in Raman scattering in terms of microscopic models, it is useful to consider the Hamiltonian of the magnetic ions, say Mn^{2+}, interacting with one another and with either band electrons or electrons bound to donors. We designate the spin of a Mn^{2+} ion at the site $\vec{R}_i$ by $\vec{S}_i$ and the spin of the electron by $\vec{s}$. In the presence of a magnetic field $\vec{H}$, the Zeeman energies of the Mn^{2+} ion and of the electron are $g\mu_B\vec{H}\cdot\vec{S}_i$ and $g^*\mu_B\vec{H}\cdot\vec{s}$ where μ_B is the Bohr magneton and g and g* are the Lande g-factors of Mn^{2+} and the electron, respectively. In addition there are exchange interactions between Mn^{2+} ions at $\vec{R}_i$ and $\vec{R}_j$ of the form $-2J_{ij}\vec{S}_i\cdot\vec{S}_j$ and between an electron and the Mn^{2+} ions. The Hamiltonian of an electron in mutual interaction with Mn^{2+} ions is

$$H = -\alpha \sum_i \vec{S}_i\cdot\vec{s}\left|\psi(\vec{R}_i)\right|^2 + g^*\mu_B\vec{H}\cdot\vec{s} + g\mu_B\vec{H}\cdot\sum_i \vec{S}_i - \sum_{i<j} 2J_{ij}\vec{S}_i\cdot\vec{S}_j \quad . \qquad (5)$$

Here $\psi(\vec{R}_i)$ is the electronic wavefunction normalized over the primitive cell and evaluated at $\vec{R}_i$; αN_0 is the s-d exchange integral, N_0 being the number of primitive cells per unit volume.

In the rest of the paper we will consider the Mn^{2+} based II-VI DMS's of which $Cd_{1-x}Mn_xTe$ and $Cd_{1-x}Mn_xSe$ are, respectively, the zinc blende and the wurtzite prototypes.

PARAMAGNETIC PHASE

We first consider Raman transitions between Zeeman sublevels of the individual Mn^{2+} ions in an external magnetic field, the sample being in its paramagnetic phase [3]. In this phase the exchange interaction between Mn^{2+} ions is smaller than the thermal energy k_BT and the ions can be considered as being independent of one another. The $^6S_{5/2}$ ground state of the Mn^{2+} ion has a total S = 5/2, orbital angular momentum L = 0 and total angular momentum J = 5/2. In this subsection we will discuss $Cd_{1-x}Mn_xTe$ as an illustrative example. The cubic crystalline field (site symmetry T_d) splits the six-fold degenerate ground state into a Γ_8 quadruplet state at +a, and a Γ_7 doublet at -2a, where 3a is the crystal field splitting. This crystal field splitting of Mn^{2+} in CdTe is too small to be observed with the resolution of a standard Raman spectrometer and we treat the ground state of Mn^{2+} in $Cd_{1-x}Mn_xTe$ as an atomic $^6S_{5/2}$ level. The application of an external magnetic field, $\vec{H}$, results in the removal of the six-fold degeneracy of the ground state, the energy levels being $E(m_S) = g\mu_BHm_S$. Here m_S, the projection of $\vec{S}$ along $\vec{H}$, has the values -5/2,-3/2,...,+5/2. These energy levels form the Zeeman multiplet of the ground state of Mn^{2+}.

In the paramagnetic phase, Raman scattering associated with spin-flip transitions between adjacent sublevels of this multiplet has been observed by Petrou *et al*, [3]. The results in $Cd_{1-x}Mn_xTe$ are shown in Fig. 1 for x = 0.03 where two features have been identified as 'PM' and 'SF' respectively; we first discuss the line labeled 'PM'. As can be seen, a strong Stokes/

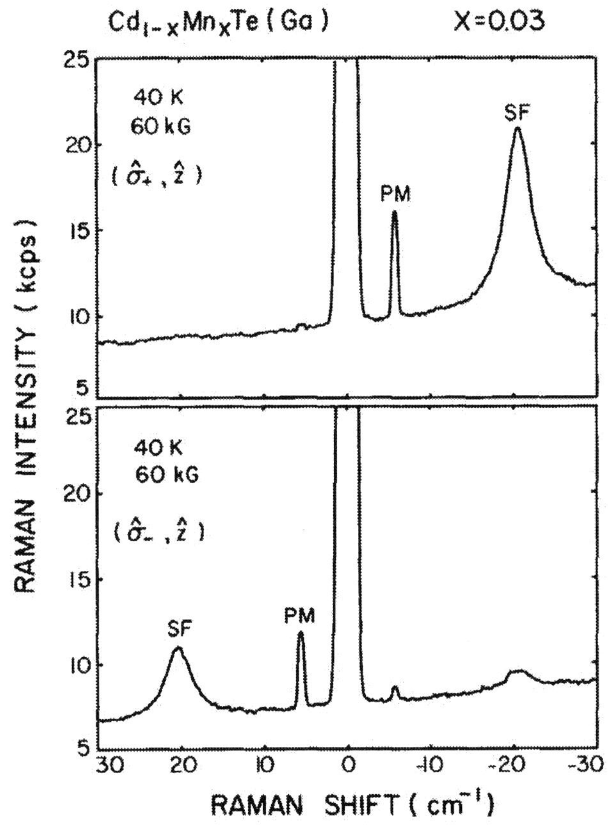

Figure 1

Raman spectra of $Cd_{1-x}Mn_xTe(Ga)$, x = 0.03, showing the $\Delta m_S = \pm 1$ transitions within the Zeeman multiplet of Mn^{2+} (PM) and the spin-flip of electrons bound to Ga donors (SF). kcps ≡ 10^3 counts/sec. [4]

anti-Stokes pair is observed with a Raman shift of $\omega_{PM} = 5.62 \pm 0.02$ cm^{-1} at room temperature and H =60 kG. Taking $\vec{H}$ and incident light parallel to $\hat{z}$, the Stokes line is observed in the $(\hat{\sigma}_+,\hat{z})$ configuration, whereas the anti-Stokes line is seen in $(\hat{\sigma}_-,\hat{z})$. When the incident light propagates at right angles to $\vec{H}(\hat{z})$, the Stokes component appears in the polarization $(\hat{z},\hat{\sigma}_-)$, while the anti-Stokes is observed in $(\hat{z},\hat{\sigma}_+)$. Within experimental error the frequency shift is linear in H. With the energy separation between adjacent sublevels of the Zeeman multiplet given by $\Delta E = g\mu_B H = \hbar\omega_{PM}$, it is found that g = 2.01 ± 0.02. The Raman line at ω_{PM} in $Cd_{1-x}Mn_xTe$ has been observed for a variety of compositions ranging from x = 0.01 to x = 0.70.

Exploiting the variation of the band gap with manganese concentration and/or temperature, it is possible to match the band gap of several samples with the energy of one of the discrete lines of a Kr^+ laser. In addition, a dye laser can also be used to achieve resonant conditions. It is found that the intensity of the ω_{PM} line increases by several orders of magnitude as the laser photon energy approaches that of band gap. The observation of this resonant enhancement in the intensity of the ω_{PM} Raman line suggests a mechanism involving interband transitions.It involves the Mn^{2+}-band electron exchange interaction described by the first term in the Hamiltonian in Eq. (5).

The term $\vec{S}_i \cdot \vec{s}$ can be written as $\vec{S}_i \cdot \vec{s} = S_i^{(z)} s^{(z)} + \frac{1}{2} S_i^{(+)} s^{(-)} + \frac{1}{2} S_i^{(-)} s^{(+)}$; here $S_i^{(\pm)}$ and $s^{(\pm)}$ are the spin raising and lowering operators for a Mn^{2+} ion and band electron, respectively, and $S_i^{(z)}$ and $s^{(z)}$ are the corresponding projections of spin along z. It can be shown that the second and the third terms in $\vec{S}_i \cdot \vec{s}$ can induce simultaneous spin-flips of the band electrons on the one hand and the Mn^{+} ions on the other, corresponding to $\Delta m_S(Mn^{2+}) = \pm 1$ and $\Delta m_J(e) = \mp 1$. All the observations discussed above and the predictions of the microscopic models considered are in accord with the general phenomenological selection rules [3]. We also note that this entire phenomenon is electron paramagnetic resonance observed as Raman shifts, i.e., it is Raman-EPR.

It is known that electrons and holes in polar crystals interact strongly with zone center longitudinal optical (LO) phonons through the Fröhlich interaction [2]. An LO phonon can be created or annihilated as a result of such an interaction. Referring to the interband mechanism responsible for the ω_{PM} line, one can visualize a fourth step in which the excited electron or hole interacts with the lattice and creates an LO phonon. Such a mechanism would result in a scattered photon with a Raman shift of $\omega_{LO} \pm \omega_{PM}$. The $\omega_{LO} + \omega_{PM}$ Stokes Raman line is expected to appear in the $(\hat{\sigma}_+, \hat{z})$ or the $(\hat{z}, \hat{\sigma}_-)$ configurations, whereas the $\omega_{LO} - \omega_{PM}$ Stokes line is allowed for $(\hat{\sigma}_-, \hat{z})$ or $(\hat{z}, \hat{\sigma}_+)$.

The new lines described above should occur only under conditions of band-gap resonance. Under such conditions we have indeed observed the new Raman lines with shifts of $\omega_{LO} \pm \omega_{PM}$ in $Cd_{1-x}Mn_xTe$ for a variety of compositions. The Raman spectra in the region of the longitudinal and transverse optical (TO) vibrational modes are shown in Fig. 2 for $Cd_{1-x}Mn_xTe$ with x = 0.10.

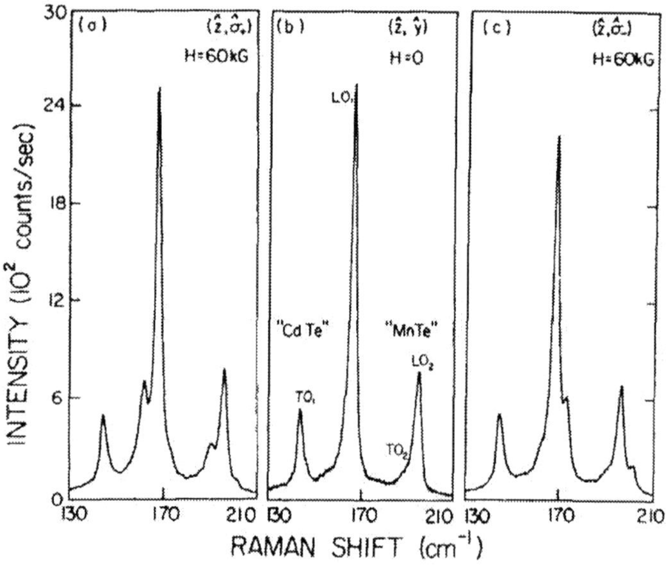

Figure 2
Raman spectra of $Cd_{1-x}Mn_xTe$, x = 0.10, showing the combination lines $\omega_{LO_1} \pm \omega_{PM}$ and $\omega_{LO_2} \pm \omega_{PM}$, where LO_1 and LO_2 are the "CdTe-like" and "MnTe-like" phonons. The sample temperature T = 120 K, H = 60 kG, and λ_L = 7525 Å; x, y, and z are along [110], [001], and [1$\bar{1}$0], respectively. (Petrou *et al.* [3].)

The zero magnetic field LO and TO phonon spectrum is shown in Fig. 2b. Here the "CdTe-like" TO and LO and the "MnTe-like" LO modes are quite distinct, while the "MnTe-like" TO appears as a shoulder to the LO. The corresponding Raman spectra, recorded in the presence of a magnetic field of 60 kG and in the $(\hat{z},\hat{\sigma}_+)$ and $(\hat{z},\hat{\sigma}_-)$ configurations, are presented in Fig. 2a and 2c, respectively. The additional Raman lines with Stokes shifts $\omega_{LO} \pm \omega_{PM}$ are clearly present with the proper polarization characteristics. The $\omega_{LO} \pm \omega_{PM}$ lines can be observed only when the exciting photon energy is strongly resonant with the band gap. There is no evidence of corresponding Raman lines associated with the TO phonons, which would have shifts of $\omega_{TO} \pm \omega_{PM}$. This supports the assumption that the Fröhlich interaction is responsible for the appearance of the new features.

The scattering amplitude for the Raman mechanism discussed above is proportional to the magnitude of the exchange coupling between Mn^{2+} ions and the band electrons which is especially strong in these alloys. Adapting Loudon's theory for optical phonons (Loudon [5]), the scattering cross-section for such a three-step process yields resonances producing a double peak in the frequency dependence of the cross section; the condition of "in resonance" results from the matching of the incident photon energy with that of an electronic excitation whereas "out resonance" occurs when the scattered photon energy equals the energy of such a transition.

On the basis of the luminescence spectra, appropriate choices of laser wavelength, sample temperature and magnetic field can be made to achieve conditions of "in resonance" or "out resonance" which can selectively enhance specific features in the Raman spectrum [6]. We discuss here, the $\omega_{LO} - \omega_{PM}$ line, i.e., the Raman shift with the creation of an LO phonon and the de-excitation of the Mn^{2+} by $\Delta m_S = -1$ involving a virtual transition at an energy $\hbar\omega_a$ for the incident radiation polarized along $\hat{z}$ and another at $\hbar\omega_b$ for scattered radiation having $\hat{\sigma}_+$ polarization. With the 7525 Å ($\hbar\omega_L$ = 1.648 eV) laser line, the "in resonance" condition is nearly fulfilled; and the manganese concentration, sample temperature and the magnetic field allow an "out resonance" for the $\omega_{LO} - \omega_{PM}$ line. The dramatic enhancement in the intensity of the $\omega_{LO_1} - \omega_{PM}$ line is illustrated in Fig. 3. The relatively broad feature at ~220 cm^{-1} is the X_+ luminescence feature attributed to the low energy exciton component which moves into the vicinity of the LO lines. In addition to the LO_1, LO_1 ± PM, LO_2, and LO_2 ± PM lines, which have been reported before (Petrou *et al.* [3]), two additional features with Raman shifts of $\omega_{LO} \pm 2\omega_{PM}$ are observed. A direct consequence of the "out resonance" conditions is the pronounced enhancement in the intensity of the LO_1 - PM line in Fig. 3 with respect to that of LO_1 + PM. In the scattering geometry and the polarization

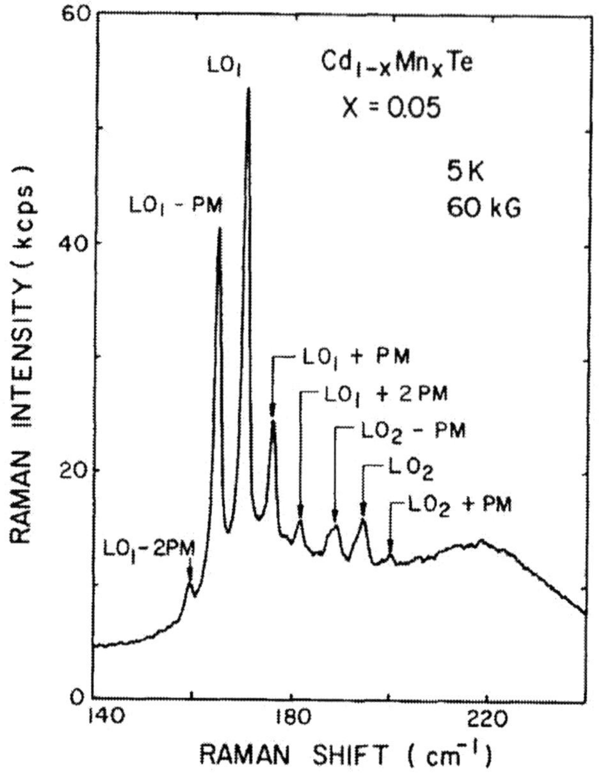

Figure 3

The Raman spectrum of $Cd_{0.95}Mn_{0.05}Te$ in the region of the LO phonons with T = 5 K, H = 60 kG, λ_L = 7525 Å, and P_L = 30 mW. Incident light polarized along $\hat{z} \| \vec{H} \| \vec{k}_S$, and scattered light unanalyzed [6].

conditions used, the preferential enhancement results from the fact that the polarization of the scattered light for LO_1 + PM is $\hat{\sigma}_-$, while that for LO_1 - PM is $\hat{\sigma}_+$, matching that of the X_+ transition. Under non-resonant conditions, for T = 5 K and H = 60 kG, the intensity of LO_1 + PM would be five times greater than that of LO_1 - PM, as calculated from the Boltzmann factor. This enhancement of LO_1 - PM becomes even more pronounced under exact "out resonance" achieved by decreasing the magnetic field to 35 kG and moving the X_+ luminescence feature under the LO_1 and the LO_1 - PM Raman lines. The resonance conditions can be controlled by varying T rather than $\vec{H}$ [6]. [6].

MAGNETICALLY ORDERED PHASE

$Cd_{1-x}Mn_xTe$ exhibits a magnetically ordered low temperature phase for x > 0.17 [1]. The transition from the paramagnetic to the magnetically ordered phase is accompanied by the appearance of a new Raman feature at low temperatures. Since this excitation is associated with magnetic order, it is attributed to a magnon. A distinct magnon feature was observed in $Cd_{1-x}Mn_xTe$ for the composition range 0.40 ≤ x ≤ 0.70. The magnon feature is absent when the incident and the scattered polarizations are parallel and appears when they are crossed, in agreement with Eq. (1) and shown in Fig. 4. This was found to be the case for several crystallographic orientations as well as for polycrystalline samples. Such a behavior, irrespective of the crystallographic orientation, is exhibited only by an excitation whose Raman tensor is antisymmetric. As the temperature is increased, the Raman shift of the magnon, ω_M, decreases, and above a characteristic Néel temperature $T_N(x)$ the feature is no longer observable. For x = 0.70, the temperature dependence of ω_M follows a Brillouin function.

The polarization features of the magnon line are those predicted for a long wavelength excitation in a magnetically ordered phase [3] and illustrated in Fig. 4. We have here the Raman-antiferromagnetic resonance (Raman-AFMR). The Brillouin function temperature dependence of ω_M yielding T_N; the behavior of ω_M in an external magnetic field; and the evolution of the Raman line at ω_{PM} into the high frequency component of the magnetically split doublet of the magnon as the sample passes from the paramagnetic into the magnetically ordered phase — these are the significant aspects of the Raman-AFMR [3,7].

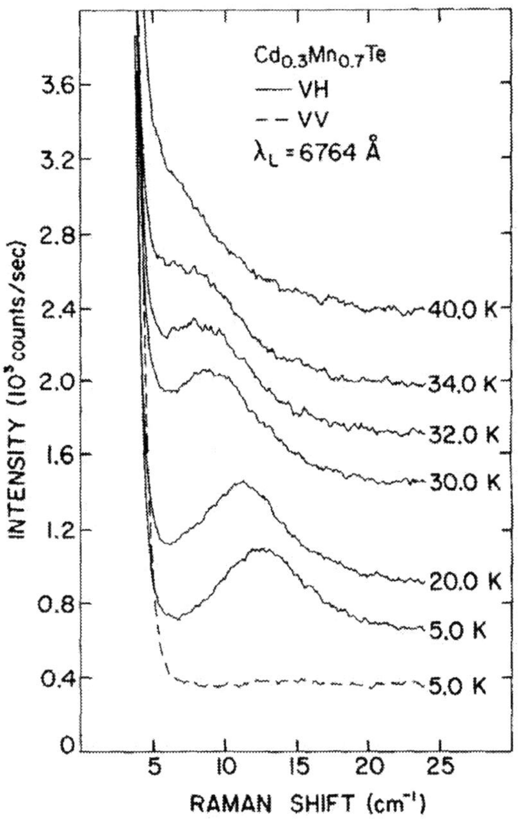

Figure 4

Temperature dependence of the low frequency Raman spectrum of $Cd_{0.3}Mn_{0.7}Te$. The intensity scale refers to the VV spectrum at 5 K and successive spectra have been displaced vertically upwards for clarity. The sample temperature is indicated for each trace. The broken curve shows the VV spectrum. All the other spectra correspond to VH polarization. λ_L = 6764 Å, the exciting laser wavelength; V ≡ Vertical; H ≡ Horizontal. The scattering plane is horizontal [7].

SPIN-FLIP RAMAN SCATTERING

When detectable, spin-flip Raman scattering by free carriers or carriers bound to donors or acceptors provides a practical means of probing the electronic structure of semiconductors, as dramatically illustrated in DMS. The large Raman shifts associated with free and bound electrons; the finite spin splitting of the levels in zero field; the effects of Mn^{2+} concentration and the antiferromagnetic coupling between Mn^{2+} ions — these are the salient features of the spin-flip Raman scattering in the Mn-based DMS. In the wide band gap DMS, spin-flip Raman scattering reported to date is associated with electrons bound to donors.

In the Raman spectra of $Cd_{1-x}Mn_xTe(Ga)$, x = 0.03, shown in Fig. 1 for the $(\hat{\sigma}_+,\hat{z})$ and $(\hat{\sigma}_-,\hat{z})$ polarization configurations with T = 40 K and H = 60 kG, the two Stokes features labeled 'PM' and 'SF' are present only in $(\hat{\sigma}_+,\hat{z})$, while the corresponding anti-Stokes features appear only in the $(\hat{\sigma}_-,\hat{z})$ configuration. We have already discussed the 'PM' line resulting from the spin-flip transitions within the Zeeman multiplet of the Mn^{2+} 3d electrons. The observed width of the 'PM' line is instrument limited, while that of 'SF' is ~ 3 cm^{-1}.

The 'SF' feature of Fig. 1 is attributed to spin-flip Raman scattering from electrons bound to gallium donors. It has the same polarization characteristics as those of the 'PM' line appearing in the $(\hat{\sigma}_+,\hat{z})$ or $(\hat{z},\hat{\sigma}_-)$ polarizations for Stokes scattering and in $(\hat{\sigma}_-,\hat{z})$ or $(\hat{z},\hat{\sigma}_+)$ for anti-Stokes. The peak Raman shift of this spin-flip feature, $\tilde{\omega}$, exhibits a strong dependence on both temperature and magnetic field. The primary source of the spin splitting of the electronic level is the exchange coupling with the Mn^{2+} ions (first term of Eq. (5)) with the Zeeman effect (second term of Eq. (5)) making a relatively small contribution. Hence, the Raman shift should be approximately proportional to the magnetization of the Mn^{2+} ion system, which amplifies the effect of the magnetic field on the electron. A finite Raman shift is observed for zero magnetic field. This effect is attributed by Dietl and Spalek [8] to the "bound magnetic polaron (BMP)": The electron localized on a donor in a diluted magnetic crystal polarizes the magnetic ions within its orbit, creating a spin cloud that exhibits a net magnetic moment. An additional effect on the binding energy of the electron bound to the donor originates from thermodynamic fluctuations of the magnetization and the resulting spin alignment of the magnetic ions around the donor. Extensive experimental results of the present authors and their collaborators are discussed in Ref. [4] in the context of the earlier work as well as the theory [8].

CONCLUDING REMARKS

Inelastic light scattering — Raman and Brillouin spectroscopy — continues to be a fruitful technique in exploring and delineating novel aspects of DMS. Extension to DMS with magnetic ions other than Mn (e.g. Fe, Co); energy levels of Mn^{2+} pairs; Brillouin scattering and the elastic constants determination; application to DMS superlattices [9] – these are illustrative examples of work currently in progress in the research program of the authors.

ACKNOWLEDGMENTS

The authors acknowledge support from the National Science Foundation (Grant No. DMR-8403325 and Grant No. DMR-8520866).

REFERENCES

1. R. R. Galazka, S. Nagata and P. H. Keesom, Phys. Rev. B22, 3344 (1980).

2. See W. Hayes and R. Loudon, Scattering of Light by Crystals (John Wiley & Sons, New York, 1978).

3. A. Petrou, D. L. Peterson, S. Venugopalan, R. R. Galazka, A. K. Ramdas and S. Rodriguez, Phys. Rev. B27, 3471 (1983).

4. D. L. Peterson, D. U. Bartholomew, U. Debska, A. K. Ramdas and S. Rodriguez, Phys. Rev. B32, 323 (1985).

5. R. Loudon, Adv. Phys. 13, 423 (1964).

6. D. L. Peterson, D. U. Bartholomew, A. K. Ramdas and S. Rodriguez, Phys. Rev. B31, 7932 (1985).

7. S. Venugopalan, A. Petrou, R. R. Galazka, A. K. Ramdas and S. Rodriguez, Phys. Rev. B25, 2681 (1982).

8. T. Dietl and J. Spalek, Phys. Rev. B28, 1548 (1983).

9. S. Venugopalan, L. A. Kolodziejski, R. L. Gunshor, and A. K. Ramdas, Appl. Phys. Lett. 45, 974 (1984).

MAGNETIC CONTRIBUTION TO THE ENERGY GAP OF $ZN_{1-x}MN_xTE$

J.A.GAJ*, A. GOLNIK*, J.P.LASCARAY**, D.COQUILLAT** AND M.C.DEJARDINS-DERUELLE**
*Institute of Experimental Physics, University of Warsaw, Hoza 69, 00-681 Warsaw, Poland
**Groupe d'Etude des Semiconducteurs (LA 357), USTL, place E.Bataillon, 34060 - Montpellier-Cedex, France.

ABSTRACT

Systematic measurements of energy gap and magnetic susceptibility of $Zn_{1-x}Mn_xTe$ in a composition range from 0 to 0.703 Mn mole fraction were performed as a function of temperature. The results show that the magnetic contribution to the energy gap variation is proportional to a product of temperature and susceptibility.

INTRODUCTION

ZnMnTe belongs to a class of semiconductor alloys called semimagnetic semiconductors (SMSC) or diluted magnetic semiconductors (DMS) [1]. An important advantage of those materials is a possibility to control the amount of magnetic component (Mn in ZnMnTe) and thus to vary magnetic properties of the alloy, from paramagnetic to spin glass or disordered antiferromagnetic phases. The principal physical interaction of interest in SMSC is magnetic ion-carrier exchange. It causes giant magnetooptical effects studied extensively in those compounds.

One of the properties distinguishing SMSC from nonmagnetic semiconductors is behaviour of their energy gap with temperature and composition. Magnetic origins of energy gap variation have been discussed by many authors.

Recent papers assume generally the carrier-ion exchange interaction as a direct cause of the Eg(T) behaviour observed in semiconductors with magnetic properties. The basic ideas of that approach have been coherently presented in a paper by F.Rys, J.C.Helman and W.Baltensperger [2].

Attempts to analyse the behaviour of the energy gap in SMSC have been done by J.Diouri et al. [3] in $Cd_{1-x}Mn_xTe$ and by T.Donofrio et al. [4] in quaternary $Cd_xZn_yMn_zTe$. In recent papers R.B. Bylsma et al [5] and independently J.Gaj and A.Golnik [6] propose a simple model in which magnetic contribution to the energy gap is found to be proportional to a product of temperature and magnetic susceptibility. This paper aims at a quantitative analysis of influence of the magnetic degrees of freedom of the system on the energy gap of $Zn_{1-x}Mn_xTe$.

EXPERIMENT

$Zn_{1-x}Mn_xTe$ single crystals have been grown by Bridgman technique al CNRS Laboratoire de Physique des Solides, Meudon, France. Six composition values in the range of x from 0 to 0.703 were used.

Mat. Res. Soc. Symp. Proc. Vol. 89. © 1987 Materials Research Society

The energy gap was measured by a polarization modulation technique described in detail by M.C.Dejardins-Deruelle et al. [7]. Energy gap was identified with the position of the principal excitonic reflectivity structure in a magnetic field producing a small Zeeman splitting sufficient to reveal the structure in polarization measurements.

A part of the data on the energy gap comes from Ref.7. To widen the composition range, additional measurements were performed on samples of x=0, 0.02 and 0.17.

For x=0 where the Zeeman splitting was very small, standard reflectivity measurements were used to determine the exciton structure energy.

Susceptibility measurements were done at Laboratoire Louis Neel, Grenoble, France, on the samples used previously for optical experiments. This is particularly important since otherwise possible inhomogeneity of the material would make the comparison ambiguous. An extraction method was used in the measurements. Magnetic field similar to that of optical measurements (of order of 1T) was applied.

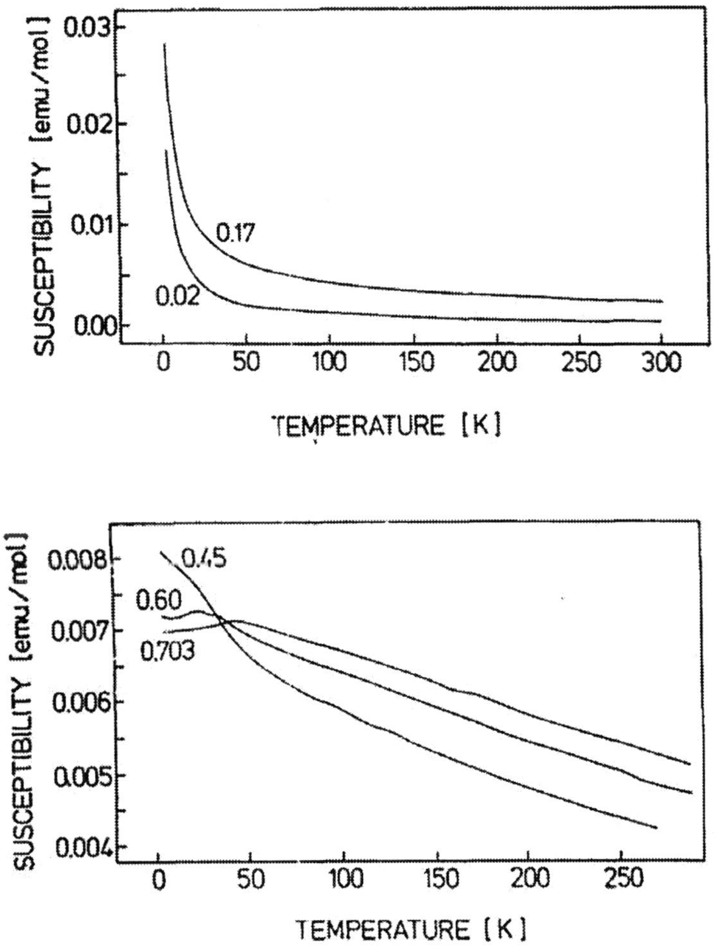

Figure 1. Susceptibility of $Zn_{1-x}Mn_xTe$ vs temperature

RESULTS AND DISCUSSION

Figure 1 shows the results of the susceptibility measurements. The only available literature data comparable with our results have been obtained by McAlister et al. [10] for x=0.17. There is a certain difference in the obtained values, originating possibly from an uncertainty in the determination of composition.

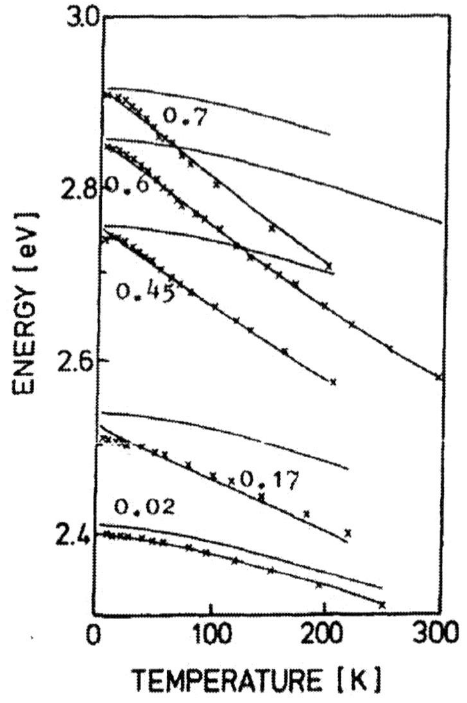

Figure 2. Energy gap of $Zn_{1-x}Mn_xTe$ alloys of indicated compositions versus temperature (experimental points). Lower lines - model described in text, upper lines were calculated without magnetic contribution.

The measured energy gap values are plotted versus temperature in Figure 2. As pointed out previously [6] the energy gap in ZnMnTe exhibits similarly to other semimagnetic semiconductors a behaviour different from that of nonmagnetic materials: even in the low temperature range the variation of the gap is fairly strong; a kind of a bump is observed for the highest Mn concentrations. Moreover, the slope of the linear high temperature part of the $E_g(T)$ curves varies strongly with composition.

Theoretical curves have been obtained using a model developed in Ref.6 based on ion-carrier exchange interaction calculated up to the second order of perturbation by F.Rys et al. [2]. Following assumptions have been made:

1. Magnetization of the Mn^{++} ion system vanishes (zero external field).

2. During the optical transition (creation of an exciton) the magnetic system remains frozen.

Describing the exchange interaction between a carrier of momentum j and magnetic ions of spins S_i by a Heisenberg hamiltonian:

$$H = -\Omega \sum_i j\, S_i\, J(r - R_i)$$

where $\Omega = V/N$ is the volume per cation, following expression for a magnetic correction to the carrier energy has been obtained:

$$E(k,T) - E_o(k,T) = \sum_q \frac{1}{4} |J_q|^2 \Gamma_q / [E_o(k,T) - E_o(k+q,T)]$$

where $J_q = \int d_3r\, J(r)\, e^{-iqr}$

and $\Gamma_q = N^{-1} \times \sum_j \langle S_o \cdot S_j \rangle_T \exp(iq \cdot R_j)$

Following further approximations have been introduced:

3. The Fourier transform J_q of the exchange coupling is significant in a limited range of wavevector q.

4. The correlation function Γ_q may be approximated by its value at k=0.

5. The contribution of the conduction band can be neglected.

Using the above assumptions, it has been shown in Ref.6 that the magnetic contribution to the variation of the energy gap is proportional to the product of temperature and magnetic susceptibility. The whole expression obtained includes also a nonmagnetic contribution, proportional to the temperature variation of the energy gap in the nonmagnetic host crystal (here ZnTe):

$$E_g(x,T) = A(x) + (1 + \alpha x)\, [E_g(0,T) - E_g(0,0)] + bT\chi(x,T) \qquad (1)$$

where values of parameters α and b are common for all the compositions studied whereas A(x), fitted independently for each x value, represents composition variation of the energy gap, extrapolated to T=0.

Equation 1 was fitted to the experimental data using parameters shown in Table I. Besides the values extracted directly from literature, an estimate of the hole effective mass has been made similarly as in Ref.6 by averaging over the light- and heavy hole bands and over a suitable energy range (comparing Refs. 5 and 6 an averaging law $m_h = 0.6m_{hh} + 0.4m_{lh}$ can be found for a parabolic case). The fitting procedure yielded values of parameters a, b and A(x) collected also in Table I. The fit is shown in Figure 2.

While the overall accuracy of the fit is quite good, some discrepancies show in the low temperature region. Apparently the approximations of the model are no longer accurate at low temperatures. This may apply both to the rigidity of the magnetic system and to the wavevector dependence of functions of interest (assumptions 2 and 4 respectively). Most probably the assumption 4 is poorly satisfied at low x values where there exists no clear tendency to an antiferromagnetic order.

Table I. Parameters used for and obtained from the fit

	parameters	values	remarks
used	$N_0\beta$ [eV]	1.09	after Ref. 10
	m_h/m_0	0.5	see text
	N_0 [cm^{-3}]	$1.68\cdot10^{22}$	after Ref. 9
obtained	α	0.2	
	b [ev Gs^2 erg^{-1} K^{-1}]	-0.123	
	A(x)	2.408	x = 0.02
		2.536	x = 0.17
		2.755	x = 0.45
		2.856	x = 0.60
		2.915	x = 0.703
	q_c [cm^{-1}]	$9.6\cdot10^7$	

It is worth noticing that setting parameter α to 0 a very small modification of the parameter b is obtained (-0.129 for α=0 compared to -0.123 for the optimum value α=0.2). That means that essentially there is no need to introduce a nonmagnetic contribution to the variation of the $E_0(T)$ behaviour with x. This was not the case in $Cd_{1-x}Mn_xTe$ [6]. Clearly an Mn^{++} ion is more similar to the host lattice cation in ZnMnTe than in CdMnTe as can be noticed e.g. from a much stronger $E_0(x)$ dependence in the latter.

The obtained value of the parameter b can be related to an integral of the square of the exchange coupling constant J_q [6]

$$\int d_3q|J_q|^2/[E_0(0) - E_0(q)] =$$

$$= b(2\pi)^3\, N_0N_A(g\mu_B)^2/[j(j+1)k_B] = 2.0\times10^{23}\ eVcm^{-3}$$

and, assuming a primitive cutoff model, can yield a cutoff wavevector value

$$q_c = 9.6\times10^7\ cm^{-1}$$

close to the wavevector value at the L point of the Brillouin zone.

It is a fairly large value keeping in mind the approximations made but it is still considerably smaller than the wavevector value corresponding to the peak of the correlation function Γ_k corresponding to an antiferromagnetic order [9].

ACKNOWLEDGEMENT

Authors wish to thank Dr J.Deportes for performing magnetic susceptibility measurements.

REFERENCES

1. N.B. Brandt and V.V. Moshalkov, Adv. in Phys. 33, 193 (1984) and references cited therein.

2. F. Rys, J.S. Helman and W. Baltensperger, Phys. Kondens. Mater. 6, 105 (1967).

3. J. Diouri, J.P. Lascaray and M. El Amrani, Phys. Rev. B 31, 7995 (1985).

4. T. Donofrio, G. Lamarche and J.C. Woolley, J. Appl. Phys. 57, 1932 (1985).

5. R.B. Bylsma, W.M. Becker, J. Kossut and M. Debska, Phys. Rev. B 33, 8207 (1986).

6. J.A. Gaj and A. Golnik, to be published in Acta Phys. Pol.

7. M.C. Desjardins-Deruelle, J.P. Lascaray, D. Coquillat and R. Triboulet, phys. stat. sol. (b) 135, 227 (1986).

8. S.P. McAlister, J.K. Furdyna and W. Giriat, Phys. Rev. B 29,1310 (1984).

9. T.M. Holden, G. Doling,V.F. Sears, J.K. Furdyna and W. Giriat, Phys. Rev. B 26, 5074 (1982).

10. A. Twardowski, P. Swiderski, M. von Ortenberg and R. Pauthenet, Sol. State Comm. 50, 509 (1984).

11. C. Hermann and C. Weisbuch in Optical Orientation, ed. F. Meier and .P. Zakharchenya, Elsevier Sc. Publishers BV 1984 p. 491

ION - CARRIER EXCHANGE INTERACTION IN $Cd_{1-x}Mn_xS$

M.NAWROCKI[a,b], J.P. LASCARAY[b], D. COQUILLAT[b] AND M. DEMIANIUK[c]
[a] Institute of Experimental Physics, University of Warsaw, Hoya 69, 00-681 Warsaw, Poland
[b] Groupe d'Etude des Semiconducteurs, USTL, place E.Bataillon, 34060 - Montpellier - Cedex, France
[c] Institute of Technical Physics, WAT, Warsaw, Poland

ABSTRACT

Magnetoreflectivity measurements in Faraday and Voigt geometry in the free exciton region (A, B and C excitons are visible) were performed on $Cd_{0.87}Mn_{0.13}S$ at T=1.6 K for magnetic field up to 5.5 T. An analysis of the results in terms of a mean field model for wurtzite type crystal enabled to determine band parameters Δ_1, Δ_2, Δ_3 and exchange integral for valence band $N_0\beta = -1.8 \pm 0.08$ eV

INTRODUCTION

$Cd_{1-x}Mn_xS$ is the second investigated semimagnetic semiconductor of wurtzite structure, after $Cd_{1-x}Mn_xSe$. Since 1981 measurements of SFRS [1-3] and magnetoreflectivity [4,5] were published giving exchange integrals for electrons and holes. The reported integral values for the valence band were 2 to 3.5 times higher than in other semimagnetic semiconductors. In this paper we present magnetoreflectivity data for $Cd_{1-x}Mn_xS$ in the fundamental absorption region as well as the magnetization data for the same sample. The obtained experimental results are discussed on the basis of a full wurtzite type hamiltonian including the exchange interaction between Mn^{2+} ions and band electrons. As a result it was possible to evaluate the band parameters for Γ point of the Brillouin zone. Using our magnetization data and the exchange interaction constant for the conduction band measured earlier by SFRS it was also possible to obtain the exchange interaction constants for the valence band.

EXPERIMENT

The $Cd_{0.87}Mn_{0.13}S$ crystals were grown by the high-pressure Bridgman method. The single crystals were cut into plates, then polished and etched in a 30% HCl solution. The samples were put in a superconducting magnet and immersed in superfluid helium. The temperature measured by the vapor pressure was about 1.4 K. The maximum field strength of the magnet was 5.5 T.

The magnetoreflectivity measurements were performed using standard experimental setup in both Faraday and Voigt geometry on samples with the c axis either normal or parallel to their surfaces. The c axis of the sample was parallel to the magnetic field direction in both configurations. The magnetization measurements were carried out by an extraction method in magnetic field up to 7 T and temperature 2 K at the Louis Neel Laboratory (Grenoble, France). The manganese mole fraction was checked by a microprobe analysis.

Mat. Res. Soc. Symp. Proc. Vol. 89. © 1987 Materials Research Society

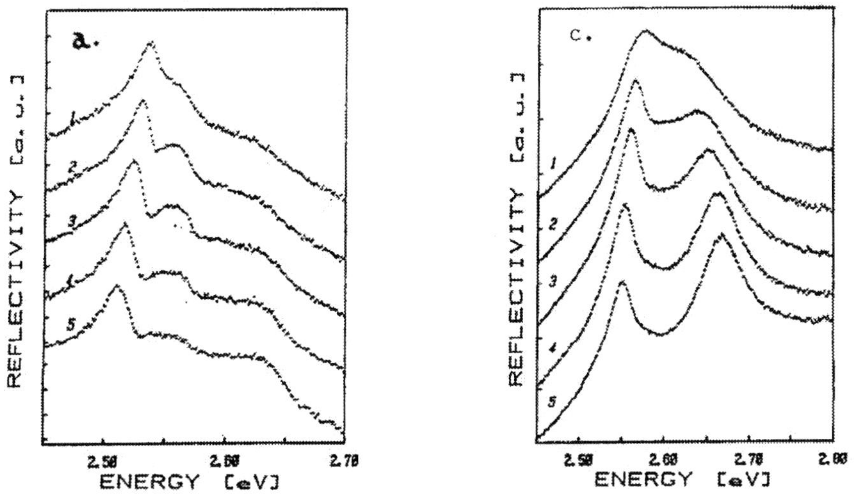

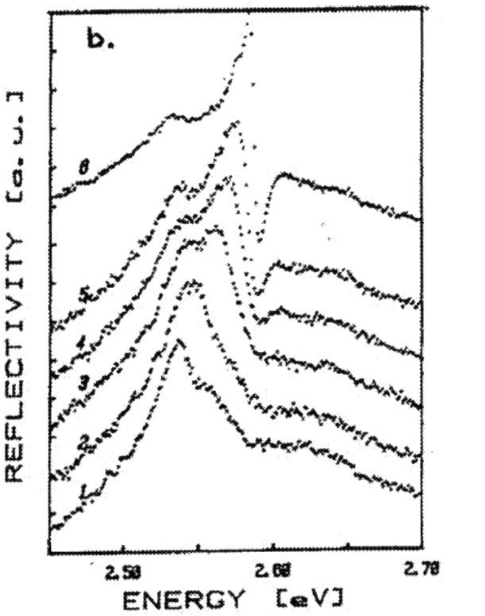

Fig.1. Magnetoreflectivity of $Cd_{0.87}Mn_{0.13}S$.
a. Faraday configuration, σ^+ polarization (1-H=0; 2-H=0.5; 3-H=1; 4-H=2; 5-H=5.5T),
b. Faraday configuration, σ^- polarization (1-H=0; 2-H=0.5; 3-H=1; 4-H=1.5; 5-H=2; 6-H=5.5T),
c. Voigt configuration, E||H (1-H=0; 2-H=1; 3-H=1.5; 4-H=3; 5-H=5.5).

RESULTS

The magnetoreflectivity results are presented in Fig. 1. An important shift of the excitonic transition energy is visible. However because of the large width of the reflectivity structures and a high background level not all transitions are visible and not for all visible transitions the energy can be determined, especially for zero and low magnetic fields. In Faraday configuration for σ^+ polarization the excitons A,B and C are visible. For the A exciton the energy can be well determined. For the C exciton it is difficult but still possible for magnetic field high enough. For σ^- polarization only the A exciton position is well defined. The B exciton is visible (e.g. on the low energy side of the A exciton for 5.5 T) but the determination of the inflection point position is impossible. For Voigt configuration two peaks can be seen. Each of them is a superposition of two π maxima. The inflection point has been attributed to the higher energy π component of B and C exciton respectively.

The magnetic field dependence of the position of reflectivity peaks is shown in Fig.2. The splitting of exciton levels vary nonlinearly with magnetic field and for higher H values the saturation is almost reached. As it was said earlier the position of σ^- component of B exciton cannot be well defined. However from the comparison of σ+ and σ^- spectra it can be seen that σ+ component has a lower energy value. That proves that the gravity center of σ^+ and σ^- components of B exciton shifts with magnetic field.

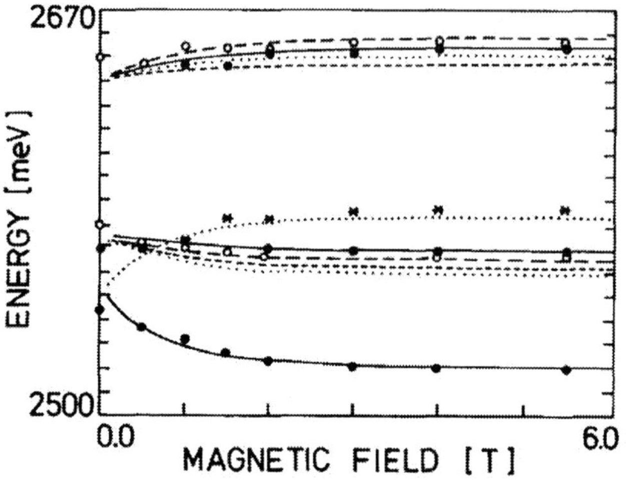

Fig.2. Energy of the free exciton lines vs. magnetic field in $Cd_{0.87}Mn_{0.13}S$ Experimental data: ● – σ^+; ✱ – σ^-; o – π Fitted curves: —— σ^+; .... σ^-; – – – π

The magnetization measurements for the sample are presented in Fig.3. The magnetization data were fitted to the modified Brillouin function [6]

$$M = M_s B_{5/2} \quad (1)$$

where M_s is fitting parameter, smaller than the theoretical saturation magnetization for the Mn^{2+} spins, $T+T_0$ is an effective temperature and $B_{5/2}$ is the Brillouin function for spin 5/2.

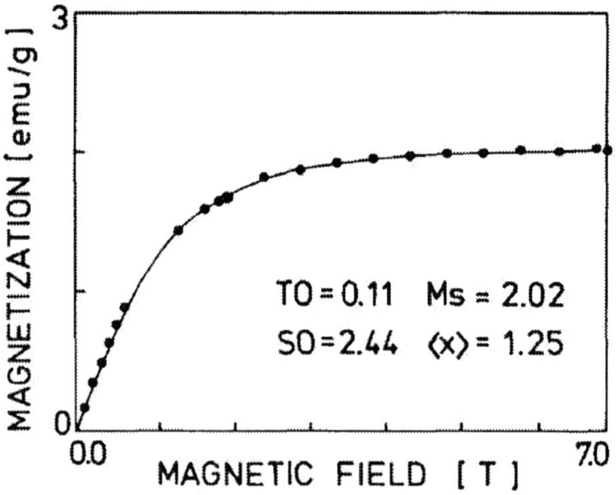

Fig.3. Magnetization vs. magnetic field for $Cd_{0.87}Mn_{0.13}S$.

THEORETICAL BACKGROUND

The theoretical description of the observed structures is possible using the complete Hamiltonian obtained by adding an exchange term H_{ex} to the original wurtzite Hamiltonian H_{wur} [4,7,8]. In what follows we shall use the notation of Ref.8. Taking the exchange term in the mean field approximation one gets the conduction band split into two subbands, corresponding to the two spin orientations of the electron. The splitting is given by $2A = N_0\alpha\, x\, \langle S_z \rangle$, where N_0 is the number of unit cells per unit volume, α is the exchange integral for the s-like conduction band electrons $\alpha = \langle S|J|S \rangle$, x denotes the manganese mole fraction, $\langle S_z \rangle$ is the thermal average of the Mn^{2+} spin component along magnetization. For the valence band in the special case where the external magnetic field is parallel to the c axis of the crystal (this was the geometry used in our experiments) the eigenvalue problem can be solved analytically producing six energies

$$
\begin{aligned}
&A\colon\ E_{(3/2,\pm 3/2)} = \Delta_1 + \Delta_2 \pm B_x,\\
&B\colon\ E_{(3/2,\pm 1/2)} = (\Delta_1 - \Delta_2)/2 \pm (B_z - B_x)/2 + E_{\mp},\\
&C\colon\ E_{(1/2,\pm 1/2)} = (\Delta_1 - \Delta_2)/2 \pm (B_z - B_x)/2 - E_{\mp},\\
&E_{\pm} = \sqrt{\left[\frac{(\Delta_1 - \Delta_2) \pm (B_x + B_z)}{2}\right]^2 + 2\Delta_3^2}.
\end{aligned}
\tag{2}
$$

where Δ_1, Δ_2, Δ_3 are wurtzite band structure parameters describing the crystal field and spin-orbit interaction and $B_{x,z} = N_0\beta_{x,z}\, x\, \langle S_z \rangle$ and β_x, β_z are the exchange interaction constants for the valence band electrons: $\beta_x = \langle x|J|x \rangle$, $\beta_z = \langle z|J|z \rangle$.

The linear combinations of wave functions which diagonalize the Hamiltonian are of a similar form as the ones used in the absence of exchange interaction:

$$
\begin{aligned}
&A\colon\ |\tfrac{3}{2}, \pm\tfrac{3}{2}\rangle = (1/\sqrt{2})|(X \pm iY)\uparrow, \downarrow\rangle,\\
&B\colon\ |\tfrac{3}{2}, \pm\tfrac{1}{2}\rangle = \pm(\sqrt{1 - b_\pm^2}/\sqrt{2})|(X \pm iY)\downarrow, \uparrow\rangle + b_\pm|Z\uparrow, \downarrow\rangle,\\
&C\colon\ |\tfrac{1}{2}, \pm\tfrac{1}{2}\rangle = (b_\pm/\sqrt{2})|(X \pm iY)\downarrow, \uparrow\rangle \mp \sqrt{1 - b_\pm^2}|Z\uparrow, \downarrow\rangle.
\end{aligned}
\tag{3}
$$

The wave function mixing coefficients b^2 become magnetic field dependent and are given by the following expressions:

$$
b_\pm^2 = \frac{1}{2}\left[1 - \frac{\dfrac{\Delta_1 - \Delta_2}{2} \mp \dfrac{B_x + B_z}{2}}{E_\mp}\right].
\tag{4}
$$

Taking into account selection rules for σ^+, σ^- and π polarizations one gets the following energies of allowed transitions:

$$
\begin{aligned}
&E_A^{\sigma\pm} = E_0 \mp A - \Delta_1 - \Delta_2 \pm B_x,\\
&E_B^{\sigma\pm} = E_0 \pm A - (\Delta_1 - \Delta_2)/2 \pm (B_z - B_x)/2 - E_\pm,\\
&E_C^{\sigma\pm} = E_0 \pm A - (\Delta_1 - \Delta_2)/2 \pm (B_z - B_x)/2 + E_\pm,\\
&E_B^{\pi} = E_0 \mp A - (\Delta_1 - \Delta_2)/2 \pm (B_z - B_x)/2 - E_\mp,\\
&E_C^{\pi} = E_0 \mp A - (\Delta_1 - \Delta_2)/2 \pm (B_z - B_x)/2 + E_\mp,
\end{aligned}
\tag{5}
$$

where $E_0 - \Delta_1 - \Delta_2$ is the energy of the transition for the A exciton in absence of external magnetic field.

DISCUSSION

The experimental data were fitted to the expressions (5) using the magnetization in the form (1). Because of the lack of the sufficient number of well defined transition energies the $N_0\alpha$ = 220 meV value known from the SFRS experiments [3,4] was used and the fitting parameters were E_0, Δ_1, Δ_2, Δ_3, $N_0\beta_x$ and $N_0\beta_z$. The best fit is presented in Fig.2 and the parameters obtained are listed in Table I. An alternative set of the band parameters Δa, Δb, Δ' introduced by Langer [9] is also included, where Δa is a trigonal spin-orbit splitting of the (j,m) = (3/2, ±3/2) bands from the (3/2, ±1/2) bands, Δb is an isotropic spin-orbit splitting of the j = 3/2 band from the j = 1/2 band and Δ' is a crystal field splitting of the X,Y bands from the Z band. For the comparison data for CdS [9] are listed too.

TABLE I*)

	E_{A0}	Δ_1	Δ_2	Δ_3	Δa	Δb	Δ'	b^{2}**
CdS	2553	28.4	20.9	20.7	-0.4	48	28.2	0.44
CdMnS	2552±2	32±2	26±1	24±1	6±2	75±3	29±2	0.48

*) all values in meV **) in absence of the magnetic field

The difference between the two exchange interaction values for valence band $N_0\beta_x$= -1.79 ± 0.08 eV and $N_0\beta_z$= -1.81 ± 0.08 eV is in our opinion non significant and we take $N_0\beta$= -1.8 ± 0.1 eV as a result. This value is more realistic than obtained earlier -5.6 ± 0.8 eV for x=0.001 [5] and -2.7 ± 0.4 eV for x=0.005 [6]. An important difference in the band parameters between CdS and CdMnS is observed. The comparison of the parameters a, b and Δ' shows that the main effect is an increase in the spin-orbit interaction. It may result from the presence of Mn^{2+} d levels in the energy region of valence band states. The fact that a change in the energy gap value was not observed is not surprising. It was reported earlier [10] that the variation of exciton energy with composition in $Cd_{1-x}Mn_xS$ is non-linear, and for small Mn content even a decrease of energy gap can be observed.

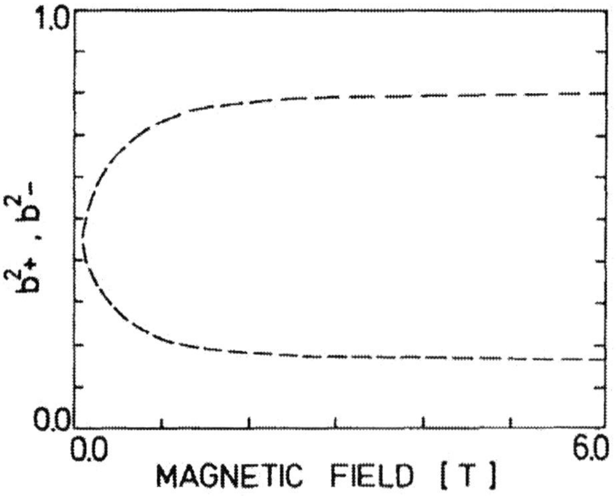

Fig.4. Calculated value of the wave function mixing coefficients vs. magnetic field.

Using the obtained parameter values the variation of the wave function mixing coefficients b^{*} with magnetic field can be computed. The result is shown in Fig.4. This effect, characteristic for semimagnetic semiconductors of wurtzite structure must be taken into account in analysis of effects influenced by the oscillator strength (e.g. birefringence in magnetic field).

It should be stressed that magnetic field dependence of wave function mixing coefficient, the change of gravity center of σ^{+} and σ^{-} transition energy for B and C excitons and correct values and sequence of transition energies cannot be obtained in the framework of the quasi-cubic model [11], which has been widely used in the case of CdS.

ACKNOWLEDGMENTS

The authors would like to thank Dr. J.Deportes for magnetization measurements, Dr. J.A.Gaj for helpful discussion and critical reading of the manuscript, and Dr. A.Golnik for valuable suggestions on fitting procedure.

REFERENCES

1. D.L.Alov, S.I.Gubarev and V.B.Timofeev
Zh.Eksp.Teor.Fiz. 84, 1806 (1983)

2. D.Heiman, Y.Shapira and S.Foner
Solid State Commun. 45, 899 (1983)

3. M.Nawrocki, R.Planel, F.Mollot and M.J.Kozielski
phys.stat.sol.(b) 123, 99 (1984)

4. S.I.Gubarev
Zh.Eksp.Teor.Fiz. 80, 1174 (1981)

5. A.G.Abramishvili, S.I.Gubarev, A.V.Komarov and S.M.Ryabchenko
Fiz.Tverd.Tela 26, 1095 (1984)

6. J.A.Gaj, R.Planel and G.Fishman
Solid State Commun. 29, 435 (1979)

7. R.L.Aggarwal, S.N.Jasperson, J.Stankiewicz, S.Foner, B.Khazai and A.Wold
Phys.Rev.B 28, 6907 (1983)

8. M.Arciszewska, M.Nawrocki
J.Phys.Chem.Solids 47, 309 (1986)

9. D.W.Langer, R.N.Euwema, Koh Era, Takao Koda
Phys.Rev.B 2, 4005 (1970)

10. M.Ikeda, K.Itoh and H.Sato
J.Phys.Soc.Japan 25, 455 (1968)

11. J.J.Hopfield
J.Phys.Chem.Solids 15, 97 (1970)

LOW-TEMPERATURE MAGNETIC SPECTROSCOPY OF DILUTE MAGNETIC SYSTEMS

D. D. AWSCHALOM AND J. WARNOCK
IBM Thomas J. Watson Research Center
P.O. Box 218, Yorktown Heights, NY 10598

ABSTRACT

Optically induced magnetization has been observed in a 10 micron diameter sample of $Cd_{.8}Mn_{.2}Te$. A newly developed integrated SQUID magnetic spectrometer was used to study both the magnitude and the picosecond dynamics of the magnetic response as a function of the energy and polarization of the optical excitation. The dramatic energy dependence of the response provides an understanding of the way in which the overall sample magnetization changes upon illumination, and the subsequent relaxation through the spin lattice interaction.

INTRODUCTION

Dilute magnetic semiconductors (DMS) represent a novel class of semiconducting materials, combining the qualities of conventional semiconductors with interesting magnetic properties which can be tuned in a well controlled manner[1]. Many interesting phenomena have been observed, arising from the strong spin coupling between the carriers and the magnetic ions in the lattice. The magnetic ions may have a profound effect on the carriers, resulting in giant spin splittings of the bands, large Faraday rotations, and strongly temperature dependent recombination and polarization effects. It is also known that carriers introduced through doping[2] or through optical excitation[3-5] can strongly influence the behavior of the magnetic ions. Changes in carrier density have been observed to reverse the sign of the effective exchange interaction between ions[2,6], and optically excited carriers can induce a finite magnetization in zero external field[4,5]. These carrier dependent magnetization effects are interesting in themselves, but they also provide a new method of probing the magnetic properties of the dilute magnetic system.

We have suceeded in directly observing the magnetism induced by optically excited carriers in $Cd_{1-x}Mn_xTe$, and have used this effect to probe the energetics and spin dynamics of this dilute magnetic system. Our methods offer a new type of magnetic "spectroscopy", in which the magnetic response is measured as the energy of the incident light is varied. The intensity, polarization and the sample temperature may also be changed, providing more information on the nature of the induced magnetism. Furthermore, the use of picosecond optical pulses extends the technique to ultrafast timescales. Dynamical studies as a function of photon energy then yield detailed information on the microscopic aspects of the magnetic system, providing a more complete understanding of the spin mechanics and relaxation rates.

Using this magnetic technique, we have addressed problems concerning magnetic polaron formation and magnetic ion spin relaxation in a new way. The magnetic polaron is formed as the strong carrier/ion spin-spin exchange interaction aligns the ion spins in the neighborhood of a localized carrier. Although the time scale for polaron formation has been measured through other time resolved experiments[7-9], the details of the spin alignment process are not well understood. In general, ion spin relaxation mechanisms are poorly understood, and the relaxation times have not been measured. Our time resolved magnetization data thus sheds new light on these problems.

Mat. Res. Soc. Symp. Proc. Vol. 89. © 1987 Materials Research Society

EXPERIMENT

The ultrafast light pulses are obtained from a synchronously pumped dye laser, producing a linearly polarized train of energy tunable pulses of 4 picosecond duration at a 76 MHz repetition rate. The train of pulses is split into pump and probe beams of intensity ratio 10:1. The probe beam passes through a variable delay before being recombined with the pump beam. The beams are circularly polarized using a quarter-wave plate, then conducted to the sample at cryogenic temperature through a polarization preserving optical fiber. The light is then incident on the sample in the direction normal to the plane of the SQUID pick-up loop. The optically induced magnetization is detected using a planar thin film dc SQUID chip[10], measuring directly the changes in magnetization of the small 10 micron size sample. The magnetic flux is sensed by two separate square pick-up coils, each 17.5 microns on the diagonal and wound in opposite senses in order to cancel the effects of stray fields. The sample is placed in a 10 micron depression in one of the loops on the integrated circuit. The sensitivity of the SQUID used for these experiments is limited by flux noise, and found to be $3 \times 10^{-6}\phi_0$ at dc, where $\phi_0 = 2.07 \times 10^{-7}$ gauss-cm^2 is the flux quantum. A detailed description of the optical configuration and the SQUID susceptometer is provided elsewhere[11].

The sample studied was a small 10x10x1 micron single crystal platelet of $Cd_{1-x}Mn_xTe$ with x=.2. The sample composition and band gap[12] were estimated from microprobe analysis and Gandolfi camera measurements, and the sample was further characterized by low-field rf SQUID magnetic susceptibility data. For this value of x, the material is paramagnetic over the temperature range of interest, and has a band-gap energy such that the system can easily be probed both above and below the gap. In addition, $Cd_{1-x}Mn_xTe$ is a cubic semiconductor whose optical properties have been extensively characterized.

RESULTS AND DISCUSSION

The sample's time-averaged magnetic response at two different temperatures is displayed in Fig. 1, where the optically induced magnetism is shown as a function of photon energy at constant intensity. The data were obtained by chopping the laser beam at 197 Hz and detecting the resulting SQUID signal with a lock-in amplifier. The signal was recorded as the excitation energy was tuned by computer. Three distinct regions of differing magnetic response are clearly defined in these spectra. At the lowest energies, well below the band edge, the signal is negligible as is expected, since the sample is relatively transparent to light at this wavelength. In the second region, just below the band edge, the signal changes rapidly, reaching a sharp maximum, then decreasing as the photon energy is increased. In the third region, at energies well above the gap, the response is relatively constant. In addition, the induced magnetism is strongly temperature dependent, decreasing rapidly as the temperature is raised.

We may draw several significant conclusions from this static magnetic data. The magnetization of the sample is definitely changed when illuminated with light of sufficient energy. Furthermore, the induced magnetism is independent of photon energy when exciting well above the gap, but increases rapidly in the band edge region. Assuming a constant quantum efficiency, we conclude that there is a spin alignment mechanism which is only important when pumping near the band edge. We attribute this effect to the formation of magnetic polarons, whose large magnetic moments are oriented perpendicularly to the SQUID loop. It has been shown[13] that a considerable spin polarization of the magnetic polarons can be achieved when exciting with circularly polarized light at an energy slightly below the band gap. On excitation with circularly polarized light at this energy, the polarized electrons and holes tend to be created in regions of the crystal where microscopic thermodynamic fluctuations in the magnetization provide strong local exchange fields. Carrier spin-flip is prohibited in these regions as a result of the strong local field, and the resulting magnetic polaron tends to be oriented in the direction described by the initial carrier spin polarization. Not only do the magnetic ions produce a strong exchange field acting on the carriers, but also the localized carrier produces a strong field which aligns the

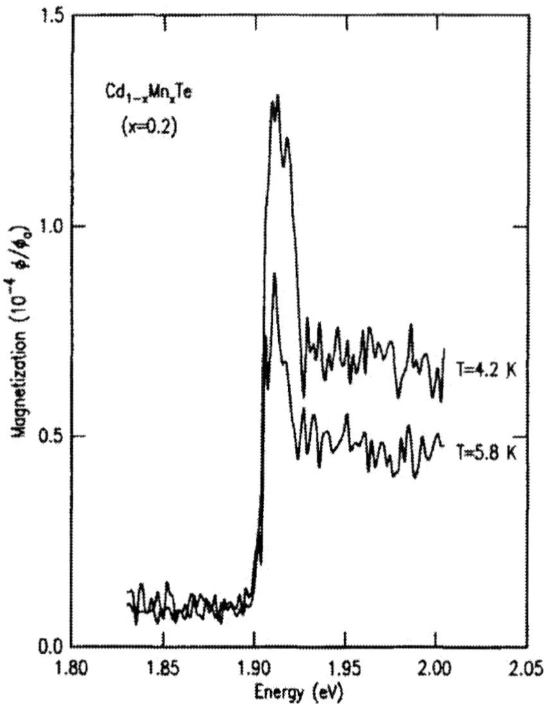

Figure 1. Time-averaged magnetic spectroscopy of a 10 micron diameter sample at two temperatures.

neighboring magnetic ions as the polaron is formed. Since the magnetism of the entire sample is observed to increase as a result of the formation of these oriented polarons, the magnetic ions must align themselves through spin lattice relaxation in the exchange field of the carrier. A spin diffusion process would not produce the observed results. Thus when pumping at this energy, the magnetism is induced through the diagonal part of the spin exchange interaction, Js_zS_z in the standard notation. Locally, this provides an effective field on the magnetic ions, which then align themselves through spin lattice relaxation, increasing the overall magnetization of the sample.

The above spin lattice relaxation process is distinct from the spin flip exchange scattering mechanism which is responsible for the signal when pumping at higher energies[4], and which proceeds through the non-diagonal parts of the exchange interaction, Js_+S_- and Js_-S_+. In a spin flip scattering event, the z component of the carrier spin is decreased (increased) while that of the ion is increased (decreased) by the same amount. Thus the carrier spin polarization is transferred to the Mn ions, resulting in the observed energy independent magnetization. This process is estimated to occur on a time scale of about 1 psec[4], so that the carriers are depolarized by the time they are eventually trapped. The polarons that will then form do not contribute to the net magnetization because of their random orientation.

The absolute magnitude of the magnetization can be calculated using the SQUID calibration data of Ketchen et al[10]. It was found that a tin sphere of radius r=2.5 microns excluded an amount of flux $\delta\phi = 0.7\phi_0$ on cooling through the superconducting transition in an external field of H=5 Oe. Thus, taking the demagnetization factor into account, we find the calibration factor F,

$$F = \left(\frac{3}{8\pi}\right)\frac{H\left(\frac{4\pi}{3}r^3\right)}{\delta\phi} = 5.58 \times 10^{-11}\frac{\text{emu}}{\phi_0} \quad (1)$$

This allows the sample magnetization to be obtained from the observed magnetic signal. Thus the measured signal, $\phi_m = 1.4 \times 10^{-4}\phi_0$ at the peak of the response can be converted to an equivalent number N_{Mn} of totally aligned Mn spins, where each spin has a moment of $Jg\mu_B$ where μ_B is the Bohr magneton, J=5/2, and g=2. Then,

$$N_{Mn} = \frac{F\phi_m}{Jg\mu_B} = 1.7 \times 10^5 \text{ spins} \tag{2}$$

This number is extremely small compared to the ~ 10^{13} spins required in magnetic resonance spectroscopy. Moreover, we may compare this number with an estimate of the expected signal based on the number N_p of oriented polarons formed by excitation with polarized light. We start with

$$N_p = \frac{\varepsilon\gamma P\tau}{E} \tag{3}$$

Here ε is the quantum efficiency for the production of polarons, which we estimate as 0.01, based on the observation that the luminescence efficiency of $Cd_{1-x}Mn_xTe$ quantum wells is at least 100 times greater than that of the corresponding bulk material[14]. The net polaron polarization[13] γ is 0.17, the polaron lifetime[9] τ is approximately 7×10^{-10} sec, the incident power P is 5 mW, and E = 1.91 eV. Thus $N_p \sim 1.9 \times 10^4$. The number of aligned magnetic ions per polaron N_I is roughly

$$N_I = \frac{4}{3}\pi r_p^3 \bar{x} N_0 \tag{4}$$

where r_p, the polaron radius is about[13] 20 Å, and $\bar{x}N_0$ is the effective number density of Mn ions, based on susceptibility measurements[15]. Thus $N_{Mn} = N_p N_I \sim 5 \times 10^5$, which is consistent with the number actually observed. It should be noted here that the sample contains about 3×10^{11} spins, so an average of only one spin in a million is polarized.

It can easily be demonstrated that the effects described so far are not associated with the heating of the sample or of the SQUID. When pumping with linearly polarized light, we expect there to be no induced moment in the sample since the average photon angular momentum is zero. To determine the polarization dependence of the signal, an electro-optic phase retarder was used to modulate the polarization from σ_+ to σ_- at a frequency of 197 Hz. The SQUID output was recorded using a digital signal averager, triggering from the modulating signal. The dramatic polarization dependence of the signal is displayed in Fig. 2, showing that the direction of the induced magnetization is indeed determined by the spin orientation of the photo-excited carriers. This represents clear evidence that heating effects do not contribute to the photo-induced signal. A study of the signal's intensity dependence was also made, and the results are shown in Fig. 3. Some evidence of nonlinearity was observed at the highest powers, but at the powers used in these measurements, there is clearly no evidence of heating effects.

In order to further understand the microscopic aspects of polaron formation, and the mechanisms underlying the observed photoinduced magnetism, a series of time resolved magnetization experiments were carried out on the same material. Unlike the time-averaged experiment discussed above, these measurements take advantage of the picosecond nature of the optical pulses. The laser beam was split into a strong excitation beam and a much weaker probe beam. Only the probe beam was chopped, allowing the sample response to the probe pulses to be measured as a function of their time delay relative to the much stronger excitation pulses. Thus at any particular delay time, the probe beam measures the magneto-optical susceptibility χ_{op} by inducing a magnetization $\delta M = \chi_{op}\delta I$ where δI is the probe intensity. The time resolved magnetism of the sample is then obtained through this measurement of χ_{op}, since to first order, the changes observed in χ_{op} will be proportional to the pump-induced magnetization. The picosecond variation of magnetization with time for the three regions of energy are shown in Fig. 4. Curve (a) is featureless, except for the anomaly indicative of the pump pulse. As was noted earlier, no magnetic response is expected when pumping well below the band gap. On the other

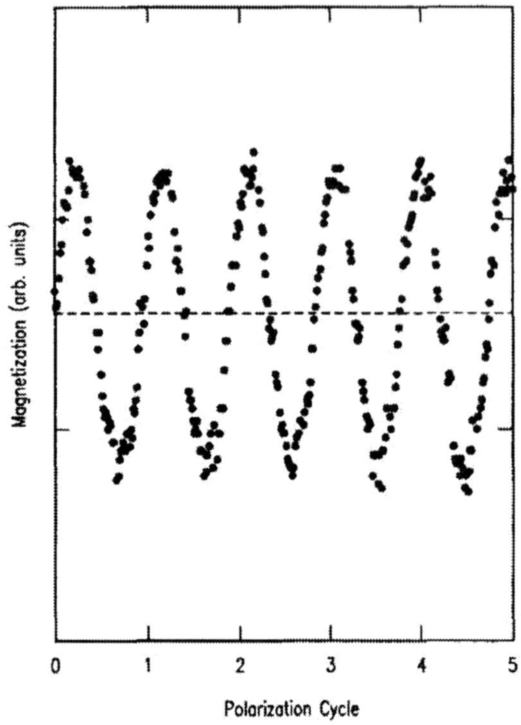

Figure 2. The induced magnetization as a function of photon polarization at E=1.915 eV. The signal was averaged for 30 minutes.

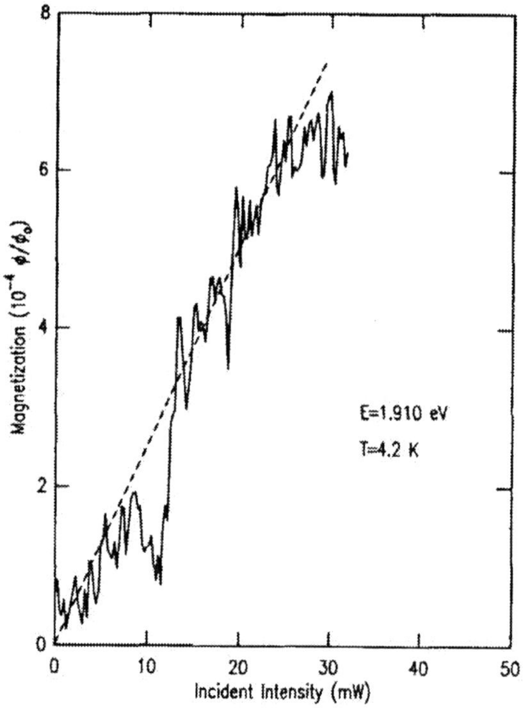

Figure 3. Magnetization as a function of laser light intensity with right-circularly polarized light.

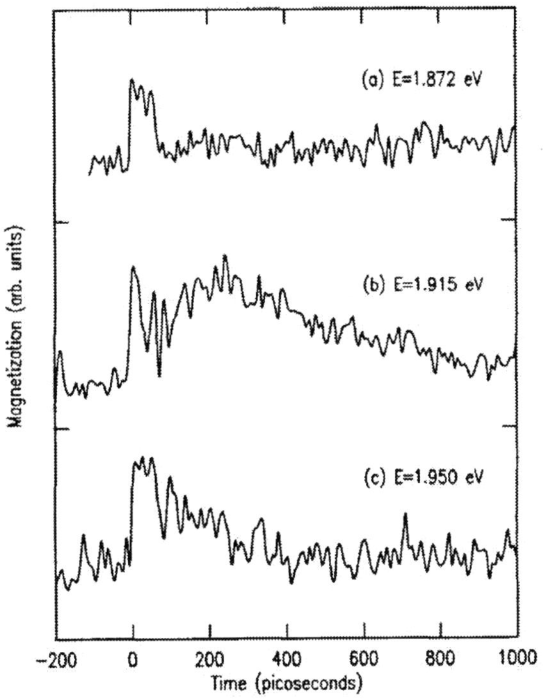

Figure 4. Picosecond time resolved magnetization for $Cd_{.8}Mn_{.2}Te$ at three different excitation energies; (a) well below the band gap energy, (b) in the band edge region, and (c) well above the band gap.

hand, when pumping at E=1.915 eV, where the response is associated with the creation of oriented magnetic polarons, the magnetization rises from zero and peaks after ~250 psec, then decays after ~1000 psec. This behavior is also quite different from that observed at higher energies, where the magnetization arises through spin flip exchange processes. On pumping at E=1.95 eV, the initial response is instantaneous within our resolution, and the magnetization decays after only about 300 psec.

Focusing first on the 1.915 eV data, we note that the initial rise time of about 250 psec is consistent with the polaron formation time as measured through time-resolved Faraday rotation experiments[8]. In addition, the data are consistent with non-magnetic measurements of this time[7,9], strengthening our conclusion that we are directly witnessing the local spin alignment as the polaron is formed. The long ~1000 psec decay is consistent with the carrier lifetime in this material[9]. The magnetic spins are kept aligned by the strong exchange field provided by the localized carriers until the electrons and holes recombine, leaving the ion spins free to relax to random orientations. Thus, at this excitation energy, we obtain a complete description of polaron formation and decay as measured through the magnetic properties.

At a higher energy, E=1.95 eV, the increase in magnetization is almost instantaneous as the excited polarized carriers transfer their polarization rapidly to the ions through spin flip scattering events. The charactistic time estimated[4] for this process is ~ 1 psec, consistent with the rapid rise observed here. This relatively uniform magnetization then relaxes through the spin lattice interaction, from which we deduce a spin lattice time of about 300 psec. It is important to note that this spin relaxation time is much shorter than the 10-100 msec measured in extremely dilute magnetic systems[16]. Furthermore, in contrast to what is observed in very dilute alloys[17], this time is independent of the field on the ion, to first order. This observation is based on the fact that the same (~300 psec) time scale is observed in the formation of the polaron, where the ion relaxes in the strong exchange field of the carrier. Similar times are also measured in the presence of an external field[8]. Finally, the temperature dependence of the signal indicates that the relaxation time decreases as the temperature is raised[4]. The above observations can be explained in terms of the spin mechanics of clusters of Mn ions, whose spins are coupled through an antiferromagnetic exchange interaction of the form,

$$\mathcal{H}_{ab} = -2J_1\vec{S}_a \cdot \vec{S}_b \tag{5}$$

where the indices a and b refer to nearest neighbor ions. Assuming a random magnetic ion distribution, there will be many clusters of various shapes and sizes. For one of the spins in such a group, the total exchange field will be anisotropic, allowing an efficient spin coupling to the lattice. Similar mechanisms have been proposed to explain spin relaxation rates in other systems[18], where at relatively modest magnetic concentrations (<10%), the relaxation rates are about 4 orders of magnitude greater than in the dilute limit.

To estimate a spin relaxation time, we start with an expression for the direct one phonon spin relaxation rate[19],

$$\frac{1}{\tau} \sim \frac{\delta^3}{\pi\rho\hbar^4 v_t^5} |\langle f|\epsilon_s|i\rangle|^2 \coth\left(\frac{\delta}{2kT}\right) \tag{6}$$

Now δ is the phonon energy, which we will take as $\approx 2.4 \times 2|J_1|$, where 2.4 is the average number of magnetic nearest neighbors for a given ion. The density ρ is 5.85 g/cm^3, and the transverse sound velocity is 1.85×10^5 cm/sec. The matrix element is between the initial and final spin states, and ϵ_s is the perturbation due to a unit strain. This matrix element will be on the order of $2J'_1$ where $2J'_1$ is the change in $2|J_1|$ per unit strain. We further assume that sample's overall spin relaxation time is determined by the relaxation time for a given spin in the cluster. We expect this assumption to hold for a large cluster in a complicated and anisotropic environment, although it is not valid for small, isolated clusters[18] where transitions which would change the overall spin of the cluster are forbidden. The exchange constant $2J_1$ is -1.2 meV, as determined through susceptibility measurements, and $2J'_1$ is 13.4 meV, calculated using the values of $2J_1$ for several compounds with different lattice parameters[20]. Thus we find $\tau \sim 120$ psec. This estimated time is consistent with the measured time, showing that the observed fast relaxation can be explained in terms of the strong spin-lattice coupling of Mn ion clusters.

CONCLUSION

In conclusion, we have directly observed optically induced magnetism in $Cd_{1-x}Mn_xTe$. By recording the magnetic response as a function of photon energy, polarization, and sample temperature, the different spin alignment processes were separately identified and studied. Effects due to magnetic polaron formation were clearly seen, showing that the overall sample magnetization increases as the polaron forms. Extending the technique to the picosecond time domain then allowed us to study in detail the underlying spin mechanics. When probing the system at an energy slightly below the bandgap, the time resolved measurements yielded a complete description of polaron formation and decay in terms of the polaron's magnetic properties. On probing the system at a higher energy, a different spin alignment mechanism was observed, and the magnetic ion spin relaxation time was determined. Thus by combining fast optical techniques with the magnetic sensitivity of the SQUID we have developed a probe of magnetic behavior which has provided us with new information on the spin dynamics of a dilute magnetic semiconductor.

ACKNOWLEDGEMENTS

We would like to thank M. Ketchen for designing the SQUID susceptometer, J. Rozen for technical assistance, F. Holtzberg for growing the samples, and N.S. Shiren for helpful discussions.

REFERENCES

1. N.B. Brandt and V.V. Moshchalkov, Adv. Phys. 33, 193 (1984).
2. T. Story, R.R. Galazka, R.B. Frankel and P.A. Wolff, Phys. Rev. Lett. 56, 777 (1986).
3. S.M. Ryabchenko, Yu.G. Semenov and O.V. Terletskii, Sov. Phys. JETP 55, 557 (1982) (Zh. Eksp. Teor. Fiz. 82, 951 (1982)).
4. H. Krenn, W. Zawadski and G. Bauer, Phys. Rev. Lett. 55, 1510 (1985).
5. D.D. Awschalom, J. Warnock and S. von Molnár, submitted for publication.
6. F. Holtzberg, T.R. McGuire, S. Methfessel and J.C. Suits, Phys. Rev. Lett. 13, 18 (1964).
7. J.H. Harris and A.V. Nurmikko, Phys. Rev. Lett. 51, 1472 (1983).
8. D.D. Awschalom, J.-M. Halbout, S. von Molnár, T. Siegrist and F. Holtzberg, Phys. Rev. Lett. 55, 1128 (1985).
9. J.J. Zayhowski, C. Jagannath, R.N. Kershaw, D. Ridgely, K. Dwight and A. Wold, Solid State Commun. 55, 941 (1985), and J.J. Zayhowski, PhD thesis, MIT (1986).
10. M.B. Ketchen, T. Kopley and H. Ling, Appl. Phys. Lett. 44, 1008 (1984).
11. D.D. Awschalom, J. Warnock, J.R. Rozen and M.B. Ketchen, to be published in Proc. 31st Conference on Magnetism and Magnetic Materials, J. Appl. Phys., April 1987.
12. A. Twardowski, M. Nawrocki and J. Ginter, Phys. Stat. Sol. (b) 96, 497 (1979).
13. J. Warnock, R.N. Kershaw, D. Ridgely, K. Dwight, A. Wold and R.R. Galazka, J. Lum. 34, 25 (1985).
14. L.A. Kolodziejski, T.C. Bonsett, R.L. Gunshor, S. Datta, R.B. Bylsma, W.M. Becker and N. Otsuka, Appl. Phys. Lett. 45, 440 (1984).
15. J.A. Gaj, R. Planel and G. Fishman, Solid State Commun. 29, 435 (1979).
16. G.R. Wagner, J. Murphy and J.G. Castle,Jr., Phys. Rev. B 8, 3103 (1973).
17. A. Abragam and B. Bleaney in "Electron Paramagnetic Resonance of Transition Ions", Ch. 10, (Clarendon Press, Oxford 1970).
18. E.A. Harris and K.S. Yngvesson, J. Phys. C (Proc. Phys. Soc.) 1, 990 (1968).
19. Ref. 18, eq. 12 in a slightly modified form.
20. A. Lewicki, J. Spalek, J.K. Furdyna, and R.R. Galazka, J. Magn. Magn. Mat. 54-57, 1221 (1986).

TRANSIENT SPECTROSCOPY AND RELATED OPTICAL STUDIES IN DILUTED MAGNETIC SEMICONDUCTOR SUPERLATTICES

A. V. NURMIKKO*, L. A. KOLODZIEJSKI**, AND R. L. GUNSHOR**
*Division of Engineering, Brown University, Providence, RI 02912
**School of Electrical Engineering, Purdue University, West Lafayette, IN 47907

ABSTRACT

Recent successes in molecular beam (MBE) epitaxial growth of diluted magnetic semiconductor (DMS) artificial microstructures are now generating structures in which optically excited lower dimensional electronic and magnetic phenomena can be investigated. We consider an example which illustrates the present early state of affairs: **transient magnetic polarons** in connection with localized quasi 2-dimensional (2D) electronic excitations. We also comment on recent measurements and antiferromagnetic ordering in ultrathin layers in magnetic semiconductors.

INTRODUCTION

Among DMS artificial microstructures the CdTe/(Cd,Mn)Te and ZnSe/(Zn,Mn)Se heteropairs have been the two systems studied to date to any significant degree [1]. The MBE growth of these superlattices is opening exciting new vistas for a range of new studies in connection with lower dimensional electronic and (coupled) magnetic phenomena. At the same time, the field is very young and it is important to note that characterization of the basic electronic properties of the two wide gap systems is still quite incomplete. Coincidentally, experimental information about the structural details and perfection of such strained layer structures is somewhat lacking and optical studies are also filling a diagnostic role. Nonetheless, intriguing glimpses of novel phenomena have been obtained. In this article we consider possibilities for bound magnetic polarons in a lower dimensional structure with some experimental results through steady state and time resolved spectroscopy. We also pose the question of the role of antiferromagnetic (AF) effects in ultrathin layers of a magnetic semiconductor (MS).

WHAT IS THE COLOR OF A BOUND MAGNETIC POLARON IN A DMS QUANTUM WELL?

The idea of a bound magnetic polaron (BMP) in a bulk DMS crystal associated with equilibrium electronic states (shallow donors or acceptors) or nonequilibrium excitations (bound excitons) in concert with orbitally enclosed magnetic ion spins has been the subject of a substantial number of studies in recent years. As a variation of this theme, consider the spatial inhomogeneity added to the problem by a quantum well structure, such as based on the CdTe/(Cd,Mn)Te system (the term quantum well is used to indicated that the surrounding barrier layers are thick enough to prevent significant coupling between adjacent wells in a multiple layer structure). Figure 1 shows a schematic of the electronic potential (dashed line) for the one dimensional square well in a nonmagnetic case. The spin exchange interactions which couple the Bloch-like extended electronic states and those of the Mn-ion d-electron states are now restricted to occur (for the short range superexchange interaction) within the evanescent tails of the confined electron (hole) wavefunction (dashed line in Fig. 1). The exchange contribution tends to lower the system energy if the electron wave increases its penetration into the 'magnetically active' barrier. Variational calculations in the mean field (MFA) limit have been recently performed to estimate the resulting effects [2,3]; these show how even in the absence of an external magnetic field a self-consistent solution allows a spontaneous shift of the

Mat. Res. Soc. Symp. Proc. Vol. 89. © 1987 Materials Research Society

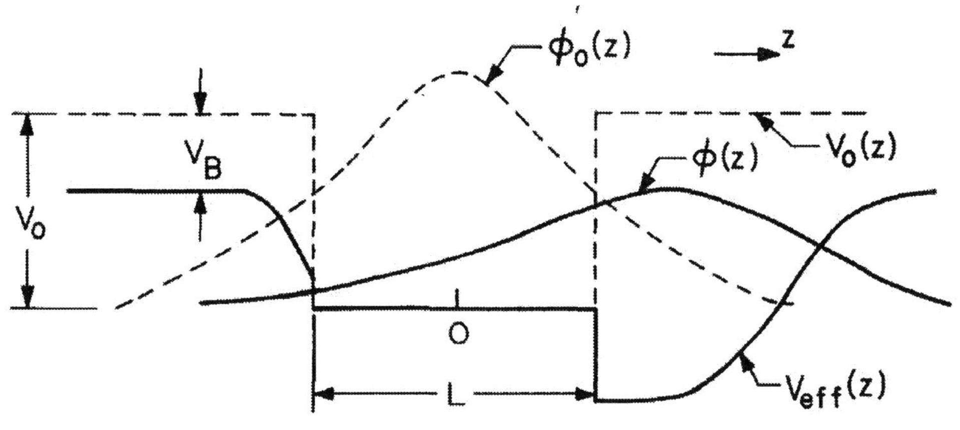

Figure 1: Schematic of a CdTe/(Cd,Mn)Te quantum well with exchange interaction (solid lines) and without (dashed lines).

wavefunction towards the barrier as an associated magnetic polarization develops within the overlapping electron density distribution. Such a BMP effect is illustrated schematically by the solid lines Fig. 1 for the wavefunction and modified effective potential of the quasi-2D electronic particle. The situation is optimized for a large effect if the confining potential V_o does not exceed the exchange contribution significantly; a condition approximated for heavy-hole like particles in the CdTe/(Cd,Mn)Te system.

Since doping of DMS superlattices and heterostructures is still at very early stages (see, however, [4]), the first experiments testing ideas such as the BMP in a lower dimensional environment have come through optical measurements of exciton recombination spectra. Theoretically, the problem as described by the one electronic scenario of a CdTe/(Cd,Mn)Te quantum well above is not fundamentally altered in the exciton case, if one assumes a small valence band offset and notes the much larger exchange coefficient which applies to a heavy-hole (heavy-hole exciton is the ground state in the strained layer QW structures). Physically, the large confinement of the electron (>300 meV) leads to a weak electron/BMP contribution and the exciton is mainly influenced by the 'magnetically attractive interface' as seen by the hole. Variational calculations, still in the MFA limit, [5,6] show significant changes in the exciton interband resonance (and in the Coulomb binding energy) which reflect BMP contributions even in zero external field. Figure 2 shows such calculated shifts (left panel) for a CdTe/(Cd,Mn)Te structure with a 60 Å quantum well width, x=0.10 in the barrier, and an assumed valence band offset of 24 meV (for other parameters, see Ref. 6). The remanent AF interactions in the Mn-ion spin system in the barrier have been accounted for in a phenomenological way [7]. The right hand panel shows how the hole envelope function is displaced from the center of the well (z_{oh}=0) even in zero external field. We also note that with sufficiently high external fields and different quantum well parameters it is possible to envision a situation in which the field actually inverts the heavy-hole well (i.e. a transition to a type II superlattice).

Such predictions for magnetically induced exciton 'interface localization' are borne out in luminescence experiments in which anomalously large spectral shifts of the exciton (interband) energy have been measured in external fields [8]. In Fig. 3, a large Zeeman effect is shown for a CdTe/(Cd,Mn)Te MQW structure with 71 Å CdTe well width and x=0.24 in a field perpendicular to layer plane; note the striking anisotropy as the field is turned along the plane. The anisotropy is

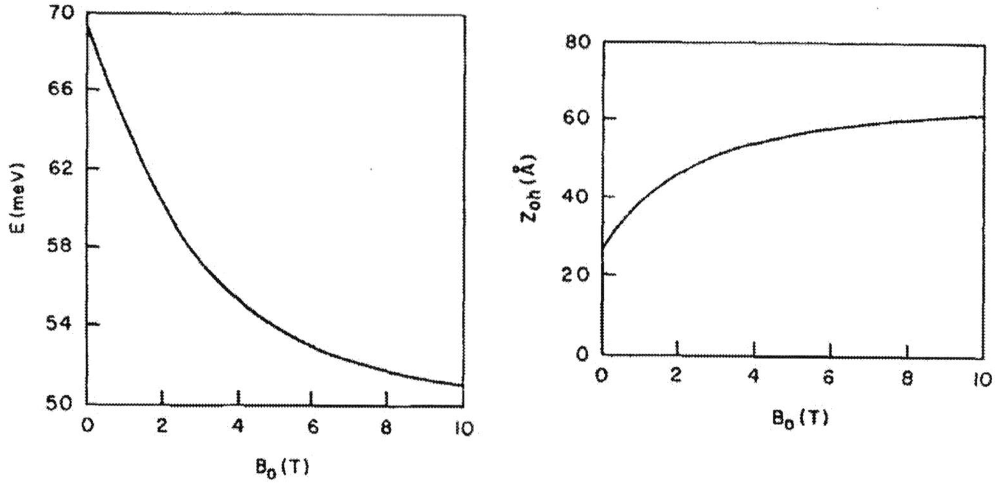

Figure 2: Left panel: calculated change in exciton interband energy of a CdTe/(Cd,Mn)Te quantum well (relative to CdTe bandedge) in external magnetic field; Right panel: change in the heavy-hole position in the well (z=0 refers to position at well center).

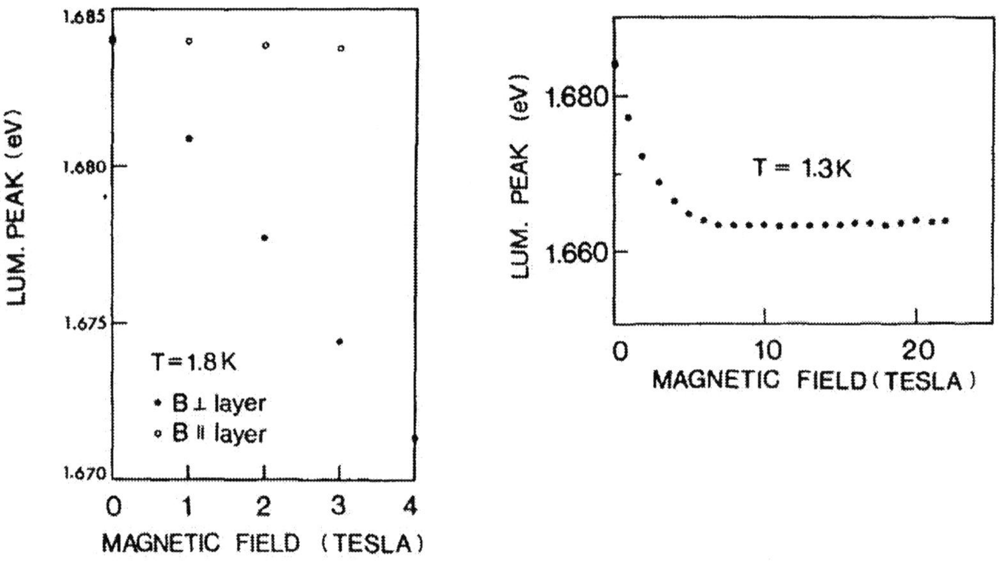

Figure 3: Experimentally observed heavy-hole exciton shifts for a CdTe/(Cd,Mn)Te MQW described in text in low and high fields.

expected for the nearly 2D heavy hole states given the usual description of the Heisenberg exchange (which now reduces to an Ising-like term). At higher magnetic fields evident saturation of the spin system is witnessed, together with a significant narrowing of the PL linewidth [9]. Experimental results such as these, which chronologically preceded the calculations outlined above, first suggested interface localization of excitons as a special characteristic in the CdTe/(Cd,Mn)Te quantum well structure. With the good agreement of the variation calculations one is tempted to attribute the experimental behavior entirely to the BMP phenomena; however, there are strong indications that strain induced random local potential fluctuations at the (111) oriented interfaces provide initial localization for the hole part of the exciton (so as to arrest its motion along the quantum well layer plane) [10].

An exciton/BMP complex offers in principle the additional advantage of time-resolved experiments for the study of its formation although the situation is inherently more complicated (specifically because of generally unknown details of the exciton localization which precedes the BMP formation). The BMP kinetics have been demonstrated recently in bulk (Cd,Mn)Se and (Cd,Mn)Te by time resolved optical methods which all derive their information from the exciton energy renormalization as the spin cloud is polarized in real time [11,12,13]. Additionally, direct time-dependent magnetization measurements have now been carried out in an elegant arrangement by Awschalom et al. [14]. In analog with the all-optical experiments, time-resolved photoluminescence has also been measured form CdTe/(Cd,Mn)Te MQW's, both from the standpoint of characterizing the structures in terms of recombination rates [15] and the dynamical details of the exciton localization and possible BMP formation. The observed time-dependent recombination spectra show distinct dynamical redshifts which are strongly influenced by the presence of external magnetic fields and temperature.

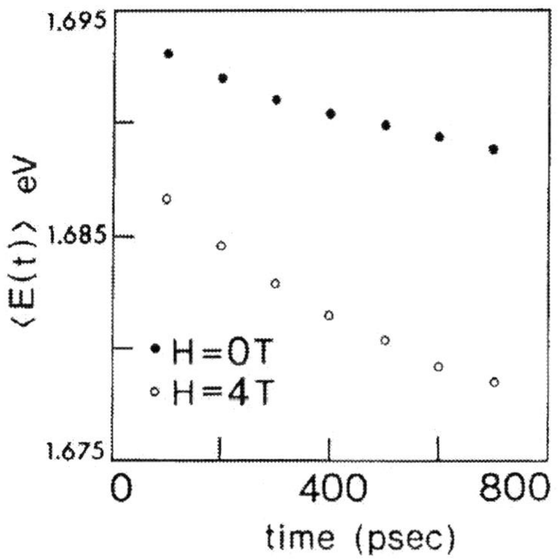

Figure 4: Transient spectral shifts in exciton luminescence of the CdTe/(Cd,Mn)Te MQW structure at T=2K with and without an applied field.

As an illustration of the time dependent spectroscopy, Fig. 4 shows the rate of change of mean exciton (interband) energy for the CdTe/(Cd,Mn)Te MQW structure already discussed in connection with Fig. 3. The data is taken from time resolved spectra following picosecond excitation in zero and 4 Tesla applied fields at T=1.6 K. The photon energy of excitation was into the low energy side of the absorption edge. As is evident from the data, the exciton spectral shifts increase in an applied field (in z-direction). With increasing temperature all the overall shifts decrease precipitously. Similar data have been observe by us on several MQW structures with distinct variations produced by changes in the quantum well dimensions and the Mn-ion concentration (including finite presence of the ion in the well itself). Qualitatively, these trends are compatible with a picture of a finite BMP contribution as follows: given the finite formation time of the complex (typically between 0.1 and 1 nsec in bulk material for B=0) and the relatively short lifetimes in the present day MQW's (typically less than 500 psec with a finite nonradiative component), the formation of a full BMP before recombination shuts off the exchange process is unlikely. Application of the external field facilitates the formation accelerating the process as seen in Fig. 3. With the presence of a finite Mn-ion concentration in the well, the BMP effect picks up an additional contribution from the spin polarization evolving in the well. In a quantum well the time dependent shifts are associated with the initial interface localization and the subsequent polaron development. It is also important to note that going beyond the mean field approximation, finite contributions to an instantaneous magnetization in the hole orbit are expected from statistical fluctuations within a thermodynamically random system of Mn-ion spins.

COMMENT ON THE ROLE OF ANTIFERROMAGNETIC INTERACTIONS IN SUPERLATTICE STRUCTURES

Experimental investigations on DMS multiple quantum well structures such as those described above have generally been performed with Mn-ion concentrations in the (Cd,Mn)Te or (Zn,Mn)Se barriers which exceed x=0.20. For the CdTe/(Cd,Mn)Te and ZnSe/(Zn,Mn)Se systems relatively high concentrations are needed in order to provide sufficient electron/hole confinement for quasi 2D effects to be significant. This raises questions about the influence of antiferromagnetic coupling of the Mn-ion spins which in bulk material severely restrict BMP effects. Qualitatively, in the CdTe/(Cd,Mn)Te (111) oriented structures, large exciton Zeeman effects with apparent BMP contributions have been seen by us in structures even with x=0.40. From a microscopic viewpoint, while the heterointerfaces in these quantum well structures are not completely understood, electron microscopy suggests some interfacial disorder. Finite disorder arising during MBE growth on a strained interface could manifest itself as an effective concentration gradient in the Mn-ion concentration across the heterointerface on the scale of a unit cell. The Mn spin-spin interaction is dominated by short range superexchange through the Te ions; thus departure from bulk antiferromagnetic behavior within a (Cd,Mn)Te barrier would be confined to such a thin transition region. This can be of importance in connection with the BMP effects considered above through the interface localized excitons, given the rather small Bohr radius of a heavy hole-like particle (13 Å).

A relevant connection can be made to recent work on ultrathin layers of the magnetic semiconductor MnSe [16] (in a cubic zincblende ZnSe matrix) which has in fact shown dramatic departures from anticipated AF ordering in structures which approximate the monolayer limit [17]. Apart from weaker AF interactions in a monolayer of MnSe through reduction of dimensionality (reduction in nearest number Mn ions), finite interface disorder on a submonolayer scale (perhaps in the form of small islands during incomplete 2D growth) has been suggested as being responsible for the absence of an AF phase transition [18]. Results of this work, which are still not completely understood, are nonetheless encouraging from the standpoint of extension of work on BMP effects towards an electronically and magnetically 2D limit in a magnetic semiconductor such as MnSe or MnTe.

The authors wish to acknowledge the participating contributions by X.-C. Zhang, S.-K. Chang, J.-W. Wu, J. J. Quinn, W. Goltsos, Donghan Lee, and S. Datta. This work was supported by Office of Naval Research contracts N00014-83-K0638 (Brown) and N00014-82-K0563 (Purdue).

REFERENCES

[1] for review see L. A. Kolodziejski, R. L. Gunshor, N. Otsuka, S. Datta, W. M. Becker, and A. V. Nurmikko, *IEEE J. Quant. Electron.* QE-22, 1666 (1986); A. V. Nurmikko, R. L. Gunshor and L. A. Kolodziejski, *IEEE J. Quant. Electron.* QE-22, 1785 (1986).

[2] C. E. T. Goncalves da Silva, *Phys. Rev.* **B32**, 6962 (1985).

[3] Ji-Wei Wu, A. V. Nurmikko, and J. J. Quinn, *Solid State Commun.* **57**, 853 (1986).

[4] J. F. Schetzina, these Proceedings.

[5] C. E. T. Goncalves da Silva, *Phys. Rev.* **B33**, 2923 (1986).

[6] Ji-Wei Wu, A. V. Nurmikko, and J. J. Quinn, *Phys. Rev.* **B34**, 1084 (1986).

[7] J. A. Gaj. R. Planel, and G. Fishman, *Solid State Commun.* **29**, 435 (1979).

[8] X.-C. Zhang. S.-K. Chang. A. V. Nurmikko, L. A. Kolodziejski, R. L. Gunshor and S. Datta, *Phys. Rev.* **B31**, 4056 (1985); A. Petrou, J. Warnock, R. N. Bicknell, N. c. Gilles-Taylor, and J. F. Schetzina, *Appl. Phys. Lett.* **48**, 692 (1985).

[9] X.-C. Zhang, S.-K. Chang, A. V. Nurmikko, D. Heiman, L. A. Kolodziejski, R. L. Gunshor, and S. Datta, *Solid State Commun.* **56**, 255 (1985).

[10] S.-K. Chang, A. V. Nurmikko, L. A. Kolodziejski, and R. L. Gunshor, *Phys. Rev.* **B33**, 2589 (1986).

[11] J. Harris and A. V. Nurmikko, *Phys. Rev. Lett.* **51**, 1472 (1983).

[12] D. D. Awschalom, J.-M. Halbout, S. von Molnar, T. Siegrist, and F. Holtzberg, *Phys. Rev. Lett.* **55**, 1128 (1985).

[13] J. J. Zayhowski, C. Jagannath, R. N. Kershaw, D. Ridgley, K. Dwight, and A. Wold, *Solid State Commun.*, **55**, 941 (1985).

[14] D. D. Awschalom and J. Warnock, these Proceedings and to be published.

[15] X.-C. Zhang, S.-K. Chang, A. V. Nurmikko, L. A. Kolodziejski, R. L. Gunshor, and S. Datta, *Appl. Phys. Lett.* **47**, 59 (1985).

[16] L. A. Kolodziejski, R. L. Gunshor, N. Otsuka, B. P. Gu, Y. Hefetz, and A. V. Nurmikko, *Appl. Phys. Lett.* **48**, 1482 (1986).

[17] A. V. Nurmikko, D. Lee, Y. Hefetz, L. A. Kolodziejski, and R. L. Gunshor, Proc. 18th Int. Conf. Physics of Semiconductors, Stockholm (1986).

[18] A. V. Nurmikko, D. Lee, Y. Hefetz, L. A. Kolodziejski, and R. L. Gunshor, Proc. 18th Int. Conf. Semiconductor Physics, Stockholm (1986); D. Lee, S.-K. Chang, H. Nakata, A. V. Nurmikko, L. A. Kolodziejski, and R. L. Gunshor, Proc. of Symposium D, MRS Fall meeting (1986).

TIME-RESOLVED PHOTOLUMINESCENCE OF CD(1-x)MN(x)SE AND CD(1-x)MN(x)TE AS A FUNCTION OF TEMPERATURE

J.J. Zayhowski,* R. N. Kershaw,** D. Ridgley,** K. Dwight, **A. Wold, ** R. R. Galazka,*** and W. Giriat*****
*MIT Lincoln Laboratory, Lexington,Massachusetts 02173-0073
**Department of Chemistry, Brown University, Providence, Rhode Island 02912
***Institute of Physics, Polish Academy of Science, Warsaw, Poland
****Centro de Fisica, IVIC, Apartado 1827, Caracas 1010A, Venevuela

ABSTRACT

The characteristics of the photoluminescence of Cd(1-x)Mn(x)Se (x = 0.05, 0.10, 0.20, 0.30) and Cd(1-x)Mn(x)Te (x = 0.20, 0.30, 0.45) change considerably as the sample temperature is reduced below the exciton-magnetic polaron (EMP) threshold temperature. At low temperatures the formation of magnetic polarons has large effects on the luminescence energy, the radiative lifetime, the radiative efficiency, and the spectral half-width of the luminescence. It is also observed that the formation time of the EMP increases almost linearly with temperature. All of these effects are explained with a simple model for the EMP.

Introduction

Cd(1-x)Mn(x)Se and Cd(1-x)Mn(x)Te belong to a group of materials known as semimagnetic semiconductors (or dilute magnetic semiconductors). In these materials there is a sizable exchange interaction between the band electrons and localized magnetic ions. This interaction leads to many novel spin dependent phenomena including magnetic polarons [1-3]. A magnetic polaron forms when a carrier becomes localized. The spin of the carrier induces a net spin alignment of the magnetic ions in its vicinity. These magnetic ions then create a potential hole for the carrier, and it becomes further localized. This process continues until the cost of carrier localization in kinetic energy balances the change in potential. As the magnetic polarons form many properties of the excitons associated with them change. At low temperatures the luminescence spectra of Cd(1-x)Mn(x)Se and Cd(1-x)Mn(x)Te are dominated by exciton-magnetic polaron (EMP) complexes, making time-resolved photoluminescence an ideal tool for studying their evolution.

Time-resolved studies of magnetic polarons have utilized the pump-probe technique [4] and, more recently, time-resolved Faraday rotation [5]. This paper reports the results of the most complete time-resolved photoluminescence studies of magnetic polarons in Cd(1-x)Mn(x)Se and Cd(1-x)Mn(x)Te to date [6].

The Exciton-Magnetic Polaron Model

In most of the semimagnetic semiconductors, including Cd(1-x)Mn(x)Se and Cd(1-x)Mn(x)Te, the exchange constant is larger for the hole than for the electron. In addition, the hole mass is considerably larger, making it easier to localize. As a result, magnetic polaron effects are much stronger for the hole than for the electron, and EMPs consist of a strongly localized hole-magnetic polaron and an electron with an extended wavefunction, as illustrated in Figure 1. In Cd(1-x)Mn(x)Se and Cd(1-x)Mn(x)Te the hole wavefunction in a magnetic polaron may be as small as 15 A, while the electron wavefunction is on the order of 50 A, determined largely by the Coulomb interaction with the hole [6]. This simple model will allow us to understand all of the experimental results.

Mat. Res. Soc. Symp. Proc. Vol. 89. ©1987 Materials Research Society

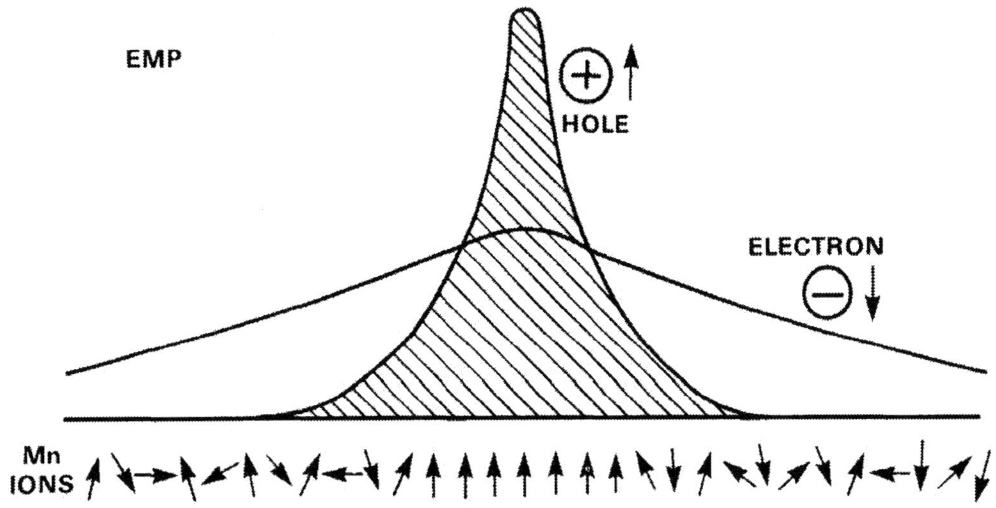

Fig. 1 Model for the exciton-magnetic polaron complex

Experimental Technique

The experimental results were obtained with highly compensated single crystals of Cd(1-x)Mn(x)Se and Cd(1-x)Mn(x)Te grown by the Bridgman technique. The samples were cooled in helium vapor under vacuum. Above bandgap optical excitation is provided by a picosecond dye laser. No dependence on laser wavelength is observed. The pump intensity is maintained at about 0.2 mW, focused onto a 50 micron spot. No intensity-dependent effects are observed at this power level. The hexagonal c-axis of the Cd(1-x)Mn(x)Se crystals are oriented parallel to the incident beam. Luminescence is collected in the back-scattering geometry, passed through a spectrometer, and time-resolved with a synchronous optical streak camera. The experimental resolution is 80 ps and 1 meV. Details of the experiment are provided elsewhere [6,7].

Experimental Results

A sample by sample summary of the experimental results for the Cd(1-x)Mn(x)Se (x = 0.05, 0.10, 0.20, and 0.30) and Cd(1-x)Mn(x)Te (x = 0.20 and 0.30) samples is given in Fig. 2 and 3. Although the Cd(0.55)Mn(0.45)Te sample showed clear indications of magnetic polaron formation, its luminescence was dominated by an intramanganese transition and could not be time-resolved [6]. The interpretation of these figures is as follows:

ENERGY: The dashed line corresponds to the energy of a free exciton. It is obtained from a linear fit to experimental values. For most of the samples a 1.8K value was available from reflectivity studies [8]. Other points were taken from the literature [8-10]. These curves are shifted by the amounts shown in parentheses. Solid circles mark the peaks of the time-averaged luminescence spectra; open circles mark the half-maxima. (At low temperatures the time-averaged luminescence spectrum of the Cd(0.90)Mn(0.10)Se sample is double peaked [6,7].) The vertical lines cover the energy range over which the peak of the luminescence spectrum moves as the magnetic polaron evolves.
FWHM: The full widths at half maximum of the instantaneous luminescence spectra after they have shifted to their final observable positions.
LIFETIME: The instantaneous lifetimes of the EMPs are obtained by fitting the luminescence intensities to an equation of the form: [6,7].

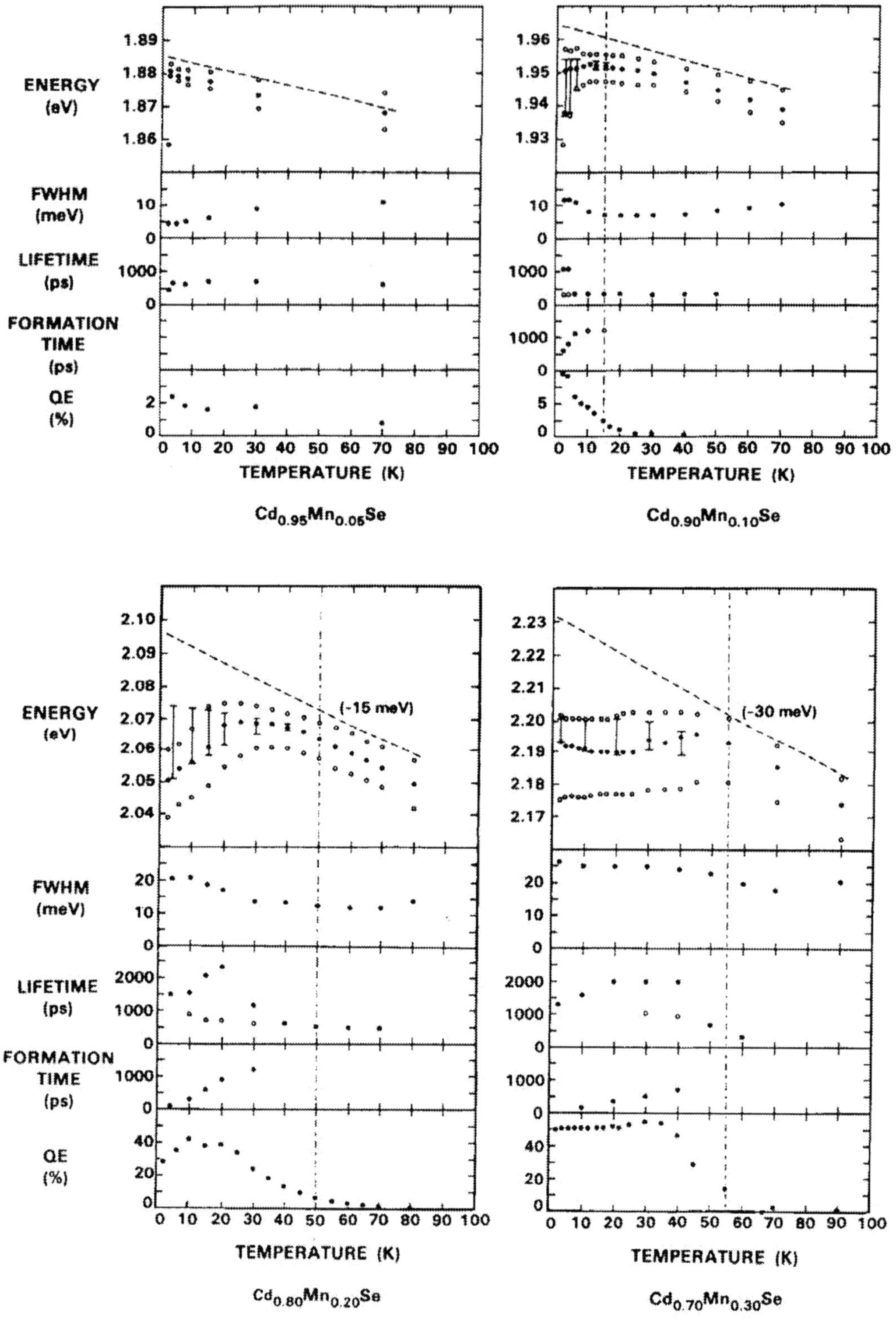

Fig. 2 Summary of time-resolved photoluminescence studies of Cd(1-x)Mn(x)Se

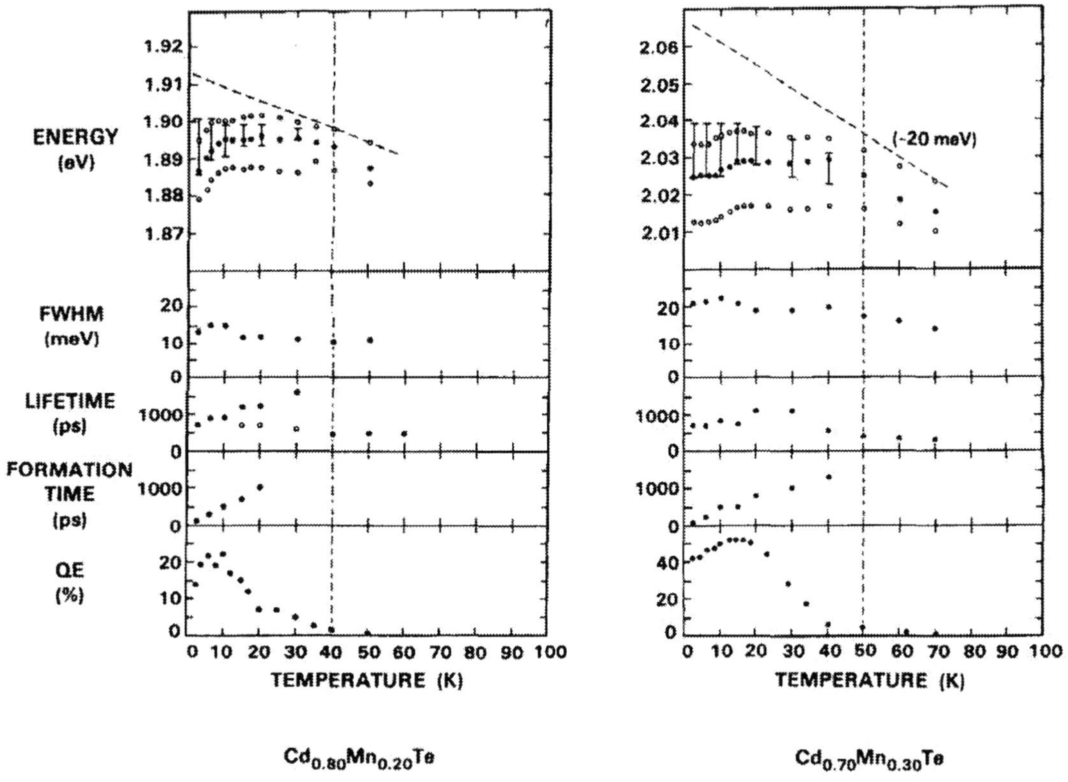

Fig. 3 Summary of time-resolved photoluminescence studies of Cd(1-x)Mn(x)Te

$$I(t) = A \exp(-t/a) + B \exp(-t/b), \qquad (1)$$

where A,a,B and b are adjustable parameters. The instantaneous lifetime is then given by

$$\tau(t) = -d/dt \ln(I(t)). \qquad (2)$$

The solid circles correspond to the instantaneous lifetimes of the EMPs after the formation of magnetic polarons, or the exciton lifetimes if no magnetic polaron formation is observed. The open circles denote the instantaneous lifetimes of the EMPs at t = 0, if different than above.

FORMATION TIME: The EMP formation time, τ_f, is obtained by fitting the peaks of the instantaneous luminescence spectra to an equation of the form: [6,7,12]

$$E(t) = E(\infty) + [E(0) - E(\infty)] \exp(t/\tau_f). \qquad (3)$$

QE: The quantum efficiencies of the radiative processes are obtained from an estimate of the collection efficiency of the experimental system and a direct measure of the power of the collected radiation [6].

The Cd(0.95)Mn(0.05)Se sample shows no indication of magnetic polaron formation. The behavior of the luminescence spectra of the other samples can be divided into two regimes, separated by an "EMP threshold temperature," indicated by the vertical dashed line in Fig. 2 and 3. Above the EMP threshold temperature the following characteristics are observed:

(1) The energies of the luminescence spectra do not change with time. They decrease linearly with temperature, paralleling the free exciton energies.

(2) The half-widths of the luminescence spectra increase with temperature, consistent with thermal broadening .
(3) Luminescence is characterized by a single lifetime, which is relatively independent of temperature.
(4) The radiative quantum efficiencies of the samples are low.

Below the EMP threshold temperature the characteristics are very different:
(1) The positions of the luminescence spectra shift to lower energies with time. Within the uncertainty of the free exciton positions, the zero time ($t = 0$) positions of the luminescence spectra track the free excitons.
(2) With decreasing temperature the temporal energy shifts of the luminescence spectra increase, with the exception of the Cd(0.70)Mn(0.30)Te sample below 10 K.
(3) The EMP formation times increase approximately linearly with temperature.
(4) As a result of (1), (2) and (3), the energies of the time-averaged luminescence spectra level off, and then decrease with decreasing temperature. In the high Mn concentration samples ($x > 0.30$), at low temperatures the decrease stops and the energies begin to increase. This effect is well documented for Cd(1-x)Mn(x)Te [9].
(5) The instantaneous half-widths of the luminescence spectra increase initially, [6,7] and then remain constant. The final values of the half-widths are greater than the values above the EMP threshold temperature, decreasing as the threshold is approached.
(6) The instantaneous EMP lifetimes increase with time. The initial values are about the same as the exciton lifetimes above the EMP threshold temperature. The final values are much longer. The evolution of the lifetimes occur in a time consistent with the EMP formation times [6,7].
(7) The final lifetimes of the EMPs increase with increasing temperature.
(8) The radiative quantum efficiencies of the samples are greatly enhanced below the EMP threshold, decreasing as the threshold is approached.

With increases in Mn concentration the following trends are observed:
(1) The EMP threshold temperatures increase.
(2) The instantaneous half-widths of the luminescence spectra increase.
(3) The EMP formation times decrease.
(4) The radiative quantum efficiencies of the samples below the EMP threshold temperature increase.

Qualitatively, all of the effects observed in the Cd(1-x)Mn(x)Se and the Cd(1-x)Mn(x)Te samples are similar. The main differences observed are:
(1) The EMP threshold temperatures are lower for the Cd(1-x)Mn(x)Te system.
(2) The EMP lifetimes are shorter for the Cd(1-x)Mn(x)Te system.
(3) EMP formation times are slightly longer for the Cd(1-x)Mn(x)Te system.
(4) The differences in energy between the EMP luminescence spectra and the free excitons are less for the Cd(1-x)Mn(x)Te system.

All of the experimental results are very repeatable. The main difference observed when an experiment is repeated is a slight shift in all of the energies. This is probably due to inhomogeneities in the samples. In addition, the experimental results are consistent with the data available in the literature [4,5].

Discussion

Most of the experimental observations can be explained with the EMP model presented at the beginning of this paper, and none are inconsistent with it. As the magnetic polaron forms, the increased binding energy of the EMP causes a decrease in the energy of the exciton luminescence. The increase localization of the exciton reduces the probability of capture by nonradiative recombination centers, increasing the excitons lifetime and radiative quantum efficiency. The lifetime of the exciton before EMP formation is believed to be dominated by nonradiative processes, and to approach the radiative lifetime as the EMP forms. Similar effects have been seen in nonmagnetic systems [13]. In additions, the localization of the exciton, and especially the strong localization of the hole, increase the inhomogeneous broadening of the luminescence features, since the wavefunctions sample a smaller portion of the inhomogeneous crystals. Also, the localization of the hole changes the overlap of the electron and hole wavefunctions in real space and in phase space, affecting the radiative lifetime of the exciton. Since maximum overlap, and therefore minimum radiative lifetime, is achieved with electron and hole wavefunctions of the same dimension, the increased localization of the hole increases the exciton lifetime. Finally, the exchange interaction between the magnetic polaron and the electron correlates the spins of the electron and hole. The energetically favored electron spin state allows radiative recombination of the carriers, the opposite spin state does not. Therefore, in the presence of a magnetic polaron, thermal excitation of the electron into the higher energy state increases the radiative lifetime of the exciton.

The differences in the observations for increasing Mn concentrations are consistent with stronger magnetic interaction (due to the increase in Mn) and increased inhomogeneities resulting from the alloy concentration. The differences in the observations for the Cd(1-x)Mn(x)Se and Cd(1-x)Mn(x)Te systems are consistent with the smaller exchange constants of the latter.

The linear dependence of the EMP formation time with temperature is explained by an electron mediated spin diffusion model of magnetic polaron formation, which starts with the EMP model presented here [6,12].

Acknowledgements

The authors wish to acknowledge P.A. Wolff, E.P. Ippen and R.L. Aggrawal for stimulating discussion and support during various stages of this work. This work was performed, in part, as part of the Ph.D. thesis research of J.J. Zayhowski [6]. It was supported in part by a grant from the Joint Service Electronic Program under DAAG29-83-K-0003, and in part by the National Science Foundation through its Division of Materials Research. J.J. Zayhowski was supported by the Fannie and John Hertz Foundation.

References

1. R.R. Galazka, Proc. Int. Conf. Physics of Semiconductors, Edinburgh (1978) p.133.
2. J.A. Gaj, Proc. Int. Conf. Physics of Semiconductors, Kyoto (1980) [J. Phys. Soc. Japan Suppl. A49, 797 (1980)].
3. D. Heiman, P.A. Wolff and J. Warnock, Phys. Rev. B27, 4848 (1983).
4. J.H. Harris and A.V. Nurmikko, Phys. Rev. Lett. 51, 1472 (1983).
5. D.D. Awschalom, J.-M. Halbout, S. von Molnar, T. Siegrist and F. Holtzberg, Phys. Lett. 55, 1128 (1985).
6. J.J. Zayhowski, Ph.D. Thesis (Unpublished), Department of Electrical Engineering, M.I.T. (1986).
7. J.J. Zayhowski, C. Jagannath, R.N. Kershaw, D. Ridgley, K. Dwight, and A. Wold, Solid State Commun. 55, 941 (1985).
8. J. Warnock, private communications.
9. A. Golnik, J. Ginter and A.J. Gaj, J. Phys. C. Solid State Phys. 16, 6073 (1983).
10. J. Stankiewicz, Centro de Fisica, Instituto Venezolano de Investigaciones, preprint.
11. Additional reference available in reference 4.
12. J.J. Zayhowski, R.N. Kershaw, D. Ridgley, K. Dwight, A. Wold, R.R. Galazka and W, Giriat, submitted to Phys. Rev. B.
13. E. Cohen and M.D. Sturge, Phys. Rev. B25, 3828 (1982).

SPIN-TEXTURE IN ACCEPTOR-BOUND MAGNETIC POLARONS

ERIC D. ISAACS AND PETER A. WOLFF
Department of Physics and Francis Bitter National Magnet Laboratory
Massachusetts Institute of Technology, Cambridge, MA 02139

ABSTRACT

We outline a theory of the acceptor-bound magnetic polaron in cubic diluted magnetic semiconductors that exhibits a nonuniform magnetization. Calculations show that the manganese spins have an overall z-alignment, but are canted in an azimuthally symmetric way from that direction. This is a new phenomenon we call BMP spin-texture. Spin-texture results from valence band degeneracy, and the important spin-orbit effects associated with it. The possibility of studying BMP spin-texture via neutron scattering is discussed.

INTRODUCTION

The bound magnetic polaron (BMP) is a collective state of both magnetic and semimagnetic semiconductors (SMSC) consisting of an impurity bound carrier that induces a ferromagnetic spin cloud [1] in the surrounding magnetic ions. The simple band structure and good optical properties of SMSC has facilitated the development of theory, and comparison with experiment, for the BMP. These studies have shown that only the acceptor-BMP, and not the less-strongly coupled donor-BMP [2], exhibits the full range of BMP behavior; from the high temperature spin-fluctuation dominated regime, to the low temperature spin-aligned regime.

We outline a theory of acceptor-BMP in cubic DMS, such as CdMnTe, that includes valence band degeneracy, and the important spin-orbit effects associated with it. In the manifold of the ground state, CdTe-like acceptor wavefunctions, we show that spin-orbit coupling results in a spatially varying hole spin alignment. The hole spin has an overall z-alignment, but is canted from that direction in a well defined pattern. Luminescence experiments [3] and theory [4,4] have shown that the Mn^{2+} spins within an acceptor-BMP align with the exchange field of the hole for temperatures $T \lesssim 30° K$. The hole spin thus induces a nonuniform magnetization in the surrounding Mn^{2+} spins - the new phenomenon we call BMP spin-texture. This is in contrast to the hydrogenic acceptor-BMP model [4-6], where the simpler non-degenerate band states result in a uniform magnetization.

SPHERICAL ACCEPTOR MODEL

To analyze the acceptor-BMP problem we start with the ground state solutions to the spherical model Hamiltonian for a shallow acceptor of Baldereschi and Lipari (BL) [7]. The BL Hamiltonian can be thought of as describing a hydrogen atom with electron spin J = 3/2, perturbed by a "spin-orbit" coupling term. The spin-orbit term mixes the four, J = 3/2 band edge Bloch states such that the solutions are eigenstates of the total angular momentum $\underline{F} = \underline{L} + \underline{J}$. For example, one of the four ground state wavefunctions is

$$|F=3/2,\ F_z=3/2\rangle \ =$$

$$f_0(r)\ Y_0{}^0(\Omega)|\ J=3/2,\ J_z=3/2\rangle \ + \ g_0(r)\ \{1/\sqrt{5}\ Y_2{}^0(\Omega)|\ J=3/2,\ J_z=3/2\rangle$$

Mat. Res. Soc. Symp. Proc. Vol. 89. ©1987 Materials Research Society

$- \sqrt{2/5}\, Y_2^1(\Omega) | J=3/2, J_Z=1/2\rangle \;+\; \sqrt{2/5}\, Y_2^2(\Omega) | J=3/2, J_Z=-1/2\rangle \}$.

$f_0(r)$ and $g_0(r)$ are radial functions determined in BL, the Y_l^m are the spherical harmonics, and the $|J, J_Z\rangle$ are the valence band edge Bloch states. Note that, due to angular momentum selection rules, this set of wavefunctions diagonalizes the BMP Hamiltonian to order $1/\sqrt{N}$, where N is the number of manganese ions within an acceptor Bohr orbit. In the continuum limit (dense Mn^{2+}) the BMP Hamiltonian is exactly diagonal.

Considering the expectation value $\langle F=3/2, F_Z=3/2|\underline{J}|F=3/2, F_Z=3/2\rangle$, averaged over the unit cell (i.e., over the Bloch states), we note that all three components J_X, J_Y and J_Z are finite:

$$\langle J_X\rangle = 3/4\pi \; \{f_0(r)\, g_0(r) \;+\; 1/2\, g_0^2(r)\, (\cos^2\theta + 1)\} \;\; \cos\theta\, \sin\theta\, \cos\phi$$

$$\langle J_Y\rangle = 3/4\pi \; \{f_0(r)\, g_0(r) \;+\; 1/2\, g_0^2(r)\, (\cos^2\theta + 1)\} \;\; \cos\theta\, \sin\theta\, \sin\phi$$

and

$$\langle J_Z\rangle = 3/8\pi \; \{f_0^2(r) \;+\; f_0(r)\, g_0(r)\, (3\cos^2\theta - 1) \;+\; g_0^2(r)\, \cos^4\theta \;\} \;.$$

where the coordinates $\{r,\theta,\phi\}$ are defined with respect to the axis of quantization for F (the z-axis). These formulas imply that the hole spin is aligned along the z-axis, but is canted in an azimuthally symmetric way from that direction.

The acceptor-BMP Hamiltonian includes a hole-Mn^{2+} spin exchange interaction term of the Heisenberg form

$$H_{exch} \;=\; -I \sum_i \left[(\underline{J}\cdot\underline{S}_i)\; \delta(\underline{r} - \underline{R}_i) \right],$$

where I is an exchange integral, $\{\underline{r},\underline{J}\}$ are the coordinate and spin of the trapped hole, and $\underline{S}_i$ is the spin of a manganese ion at lattice site $\underline{R}_i$. The expectation value of the exchange Hamiltonian with respect to the BL wavefunction can be written in terms of an exchange field as

$$\langle F=3/2, F_Z=3/2|\; H_{exch}\; |F=3/2, F_Z=3/2\rangle \;=\; - \mu_B g_{Mn} \sum_i \left[(\underline{B}_{exch}\cdot\underline{S}_i)\; \delta(\underline{r} - \underline{R}_i) \right],$$

where $\underline{B}_{exch} = I\langle F=3/2, F_Z=3/2\, |\underline{J}|\, F=3/2, F_Z=3/2\rangle / \mu_B g_{Mn}$, μ_B is the Bohr magneton and $g_{Mn} = 2$. From this substitution it is clear that the Hamiltonian can be thought of as describing a set of manganese spins, within an acceptor Bohr orbit, experiencing the exchange field of a trapped hole. As a result, for $T \lesssim 30°K$, the manganese spins tend to align with the canting hole spin. This is the phenomenon of acceptor-BMP spin-texture. Figure 1 shows spin-texture for CdMnTe at T = 0°K.

HOLE-Mn^{2+} SPIN CORRELATION

The acceptor-BMP has as many as 6^N spin states, where N is the number of magnetic ions within an acceptor Bohr orbit and can be as large as 100. Many of these states are accessible to the BMP, even at low temperatures. Thus, many of the BMP properties are determined by the spin partition function. One of these, the hole-Mn^{2+} spin correlation function, is given by the ensemble average

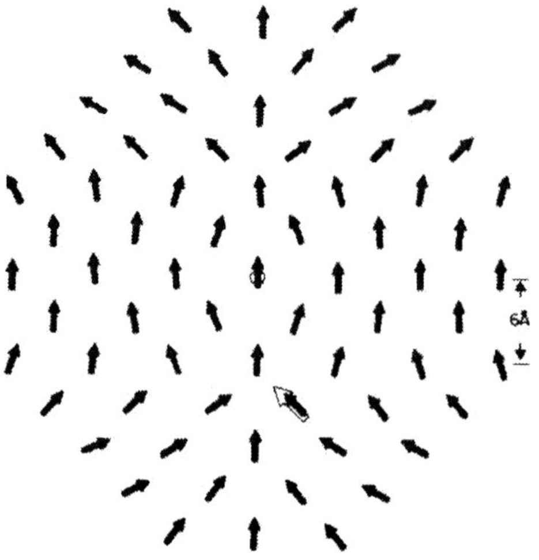

FIGURE 1. Spin-texture in an ideal acceptor-BMP. A Mn^{2+} spin has been indicated at every cation, fcc sublattice site for clarity and aesthetics. In an actual sample, only a fraction of these sites will be occupied, randomly, by a magnetic ion. Note that the texture is azimuthally symmetric about the direction of the net moment (z-axis).

$\uparrow$ - Mn^{2+} ion spin direction.

$\Uparrow$ - Hole spin direction.

O - Acceptor impurity center.

$$\langle \underline{F}\cdot\underline{S}_k \rangle = \frac{\underset{\{\underline{F},\underline{S}_k\}}{\mathrm{Tr}} \langle F,F_z| \left[(\underline{J}\cdot\underline{S}_k) e^{-H_{exch}}\right] |F,F_z\rangle}{\underset{\{\underline{F},\underline{S}_k\}}{\mathrm{Tr}} \langle F,F_z| \left[e^{-H_{exch}}\right] |F,F_z\rangle} .$$

$\langle \underline{F}\cdot\underline{S}_k \rangle$ can be evaluated in the continuum limit following Wolff [5] and is plotted in Figure 2 along the directions of a few of the neighboring cation, fcc sublattice sites for T = 5 K. We see that not only does texture exhibit a nonuniform magnetization direction, but also a nonshperically symmetric hole-Mn^{++} spin correlation.

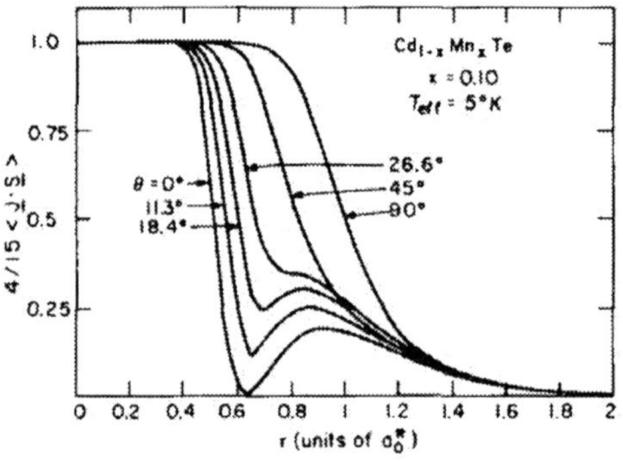

Figure 2. Hole-Mn^{2+} spin-correlation in the continuum limit, plotted as a function of distance from the impurity center and along the directions of a few of the neighboring fcc, cation sublattice sites.

NEUTRON SCATTERING

Acceptor-BMP spin-texture should be most pronounced in the SMSC materials for which there is significant valence band mixing. Such is the case for both (Cd,Mn)Te and (Zn,Mn)Te, where a degenerate valence band is perturbed by a large spin-orbit coupling. However, this is not the case for (Hg,Mn)Te, where, even with valence band degeneracy, the spin-orbit coupling is small and the BMP wavefunction is best described by a single band, hydrogenic model. The noncubic SMSC, such as (Cd,Mn)Se, are not considered here because the sizable crystal field splitting of the valence band (about 20 meV) leads to negligible mixing. In addition, they are by nature n-type and thus contain only a small number of acceptors.

Neutron scattering has been used as a tool to study the magnetic properties in SMSC; elastic scattering has been used to identify magnetic ordering [8], and inelastic scattering to determine the Mn-Mn exchange constant [9]. Elastic neutron scattering from the Mn spins could provide a striking visualization of the spatially varying hole spin density in the acceptor-BMP. At the same time it could provide us with a picture of how the size of the BMP, i.e., the spatial extent of the spin correlation, varies with temperature, magnetic field, and manganese concentration.

Neutron scattering from acceptor-BMP is difficult. For example, in a typical sample of (Cd,Mn)Te there are only about 10^{17} acceptor/cc. For a manganese molar concentration of x = 0.10 and an acceptor Bohr radius of 13 angstroms, only about 0.05% of the total number of manganese spins participate in BMP scattering. In addition, the cation elements in SMSC tend to be efficient neutron absorbers, the largest of which is Cd^{112}, which is often used as a control rod in nuclear reactors. Because of these limitations we have tried the first set of experiments using a thin (1 mm thick) sample of $Hg_{0.09}Mn_{0.01}Te$ with 2 x 10^{17} acceptors/cc. Preliminary results show no BMP effects, which we believe to be caused by the relatively weak hole-Mn^{2+} coupling (large acceptor Bohr radius) leading to small polaronic effects. We are presently considering other materials, such as (Cd,Mn)Te grown with the isotope Cd^{114}, a considerably weaker neutron absorber, and p-(Zn,Mn)Te, for future experiments.

ACKNOLWEDGEMENTS

We would like to thank Dr. Donald Heiman for his valuable insight, and Dr. Julius Hastings and Dr. Steven Shapiro at Brookhaven National Laboratories for their time and effort in running the neutron scattering experiments. EDI is supported by an AT&T Bell Laboratories Ph.D. Scholarship. This work was supported by the National Science Foundation Grants Nos. DMR-8504366 and DMR-8511789.

REFERENCES

1. T. Kasuya and A. Yanase, Rev. Mod. Phys. 40, 684 (1968).
2. D. Heiman, P.A. Wolff, and J. Warnock, Phys. Rev. B27, 4848 (1983).
3. T.H. Nhung and R. Planel, Proc. 16th Int. Conf. Phys. Semicond., Montpellier (1981); Physica 117B and 118C, 488 (1973).
4. J. Warnock and P.A. Wolff, Phys. Rev. B31, 6579 (1985);
5. P.A. Wolff, Semiconductors and Semimetals, ed. by R.K. Willardson and A.C. Beer, (Academic Press, Inc., New York, 1986).
6. A. Golnick, J.A. Gaj, M. Nawrocki, R. Planel, and C. Benoit a la Guillaume, J. Phys. Soc. Japan, Suppl. A, 49, 819 (1980).
7. A. Baldereschi and N.O. Lipari, Phys. Rev. B8, 2697 (1973).

8. T. Giebultowicz, H. Kepa, B. Buras, K. Clausen and R.R. Galazka, Solid State Commun. 40, 499 (1981); G. Dolling, T.M. Holden, V.F. Sears, J.K Furdyna, and W. Giriat, J. Appl. Phys. 53, 7644 (1982).
9. L.M. Corliss, J.M. Hastings, S.M. Shapiro, Y. Shapira, and P. Becla, Phys. Rev. B33, 608 (1986).

EFFECTS OF EXCHANGE INTERACTION IN DILUTED MAGNETIC SEMICONDUCTOR QUANTUM WELLS

JACEK KOSSUT* AND JACEK K. FURDYNA
Department of Physics, Purdue University, West Lafayette, IN 47907, U.S.A.

ABSTRACT

The presence of transition metal ions (typically Mn^{2+}) in diluted magnetic semiconductors (DMS) results in a strong spin-spin coupling between localized magnetic moments and band electrons. This leads to considerable modifications of the semiconductor band structure in the presence of strong magnetic fields, e.g., to large spin-dependent shifts of the electronic states at the band edge. This feature is of particular interest in the context of quantum wells involving DMS. Starting with the original idea of a "spin-superlattice", we concentrate on various opportunities which arise due to the tunability of the depth of the quantum wells by the magnetic field and/or temperature associated with the aforementioned spin-dependent effects. Thus, we discuss boil-off and freeze-out of electrons to and from quantum wells, selective spin tunneling across the barriers, tunable infrared emitters, enhancement of electronic g-factors in shallow non-magnetic wells surrounded by DMS barriers, the possibility of transition from a type-I to a type-II superlattice induced by the magnetic field, and quantum oscillations anomalies in DMS quantum wells.

INTRODUCTION

The unique properties of diluted magnetic semiconductors (DMS) continue to attract considerable scientific interest [1]. The majority of the studies carried out so far concern bulk crystals of DMS. With the advances in the molecular beam epitaxy and other film deposition techniques, the possibility of growth of good quality quantum well structures and superlattices involving DMS has been demonstrated [2-5]. These achievements facilitate the extension of the studies on DMS also to quasi-two-dimensional (2D) electronic systems. We shall not attempt here to give an extensive review of the results of measurements performed already on 2D structures of DMS [6], but rather concentrate on novel opportunities which these systems appear to hold in store. More specifically, we shall limit our discussion to electronic properties of DMS quantum wells, leaving such topics as the possible modifications of the magnetic properties due to their 2D character [7,8] outside the scope of this paper.

The most extensively investigated and most thoroughly understood group of DMS (i.e., semiconducting mixed crystals containing substitutional transition metal ions) are II-VI compounds with Mn^{2+} ions. Of these, $Hg_{1-x}Mn_xTe$ and $Hg_{1-x}Mn_xSe$ are narrow-gap semiconductors, and $Cd_{1-x}Mn_xTe$ and $Zn_{1-x}Mn_xSe$ are examples of semiconductors with wide energy gap. These materials crystallize in the zinc blende or in the wurtzite structures, with good quality monocrystalline samples obtainable even for sizable amount of Mn^{2+} present in the host lattice (e.g., in the case of $Cd_{1-x}Mn_xTe$ the limit of Mn^{2+} solubility is close to $x \simeq 0.7$).

There are several features which make DMS particularly attractive. First, both the lattice parameters and the energy gap change smoothly with the the Mn^{2+} mole fraction x, making it possible to "tailor" these parameters appropriately. Needless to say, this fact is of special importance in the context of the quantum wells and superlattices, since it provides the means of controlling the quantum well depths, as well as the degree of lattice mismatch (which determines internal strains) between the adjacent layers forming the structure. To illustrate this in the case the telluride DMS family $Cd_{1-x}Mn_xTe$, $Zn_{1-x}Mn_xTe$ and $Cd_{1-x}Zn_xTe$, we

Mat. Res. Soc. Symp. Proc. Vol. 89. © 1987 Materials Research Society

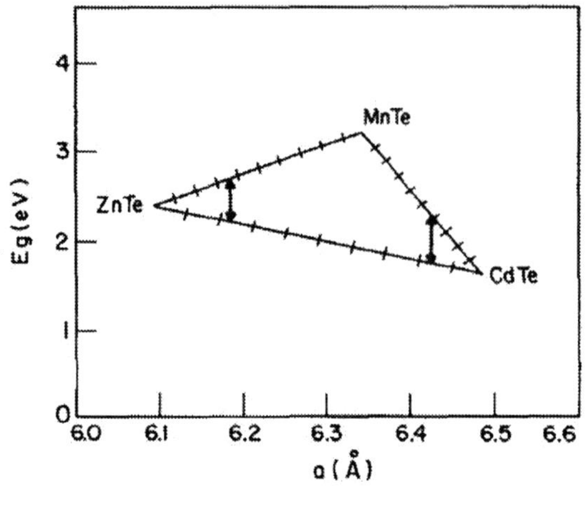

Fig.1. Energy gap as a function of the lattice constant in three ternary compounds $Cd_{1-x}Mn_xTe$, $Zn_{1-x}Mn_xTe$, and $Cd_{1-x}Zn_xTe$. The distance between two tick marks on each side of the triangle represents an increment $\Delta x = 0.1$ of the molar fraction x. The length of the thick arrows corresponds to the gap discontinuity $\Delta E_g = 0.5$ eV.

show in Fig.1 the plot of the energy gaps as afunctionof the lattice constants for these three materials. The corners of the triangle correspond to the parental binary compounds, and a point on any of the sides represents a ternary alloy of a given composition. The value of the molar fraction x correspondig to a given point on the triangle is given by its distance from the apices. A vertical line connecting two points of the triangle then determines the compositions of the two ternaries which are perfectly lattice-matched, while the length of the connecting line gives the value of the energy gap discontinuity.

Another physical property which makes the DMS extremely interesting is the fact that the magnetic moments localized on Mn^{2+} ions are strongly coupled via sp-d exchange interaction to the conduction and valence band electrons. This coupling strongly affects the energies of the band and impurity states, thus leading to significant modifications of various electronic properties of DMS. In particular, it was shown theoretically, as well as demonstrated in magnetotransport [9] and magnetooptical [10] experiments, that the exchange interaction affects quite dramatically the spin splitting of the electronic states in the presence of a strong magnetic field. The origin of this "amplification" of the spin properties is best illustrated by the expression describing the electronic g-factor in the parabolic approximation (particularly suitable for the conduction electrons in wide-gap DMS), where we can write

$$g_{eff} = g^* - \frac{\alpha M}{g_{Mn}\mu_B^2 B} \tag{1}$$

Here g^* is the g-factor determined solely by the band parameters, α is the s-d exchange constant for the conduction band, M is the magnetization, μ_B is the Bohr magneton, g_{Mn} is the g-factor of Mn^{2+} ions, and B is the external magnetic field. The second term in Eq.(1) is entirely due to the exchange interaction and, for typical values of x ($x > \sim 0.02$), exceeds the first term in Eq.(1) at low temperatures. Even in the case of narrow-gap DMS, where the values of g^* are quite sizable, the exchange-induced part of the g-factor can be the dominant contribution. Since the magnetization M is a sensitive function of temperature and magnetic field, the effective g-factor--and the resulting spin splitting--may vary over a broad range, depending on the values of these quantities. Thus the band structure of DMS is much more sensitive to B and T than in ordinary (i.e., non-magnetic) semiconductors.

The main purpose of this paper is, as mentioned above, to provide an overview (by no means exhaustive) of new phenomena expected to be found in DMS quantum wells and superlattices. Several of these have already been described in the literature [11]. In this paper we shall attempt to be specific as to the choice of materials in which the phenomena under consideration appear practical, and, whenever possible, we shall try to discuss the predicted effects in quantitative terms.

NEW OPPORTUNITIES IN DMS QUANTUM WELLS

There are three facts which are important for the effects in this section: (i) the spin splitting of electronic states in DMS may easily be greater (by as much as two orders of magnitude in the case of wide gap DMS at low temperatures) than in adjacent non-magnetic layers; (ii) the spin splitting of the electronic states in DMS layers is comparable to the ionization energy of shallow impurities; (iii) the spin splitting is considerably greater than (in the case of wide-gap DMS) or comparable to (in the case of narrow-gap DMS) the energy separation between consecutive Landau levels in the presence of the magnetic field.

The spin superlattice

Let us start with the original idea of the spin superlattice [12]. It is based on the observation that the exchange interaction with localized magnetic moments srongly modifies the spin splitting of the Landau levels in a narrow-gap DMS, but leaves the effective mass of electrons practically unaffected. For a given DMS it is then possible to choose a non-magnetic counterpart, whose energy gap and conduction electron effective mass are equal to those of the DMS. Two materials would, on the other hand, differ greatly in their values of the electronic g-factors. The pair $Hg_{1-x}Mn_xSe$ and $Hg_{1-y}Cd_ySe$ with x=0.01 and y=0.024 was chosen in Ref.12 as an example of such complementary materials. The electron travelling along the growth axis of the superlattice consisting of semiconductors chosen in this way would experience an approximately uniform potential in the absence of the magnetic field (see Fig.2a), but with the field applied to the system (Fig.2b) the potential "seen" by the electron would have the form of a sequence of square wells. Thus the presence of the field is expected to split the conduction band into a series of "minibands" separated from each other by "minigaps" which, in turn, are proportional to the difference between the spin splittings in the constituent materials. The estimate carried out in Ref.12 for the above mentioned values of x and y, (for layers 97 Å thick and for B=40 kG) rendered values of the minigaps varying between 0.6 meV at 25K and 3 meV at 1.8K, corresponding to the submillimeter range of wavelengths.

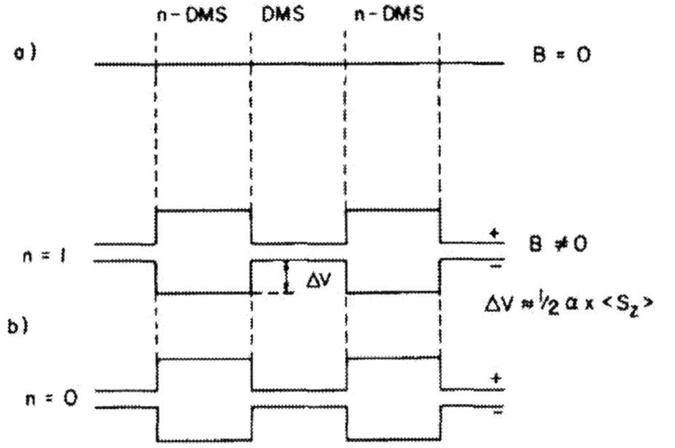

Fig.2. Schematic representation of the potential experienced by an electron in the spin-superlattice involving open narrow-gap semiconductor at zero magnetic field (a), and with the magnetic field applied (b), when the electron states are spin split (index +/-) and quantized into Landau levels (index n).

Exchange effects in shallow non-magnetic wells

Turning now to the wide-gap DMS, let us note that for materials differing only slightly in their values of the energy gap the potential wells "seen" by

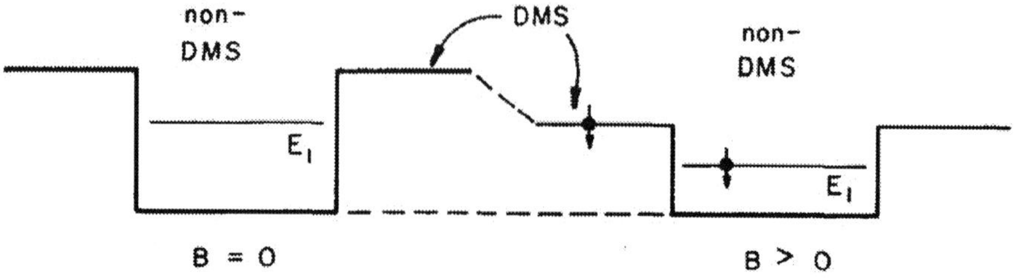

Fig.3. Shallow non-DMS quantum well between DMS barriers, showing the enhanced magnetic field dependence of the energy position of the ground (spin-down) state of an electron confined in the non-magnetic well due to variation of the well depth.

electrons can be comparable with the shifts induced by the exchange interaction in the presence of the magnetic field (see Fig.3). In such configurations one can expect that the enhancement of the g-factor will occur also for states in the non-magnetic well surrounded by DMS barriers, because their energy positions depend on the depth of the (shallow) wells. Fig.4 shows results of the calculation of the ground state spin splitting for 50Å CdTe well sandwiched between $Cd_{0.98}Mn_{0.02}Te$ layers. One can easily see that the spin splitting ΔE and, consequently, the corresponding g-factor are cosiderably greater in the case of CdTe quantum well ($g_{eff} \cong 17$ at 50kGs and 4.2K) than those in bulk CdTe ($g^* \cong 0.5$).

The phenomenon described above may also be viewed as being due to the "leakage" of the wave functions of the electrons confined in the well into DMS barriers, where it can interact with the Mn^{2+} ions via sp-d exchange. Of course, the degree of the "leakage" is the greater the closer a given electronic state is situated to the top of the well.

The use of non-rectangular wells (e.g., parabolic or triangular, as suggested in Ref.[13]) may lead to an even stronger dependence of the electron energies on the depth of the well, thus making the tuning range by the magnetic field wider than in the above simple example.

Transition from type I to type II superlattice

Consider now a superlattice consisting of alternating DMS and non-DMS layers, with DMS forming the barriers for the conduction electrons, as in the $CdTe/Cd_{1-x}Mn_xTe$ superlattice. Although at present the actual values of the band offsets in these materials are not known, it is commonly assumed that the valence band offset is much smaller than the conduction band discontinuity, and probably such that the potential wells for the holes occur in the absence of the magnetic field also in non-magnetic layers [14]. This corresponds to a type-I superlattice. This situation is depicted in Fig.5a. The localization of the electrons and the holes in the same spatial regions facilitates an efficient recombination of excitons created in the material, leading to, e.g., an intense luminescence associated with these transitions.

When the magnetic field is applied to the system, the conduction electrons will remain localized in the CdTe layers provided that the initial depth of the wells is not very small. On the other hand, the spin splitting of the valence band states can be easily made comparable to the discontinuities in this band. If the value of the magnetic field is reached such that the spin splitting of the valence band edge in the DMS barrier exceeds the zero-field band offset (Fig.5b), then one may expect a transition to type-II superlattice to occur, with a sudden decrease in the intensity of the excitonic luminescence being one of the consequences [15].

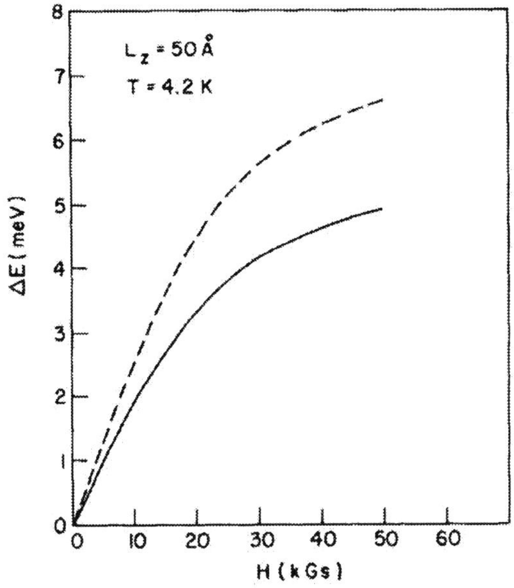

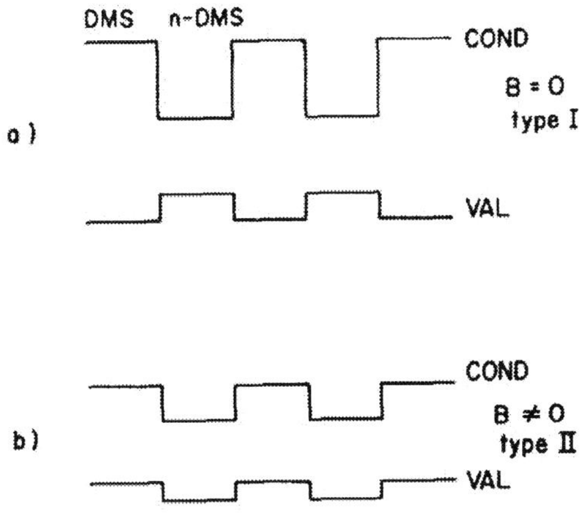

Fig.4. The spin splitting ΔE of the electron ground state in the non-magnetic 50 Å thick CdTe quantum well surrounded by $Cd_{1-x}Mn_xTe$ (x=0.02) barriers as a function of magnetic field at 4.2K. The solid line was calculated assuming that the conduction band offset is equal to the gap discontinuity. The broken line corresponds to the band offset reduced twice.

Fig.5. Schematic representation of probable band offsets in the CdTe/$Cd_{1-x}Mn_xTe$ superlattice (for clarity, strain effects are neglected). in the absence (a), and in the presence of strong magnetic field, with the band edges shown for spin-down electronic states only.

Observation of this effect is crucially dependent on the choice of the starting materials, so that the valence band discontinuity is appropriately small. Since, as mentioned above, no firm information exists about the valence band offsets in DMS, it is difficult to make predictions at this point concerning the choice of suitable materials. Possibly the required magnitude of the band offset can be achieved by a judicious use of the strain-related shifts of the valence band edges in lattice-mismatched structures.

Magnetic freeze-out and boil-off

Selectively doped structures involving DMS can be expected to show a variety of effects due to magnetic-field-induced changes in the relative position of the donor levels in the barrier layer and the confined electronic states in the well. The following examples are based on this opportunity.

Imagine, for example, a modulation-doped single quantum well consisting of non-DMS well positioned between two n-type DMS barriers, shown in Fig.6. By selecting suitable materials for the well and the barriers, one may achieve the situation that at zero magnetic field the ground electronic state E_1 in the well (whose position is determined primarily by the depth and thickness of the well) is slightly (say, 5 meV) below the donor level in the barrier. Therefore at low temperatures the donor electrons will transfer to the E_1 states, yielding the sample conducting in two dimensions. After the magnetic field is applied, one of the spin components of the donor state will shift strongly down in energy, while the

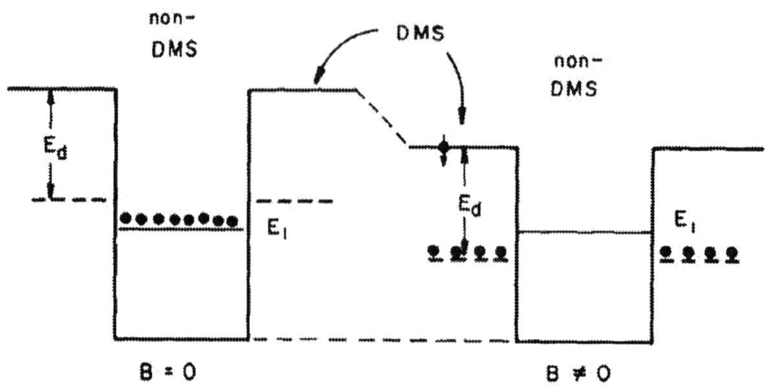

Fig.6. A non-DMS quantum well between DMS barriers, illustrating exchange-induced freeze-out. Only the ground state (spin-down) is shown for B≠0.

position of the E_1 state in the non-magnetic material will remain practically unaffected. Using realistic DMS parameters it is possible to show that in strong magnetic fields the original sequence of the donor and the E_1 levels can be reversed, so that the electrons will then "spill" back to the donor states, resulting in a sudden increase of the sample resistivity associated with this transition (the freeze-out effect).

A specific pair of materials which are well suited for observation of this effect is (lattice-matched) $Cd_{0.982}Zn_{0.018}Te$ and $Cd_{0.95}Mn_{0.05}Te$. The result of the model calculation of the number of electrons occupying the states in the well is shown in Fig.7. Of course, as can be seen in the figure, the rate of "escape" of electrons from the conducting wells depends sensitively on the temperature.

A reverse effect can be expected to take place in the structure consisting of DMS wells and selectively doped n-type non-magnetic barriers, chosen in such a manner that the donor level is now slightly below the ground state E_1 in the well at zero magnetic field. $Hg_{0.1}Cd_{0.85}Mn_{0.05}Te$ and $Cd_{0.977}Zn_{0.023}Te$ may serve as an example of such a (lattice-matched) system. Application of a magnetic field in this configuration causes the donor electrons to transfer to the well states (the boil-off effect), since now one of the spin components of the E_1 state is forced by the field to move below the donor level in the barrier layer.

Tunable resonant tunneling

The possibility of adjusting relative positions of the energy levels in the quantum well and the donor levels in the barriers suggests the feasibility of resonant tunneling of carriers through the barriers tuned by magnetic field and/or temperature. A sequential tunneling across the non-DMS barriers is depicted schematically in Fig.8. The original idea of the sequential tunneling [16] made use of the ground and first excited electron states in the quantum well. Here, the tunneling involves both spin components of the ground E_1 state spin-split by the magnetic field. The electric field applied across the quantum well structure can be varied to satisfy the resonance condition, i.e., to ensure the equality of the spin splitting of the E_1 level and the voltage drop across the barrier. Imagine now that the electrons injected to the excited states in the well lose their energy via a radiative process. Thus, it is conceivable that this, or analogous configurations involving various spin components of excited electron states, may lead to the construction of efficient infrared sources of radiation, with a definite polarization of the outgoing light [17].

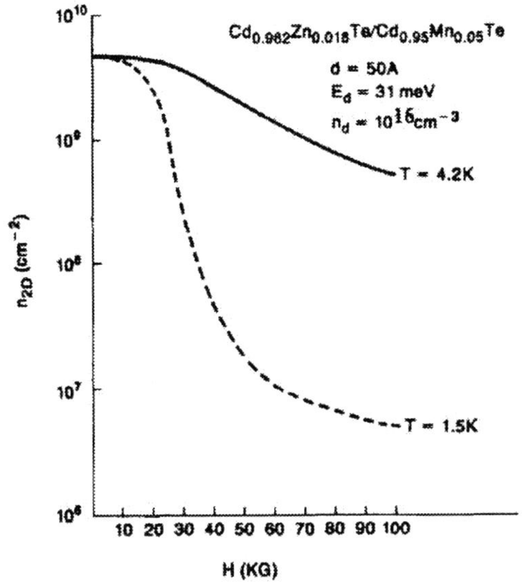

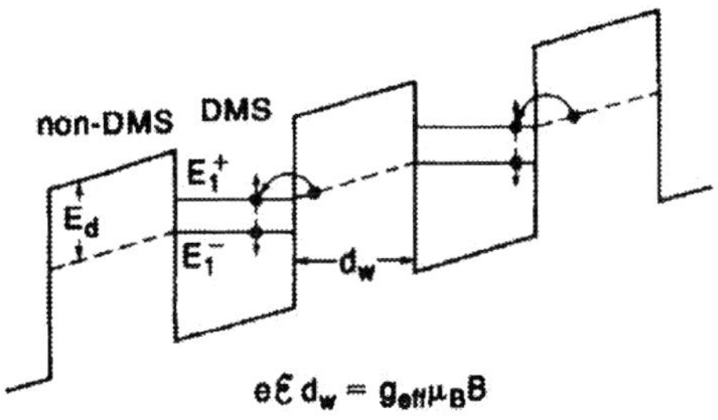

Fig.7. Calculated density of 2D electrons confined in the 50 Å thick well formed by $Cd_{0.982}Zn_{0.018}Te$ and $Cd_{0.95}Mn_{0.05}Te$ showing the freeze-out effect and its temperature dependence. The ionization energy of donors was assumed to be E_d = 31meV and their concentration $n_d = 10^{16}$ cm^{-3}.

Fig.8. DMS quantum wells between non-DMS barriers (doped n-type) in the presence of external electric and magnetic fields, showing sequential resonant tunneling between the spin-split electron states in the well.

The probability of the tunneling process will be greater for the electrons occupying the higher state (the spin-up state in our example) in the well, since their wave functions penetrate the barriers to a greater extent. This property may be used as a basis of the "spin filtering" mechanism, with only one spin species being passed through a series of quantum wells. Since the spin splitting of the quantum states in a DMS well is so large, the mechanism in question may turn out to be very efficient.

Infrared emitters

Another heterostructure which holds promise as an efficient infrared emitter consists of a wide-gap DMS layer, doped n-type, interfaced with a narrow-gap non-magnetic semiconductor [11]. As an example of materials suitable in this case one may choose $Cd_{1-x}Mn_xTe$ and InSb or $Hg_{1-x}Cd_xTe$. The g-factor of the conduction band electrons in the narrow-gap semiconductors, being due to the spin-orbit interaction, is opposite in sign to that in $Cd_{1-x}Mn_xTe$ (where it is mainly determined by the sign of the exchange constant α). Therefore, the electrons injected from donors in $Cd_{1-x}Mn_xTe$ to the non-magnetic layer by means of an external electric field will be mainly in their spin-down state, since at low temperature they are simply more numerous. Assuming that the spin is conserved during the transfer, these electrons will occupy spin-down states, which happen to be higher in energy than the spin-up states in the narrow-gap semiconductor. Thus an inverted occupation of these levels may be achieved, which may, in turn, be exploited for the process of stimulated emission.

Quantum oscillations

In the majority of the effects listed so far, the orbital quantization of the electron states into the Landau levels has been neglected. This was justified by the relatively large values of the effective masses characteristic of wide-gap DMS. Here we want to show that the enhanced spin splittings of DMS can lead to interesting consequences in the regime of quantum oscillatory phenomena. Thus we shall assume that well resolved Landau levels are formed. We shall, however, still assume that the spin splittings in the DMS far exceed the separation between consecutive Landau levels

Consider a configuration similar to that discussed above in connection with the freeze-out effect , i.e., with the parameters of the n-type DMS barriers and non-magnetic wells chosen such that the ground state E_1 in the absence of the magnetic field lies slightly below the donor level in the barrier (Fig.9a). In the presence of the field, the E_1 state splits into the Landau levels whose energy increases with increasing field strength. An oscillation of, say, magnetoresistance appears

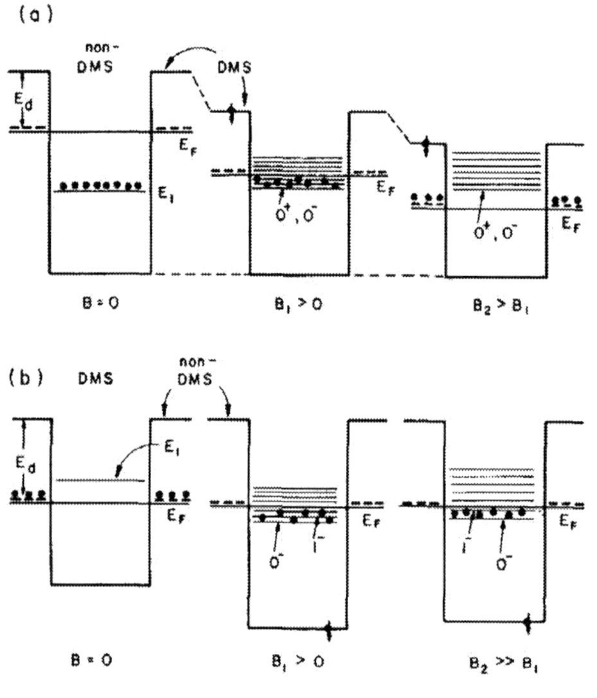

Fig.9. A non-DMS quantum well between DMS barriers (a), and DMS quantum well between non-magnetic barriers (b), illustrating the mechanism of quantum oscillations in DMS heterostructures. B_1 and B_2 denote two different values of the applied magnetic field. Note that the energy of the spin-down states in DMS wells does not continue decreasing infinitely as the field increases owing to saturation of the magnetization, as shown in the right-most part of the case (b).

whenever the energy of the Landau level coincides with the position of the Fermi level, which is determined by the donor energy in the barrier. However, in the present situation the Fermi level is not constant (as it would be in, e.g., $Ga_{1-x}Al_xAs$) but, owing to the large g-factor of the DMS donor states, moves relatively fast to lower energies. This fact will result in a modification of the usual B^{-1} periodicity of the oscillations, and will also permit one to observe the oscillation associated with the lowest Landau level, an observation not possible in non-magnetic semiconductors. Moreover, the period of the oscillations will now become a function of temperature, reflecting the temperature dependence of the donor level, and thus the Fermi energy.

Figure 9b illustrates a particularly striking anomaly of the quantum oscillations in DMS quantum wells. Consider a DMS well surrounded by n-type non-magnetic barriers, with the donor level slightly below the E_1 state at zero field, as in the discussion of the boil-off effect. Due to the initial strong downward shift of the spin-down components of the Landau ladder, originating from the spin-splitting of E_1, one may expect the oscillation related to the lowest Landau index n to appear first (i.e., in the lowest field). This is in sharp contrast with the sequence observed in the usual Shubnikov-de Haas effect, where the oscillations appear in the order corresponding to decreasing Landau indices as the field is increased. For very strong fields, when the Mn^{2+} spins saturate, the order of oscillations should eventually reverse, and the same oscillations should reappear, this time with the Landau index decreasing with growing field, i.e., in the standard sequence.

CONCLUDING REMARKS

The discussion of some of the effects given above is based on very simple band structure models. The reality of DMS quantum wells may turn out to be more complicated and may lead to additional results, not considered here. For instance, the observed enhancement of the spin splittings in CdTe/$Cd_{1-x}Mn_xTe$ superlattices is too large to be explained by the simple wave function leakage mechanism invoked above [18]. It is now believed that some electronic states in the non-DMS well have the tendency to localize immediately near the interface with the DMS barrier, leading to a more efficient interaction with the Mn^{2+} spins located on the other side of the heterointerface. A mechanism analogous to that of bound magnetic polaron is thought to be responsible for this effect [19-21].

On the other hand, the same experiments (see, e.g., Ref.19) do show that the electronic structure of a DMS quantum well does indeed depend very strongly on the magnetic field and the temperature, and thus is susceptible to manipulation by these external factors. In the case of epitaxial DMS samples the possibility arises of introducing moderately strong magnetic fields by growing a thin ferromagnetic film (appropriately designed) on the surfaces of the DMS specimen [22].

The phenomena listed in the paper are still awaiting to be demonstrated experimentally. Needless to say, their realization depends on progress being made in the (selective) doping technology of II-VI semiconductors. Advances in this field, together with recent successes in the MBE growth of DMS, permit one to expect that at least some of the effects discussed here should become implemented in the laboratory in the near future. The effort in this direction will be undoubtedly further stimulated by the fact that the 2D DMS structures appear to have the potential for device applications, as evidenced by the recent observation of stimulated emission already demonstrated in superlattices involving $Cd_{1-x}Mn_xTe$ and $Zn_{1-x}Mn_xSe$ [23-25].

ACKNOWLEDGEMENTS

The support of NSF Grants DMR-8520866 and DMR-8600014 is gratefully acknowledged.

REFERENCES

* On leave from the Institute of Physics, Polish Academy of Sciences,Warsaw, Poland.

[1] For the recent reviews of DMS see: J.K. Furdyna, J.Vac.Sci.Technol. A 4 , 2002 (1986); J.K.Furdyna, J.Appl.Phys. 53 , 7637 (1984); J.Mycielski, Progr. Cryst. Growth Character. 10 , 101 (1985); N.B.Brandt and V.V.Moshchalkov, Adv. Phys. 33 , 193 (1984).

[2] R.N.Bicknell, R.W.Yanka, N.C.Giles-Taylor, D.K.Blanks, E.L.Buckland, and J.F.Schetzina, Appl.Phys.Lett. 45 , 92 (1984).

[3] L.A.Kolodziejski, T.Sakamoto, R.L.Gunshor, and S.Datta, Appl.Phys.Lett. 44 , 799 (1984).

[4] L.A.Kolodziejski, R.L.Gunshor, T.C.Bonsett, R.Venkatasubramanian, S.Datta, R.B.Bylsma, W.M.Becker, and N.Otsuka, Appl.Phys.Lett. 47 , 169 (1985).

[5] M.Dobrowolska, Z.Yang, H.Luo, J.K.Furdyna, K.A.Harris, J.W.Cook,Jr., and J.F.Schetzina, J.Vac.Sci.Technol., to be published; see also this conference.

[6] J.K.Furdyna, J.Kossut, and A.K.Ramdas, Proc. Advanced Workshop on Optical Properties of Narrow-Gap Low Dimensional Structures, St.Andrews, Scotland, 1986 (in press).

[7] S.Venugopalan, L.A.Kolodziejski, R.L.Gunshor, and A.K.Ramdas, Appl.Phys.Lett. 45 , 974 (1984).

[8] L.A.Kolodziejski, R.L.Gunshor, N.Otsuka, B.P.Gu, Y.Hefetz, and A.V.Nurmikko, Appl.Phys.Lett. 48 , 1482 (1986).

[9] M.Jaczynski, J.Kossut, and R.R.Galazka, Phys.Status Solidi (b) 88 , 73 (1978).

[10] G.Bastard, C.Rigaux, Y.Guldner, J.Mycielski, and A.Mycielski, J.Phys. (Paris) 39 , 87 (1978).

[11] S.Datta, J.K.Furdyna, and R.L.Gunshor, Superlatt. Microstruct. 1 , 327 (1985).

[12] M. von Ortenberg, Phys.Rev.Lett. 49 , 1041 (1982).

[13] W.Pötz and D.K.Ferry, Phys.Rev. B 32 , 3863 (1985).

[14] See, e.g., D.K.Blanks, R.N.Bicknell, N.C.Giles-Taylor, and J.F.Schetzina, J.Vac.Sci.Technol. A 4 , 2120 (1986).

[15] J.A.Brum, G.Bastard, and M.Voos, Solid State Commun. 59 , 561 (1986).

[16] F.Capasso, K.Mohammed, and A.Y.Cho, Appl. Phys. Lett. 48 , 478 (1986).

[17] R.F.Kazarinov and R.A.Suris, Sov.Phys. Semicond. 5 , 707 (1971) [Fiz.Tekh.Poluprov. 5 , 797 (1971)].

[18] X.-C.Zhang, S.-K.Chang, A.V.Nurmikko, L.A.Kolodziejski, R.L.Gunshor, and S.Datta, Phys.Rev. B 31 , 4056 (1985).

[19] C.E.T. Goncalves da Silva, Phys.Rev.B 32 , 6962 (1985).

[20] J.-W.Wu, A.V.Nurmikko, and J.J.Quinn, Solid State Commun. 57 , 853 (1986).

[21] C.E.T. Goncalves da Silva, Phys.Rev. B 33 , 2923 (1986).

[22] C.Vittoria, F.J.RAchford, J.J.Krebs, and G.A.Prinz, Phys.Rev. B 30 , 3039 (1984).

[23] R.B.Bylsma, W.M.Becker, T.C.Bonsett, L.A.Kolodziejski, R.L.Gunshor, M.Yamanishi, and S.Datta, Appl. Phys. Lett. 47 , 1039 (1985).

[24] R.N.Bicknell, N.C.Giles-Taylor, J.F.Schetzina, N.G.Anderson, and W.D.Laidig, J.Vac.Sci.Technol. A 4 , 2126 (1986).

[25] E.D.Isaaks, D.Heiman, J.J.Zayhowski, R.N.Bicknell, and J.F.Schetzina, Appl.Phys.Lett. 48 , 275 (1986).

SEMIMAGNETIC LEAD SALT ALLOYS

GÜNTHER BAUER
Institut für Physik, Montanuniversität Leoben, A-8700 Leoben, Austria

ABSTRACT

The magnetic and electronic properties of pseudobinary alloys of the lead compounds (PbTe, PbSe) with Manganese, Europium and Gadolinium are reviewed. These materials have a direct minimum gap at the L-point of the Brillouin zone. The addition of Mn, Eu, or Gd to the diamagnetic hosts causes a rather weak antiferromagnetic exchange interaction. For sufficiently high carrier concentrations (10^{20}-$10^{21}cm^{-3}$) a ferromagnetic interaction occurs due to RKKY interaction. The spin-spin exchange interaction between the Bloch states and the localized 3d or 4f states is characterized by rather small exchange integrals and the limited applicability of the mean field approximation for the description of energy band parameters like g-factors. Interband transitions in PbMnTe and PbMnS lend support to a zero-magnetic field splitting ("spin-splitting") of band states which increases linearly with Mn-content up to several meV at temperatures below 4K.

INTRODUCTION

The pseudobinary alloys of the lead compounds PbTe, PbSe and PbS with Mn, Eu, Gd have been studied by various methods during the recent past [1-6]. The magnetic ions are incorporated into the cubic rock salt lattice substitutionally and are rendomly distributed in the Pb-fcc sublattice. In contrast to the diluted magnetic semiconductors like $A^{II}_{1-x}Mn_xB^{VI}$, the minimum direct gap of the semimagnetic IV-VI compounds is located at the L-point of the Brillouin zone. The band edge states are composed primarily of p-orbitals.

These compounds have attracted attention because of the fact that they permit a study of the effects of the incorporation of ions with partially filled 3d or 4f states in a cubic but non-zinc blende type structure. Of course similarities with other diluted magnetic semiconductors of II-VI compounds [7] exist: in the ternary IV-VI compounds the energy gap can be tuned as well, the effective masses change accordingly and other physical parameters as well by varying the composition. As far as the magnetic properties are concerned all materials studied so far (PbMnTe [8], PbMnSnSe [2], PbMnS [9], PbEuTe [10], PbGdTe [6]), indicate a dominant antiferromagnetic interaction. In PbMnTe at low temperatures below 1K a spin glass transition was observed [11]. On the other hand it is already known for years that the alloys of IV-VI semiconductors with MnTe exhibit ferromagnetic ordering (PbGeMnTe [12] for x>0.005), $Sn_{1-x}Mn_xTe$ [13] or $Ge_{1-x}Mn_xTe$ [14] if the materials are highly degenerate with hole concentrations of the order of 10^{20}-$10^{21}cm^{-3}$. The carrier concentration induced ferromagnetism was recently demonstrated in $[(PbTe)_{1-x}(SnTe)_x]_{1-y}[MnTe]_y$ (x=0.72, y=0.03) [15]. The transition temperature from the paramagnetic high temperature phase to the ferromagnetic low temperature phase depends on the carrier concentration, an

Mat. Res. Soc. Symp. Proc. Vol. 89. © 1987 Materials Research Society

effect interpreted as a manifestation of the Ruderman-Kittel-Kasuya-Yosida (RKKY) interaction.

An important feature of the IV-VI compounds is the fact that the addition of manganese to the diamagnetic lead salt hosts causes a rather weaker antiferromagnetic exchange interaction in comparison to the corresponding pseudobinary II-VI-compounds [8,9].

The presence of the magnetic ions Mn^{2+}, Eu^{2+}, Gd^{3+} leads in the lead salts of course also to spin-spin exchange interactions between the localized magnetic moments and the band electrons. The appropriate description of this interaction seems to be the most intriguing feature of the dilute magnetic lead salts. Magnetotransport experiments in PbMnTe and PbMnS as well as magnetooptical experiments in PbMnTe, PbMnSe and PbMnS seem to indicate that the part of the exchange interaction which is described by mean field theory alone is really small if not negligible . In PbMnTe but also in PbMnS there does not exist a magnetic field range where the magnetic field and temperature dependence of the macroscopic magnetization determines the g-factors (tensor components in the many valley system) of free carriers in the sense of a simple mean field approximation.

CRYSTAL GROWTH

Bulk IV-VI compounds are usually grown by using the Bridgman method. Isothermal annealing in an appropriate atmosphere controls the number of metal of chalcogen vacancies and thus the number of holes or electrons. Very homogeneous and high mobility material can be grown by using epitaxial methods: either the hot wall method [19] as used for the growth of PbMnTe and PbEuTe or molecular beam epitaxy (used for the growth of $Pb_{1-x}Eu_xSe_yTe_{1-y}$ and $Pb_{1-x}Eu_xSe$) [20]. For $Pb_{1-x}Mn_xTe$ the solubility limit of MnTe in PbTe is reached at about $x=0.12$ [21]. Experimentally, Escorne et al. [11] as well as Vinogradova et al. [21] found a Vegard type behaviour of the weighted average of the bond length up to this composition with an empirical rule: $a_o-0.00491\cdot x=a$ [22] where a_o denotes the lattice constant of PbTe and a that of PbMnTe, respectively.

Epitaxially grown PbMnTe samples have Mn contents less than about 3% (for substrate temperatures less than 500 ^oC).

PbEuTe has been grown both with the hot wall method (HWE) [4] as well as with molecular beam epitaxy (MBE). In the case of HWE samples with $x>0.05$ already a two phase behaviour was observed whereas MBE grown material is available in the entire range of Eu compositions up to $x=1$. $Pb_{1-x}Eu_xSe$ has been grown by MBE up to compositions of about $x=0.1$. $Pb_{1-x}Mn_xS$ has been prepared by the Bridgman method with S compositions up to 4% and $Pb_xMn_ySn_zSe$ with Mn compositions up to 5% [24]. For the control of composition usually an electron microprobe analysis is employed.

ENERGY GAPS

The replacement of lead by Mn, Eu, Gd in the tellurides, selenides or sulfides leads to a significant change of the energy gap. Data are available for $Pb_{1-x}Mn_xTe$, $Pb_{1-x}Eu_xTe$ as well as $Pb_{1-x}Mn_xSe$ and are shown in Fig.1. Especially in the case of

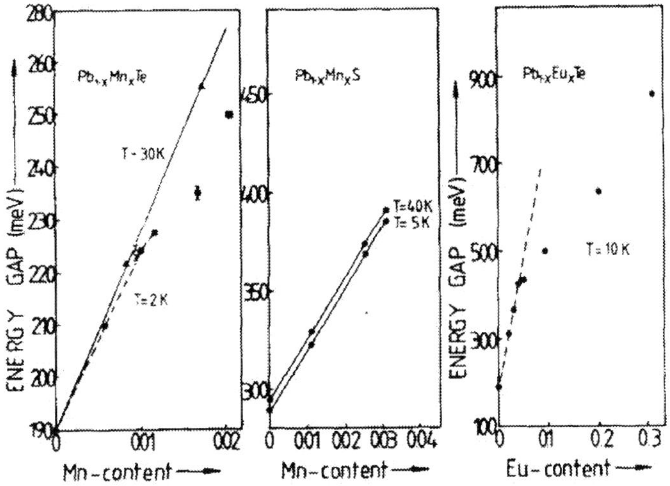

Fig.1: Energy gaps vs composition [17,35,39],[34],and [23].

PbEuTe a remarkable bowing is observed . Whereas in EuTe the f-levels form a narrow band within the forbidden gap between conduction and valence band,in PbEuTe with low Eu content the corresponding levels should be situated well below the valence band edge. Since also the coefficient dE_g/dT changes sign at x=20%, the gap does no longer behave PbTe-like above such concentrations. It is interesting to note,that also in $(Pb_{1-x}Sn_x)_{1-y}Mn_ySe$ with tin compositions up to 0.08 the replacement of the cation sites by Mn increases the gap (measurements up to Mn=0.04 [24]).

For $Pb_{1-x}Mn_xTe$ with x≈0.114,Escorne et al. [11] have concluded from galvanomagnetic measurements that at low temperatures the energy gap is already indirect, since the second valence band situated along the Σ-lines of the Brillouin zone lies higher than the valence band extrema at the L-point.

MAGNETIC PROPERTIES

Magnetic properties of the dilute magnetic lead compounds were studied by low field susceptibility, high field magnetization and electron paramagnetic resonance measurements [25]. The results of the susceptibility measurements indicate rather weak antiferromagnetic interactions with characteristic temperatures Θ around 2K. Apart from an early paper by Andrianov [26] on $Pb_{1-x}Mn_xTe$ all reports agree on this value of Θ. The magnetic field and temperature dependence of the magnetization of PbMnTe and PbMnS has been analysed by using a modified Brillouin function

$$S_z = S_o\, B_{5/2}\left(\frac{\frac{5}{2}\, g\mu_B H}{k_B\,(T+T_o)}\right) \qquad (1)$$

For both compounds, Karczewski et al. [9] and Anderson and Gorska [8] found a reasonable fit. The value of T_o increases slightly with composition x for PbMnTe at a given lattice temperature (0.7-2K) and T_o appears to decrease with lattice temperature.

Karczewski et al. [9] have also used a self consistent cluster approximation for the interpretation considering that in the rock salt lattice Mn has 12 nearest neighbours on cation sites and 6 next nearest neighbours and furthermore 66 cation neighbours separated from the central manganese ion by 2 anions. The corresponding exchange integrals denoted by J_1, J_2 and J_3, respectively have values of $J_1=J_2=0.54$ and $J_3=0.060$ K.

The reciprocal susceptibility derived from all data on $Pb_{1-x}Mn_xS$ and $Pb_{1-x}Mn_xTe$ ($x\leq0.04$) yields a small antiferromagnetic Curie temperature $|\Theta|<3K$. The values by Andrianov et al. [26] deviate from all other published data with $|\Theta|$ values as large as 70K. Also for $Pb_{1-x}Eu_xTe$ Andrianov et al. [27] obtained a peculiar dependence of the paramagnetic Curie temperature on the Eu content with a minimum value of $\Theta=100K$ for $x=0.01$.

On the other hand Bartkowski et al. [25] reported Θ values close to zero for PbMnTe using susceptibility and EPR measurements and confirming thus the rather weak antiferromagnetic exchange coupling between the Mn^{2+} ions.

An important feature, also found in other dilute magnetic semiconductors is the occurence of a spin glass behaviour in $Pb_{1-x}Mn_xTe$ at temperatures below 1K [11] inferred from a cusp of the low field susceptibility at the temperature T_g. From the monotonic decrease of the reversible part of the susceptibility below T_g, Escorne et al. [11] postulated the existence of long range exchange interactions in this narrow gap material.

One of the most fascinating magnetic properties of the semimagnetic lead salt alloys has been described recently by Story et al. [15,28]. In $[(PbTe)_{1-x}(SnTe)_x]_{1-y}[MnTe]_y$ ($x=0.72$, $y=0.03$) a ferromagnetic ordering was observed above a critical carrier (hole) concentration $p\sim3\cdot10^{20}cm^{-3}$. At transition temperatures (3-5K) which increase with hole concentration, the paramagnetic alloy becomes ferromagnetic as observed by magnetization, magnetic susceptibility and specific heat measurements. The ferromagnetic coupling between the magnetic ions is mediated through the RKKY interaction. The important property of this material is the fact that apparently the magnetic properties are determined both by the concentration of magnetic ions and by the free carrier concentration. Suski et al. [28] have shown that RKKY interaction plays indeed the dominant role for the occurence of ferromagnetism in this material by hydrostatic pressure experiments. Under pressure, the number of highly mobile carriers in the L-extrema increases at the expense of less mobile carriers in the Σ-extrema. The lighter holes in the L-extrema are more effective in the inter-ion interaction mechanism. In Fig.2 the carrier concentration dependence of the critical temperature for an alloy with an Mn-content $y=0.03$ is shown [28] together with magnetization isotherms for two temperatures for a sample with a critical temperature $T_c=4.1K$ [15]. The hole concentration is increased by annealing the PbSnMnTe alloy in a Te-rich atmosphere.

Among the dilute magnetic semiconductors only the IV-VI-compounds have so far demonstrated the possibility of generating ferromagnetic ordering by changing the carrier concentration and keeping the concentration of magnetic (Mn^{2+}) ions constant.

EXCHANGE INTERACTION

For Mn, Eu or Gd concentrations which are sufficiently small, the band structure remains PbTe like. For the description of the band extrema, situated at the L-points of the Brillouin zone, a two band model with four far bands treated in k^2 perturbation is used. The band edge functions have the symmetry [29,30]

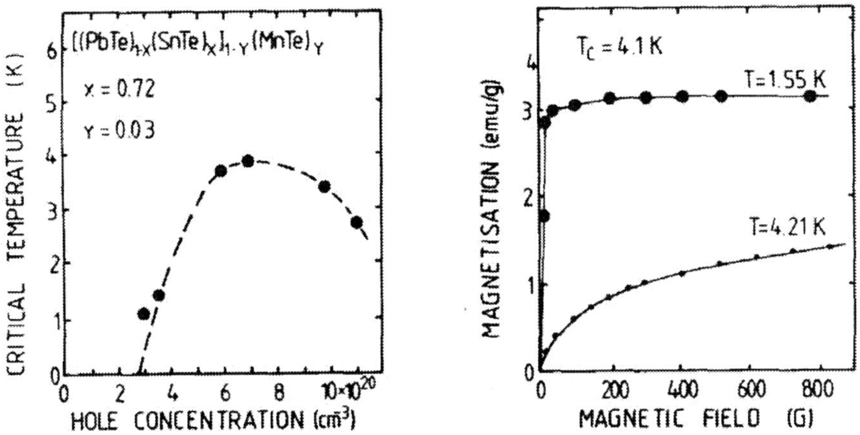

Fig.2: Paramagnetic-ferromagnetic phase transition: dependence of the critical transition temperature on hole concentration (lefthandside); the magnetization exhibits a rapid saturation for lattice temperatures below T_c with a measured saturation value close to the theoretical one for Mn content y=0.03 and S=5/2 (M_{theor}=3.1 emu/g). After [15] and [28].

$$
\begin{aligned}
V^+ &= i\cos\Theta^+ R\uparrow + \sin\Theta^+ S_+\downarrow \\
V^- &= i\cos\Theta^+ R\downarrow + \sin\Theta^+ S_-\uparrow \\
C^+ &= -\sin\Theta^- Z\uparrow - \cos\Theta^- X_+\downarrow \\
C^- &= \sin\Theta^- Z\downarrow - \cos\Theta^- X_-\uparrow
\end{aligned} \tag{2}
$$

Within the framework of the molecular field approximation, the exchange induced corrections to the (4×4) k.p Hamiltonian were calculated by Zawadzki [31].Due to the many valley band structure H_{exch} is given by:

$$H_{exch} = \frac{1}{2}\, x S_o\, B_{5/2}\left[\frac{5\mu_B H}{k_B(T+T_o)}\right] J\,(\sin\phi\sigma_x + \cos\phi\sigma_z) \tag{3}$$

J denotes the exchange integral and σ_x, σ_z are electron spin operators, ϕ denotes the angle between the magnetic field $\vec{H}$ and the [111] direction for $\vec{H}$ in a (x,z) ([$\bar{2}$11], [111] plane. The exchange matrix is of the form [31, 17, 16]

$$\begin{pmatrix} a & c & 0 & 0 \\ c & -a & 0 & 0 \\ 0 & 0 & b & d \\ 0 & 0 & d & -b \end{pmatrix} \tag{4}$$

where

$$
\begin{aligned}
a &= A\cos\phi\cos\gamma - a_1\sin\phi\sin\gamma \\
c &= -A\cos\phi\sin\gamma + a_1\sin\phi\cos\gamma \\
b &= B\cos\phi\cos\gamma - b_1\sin\phi\sin\gamma \\
d &= -B\cos\phi\sin\gamma - b_1\sin\phi\cos\gamma
\end{aligned} \tag{5}
$$

are functions of the four exchange integrals for the valence (A) and conduction band (B):

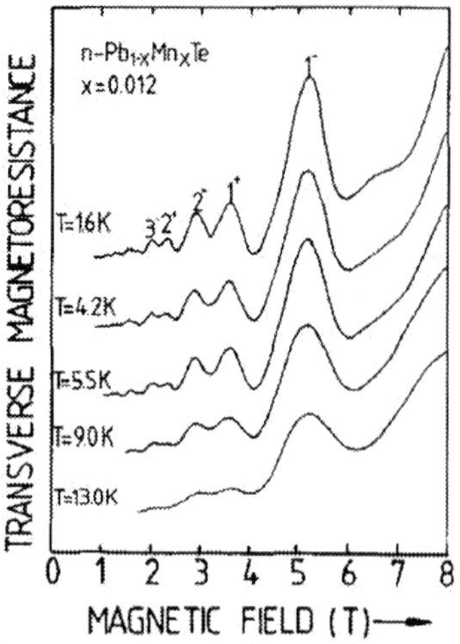

Fig.3: Temperature dependence of the Shubnikov- de Haas oscillations in n-PbMnTe [31].

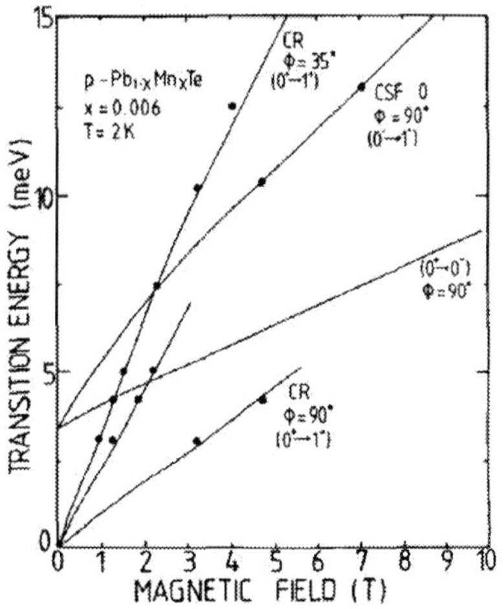

Fig.4: Magnetooptical intraband-transitions in p-PbMnTe. Full lines: model calculations.

$$A = a_1 - a_2 = \alpha \cos^2\Theta^+ - \delta \sin^2\Theta^+$$
$$B = b_1 - b_2 = \beta_\parallel \sin^2\Theta^- - \beta_\perp \cos^2\Theta^- \qquad (6)$$

where $\Theta^\pm$ are spin orbit mixing parameters defined by Bernick and Kleinman [32]. The four exchange integrals are given by:

$$\alpha = (R|J|R)/\Omega, \quad \delta = (S_\pm|J|S_\pm)/\Omega$$
$$\beta_\parallel = (X_\pm|J|X_\pm)/\Omega, \beta_\perp = (Z|J|Z)/\Omega$$

The angle γ (Eq.5) results from a coordinate transformation which is necessary for arbitraty directions of magnetic field [33]. The four exchange parameters are characteristic for the symmetry at the L-point. Of course, this molecular field approximation results in exchange induced spin splittings which reflect the temperature and magnetic field dependence of the macroscopic magnetization. According to this approximation, e.g. the longitudinal g-factors in the valence and conduction band are given by:

$$g^{eff}_{v,\ell} = g^+_\ell + 4P^2_\perp/m_o E_g + xA \times \langle S_z\rangle/\mu_B H$$

$$g^{eff}_{c,\ell} = g^-_\ell + 4P^2_\perp/m_o E_g + xB \times \langle S_z\rangle/\mu_B H \qquad (7)$$

$P_\perp$ denotes the transverse momentum matrix element and $g^\pm_\ell$ far band contributions to the longitudinal g-factors. Using magneto-optical interband transitions in Faraday and Voigt geometry, Karczewski [34] has determined the sum and difference of the exchange intetrals $|A+B|$ and $|A-B|$ in $Pb_{1-x}Mn_xS$ as a function of composition x. He found that the exchange integrals are rather small (200meV) and decrease rapidly with increasing Mn-content to about 50meV at 3%. Also for $Pb_{1-x}Mn_xTe$ ($x \leq 0.02$) the exchange induced corrections to electronic g-factors according to the molecular field approximation are rather small. E.g. an estimate from inter- and intraband magnetooptical data yields corrections (last terms of Eq.7) up to 4% to g_ℓ and up to 20% to g_t [35].

For $Pb_{1-x}Eu_xTe$ Goltsos et al. [23] have given estimates for the magnitude of the exchange integrals from interband magnetooptical data. These estimate also yield very small values for |A-B| of less than 50meV, decreasing with increasing x. These exchange induced corrections are really small, if compared to dilute magnetic II-VI compounds. Therefore also Shubnikov-de Haas measurements as a function of temperature as shown in Fig.3 do not reveal any typical anomalies as found e.g. in $Hg_{1-x}Mn_xTe$ [36].

Since PbSnMnTe exhibits a paramagnetic-ferromagnetic phase transition it should be noted, that Mycielski [37] has calculated the band structure near the L-points of the Brillouin zone for the ferromagnetic case. The resultant splitting of the conduction and valence band edges depends on the direction of the magnetization. The ellipsoids of constant energy are no longer ellipsoids of revolution but ellipsoids with three different masses and the main axes of the ellipsoids are rotated into the direction of $\langle S \rangle_{av}$. No experimental data are yet available.
In intraband magnetooptical experiments on n-PbMnTe, Niewodniczanska et al. [31] and Bauer [16] have reported the occurence of additional magnetooptical transitions due to the presence of Mn. In a $\vec{H}||\vec{E}$ geometry, apart from the cyclotron resonance, also combined spin flip resonances and spin flip resonances were observed. Similar features were observed by Karczewski and von Ortenberg in $Pb_{1-x}Mn_xS$. In this material the variation of the g-factor with Mn-composition was deduced. Far infrared transmission data involving the strip line method [16] even exhibited three additional resonances extrapolating to finite energies for $H \to 0$ ($\approx$1meV at T=2K) and tentatively ascribed to transitions involving impurity states. In p-type $Pb_{1-x}Mn_xTe$, which has at $T \sim 4K$ mobilities of the order of 30.000 cm^2/Vs, magnetooptical data in $\vec{H}||\vec{E}$ geometry also revealed, apart from the cyclotron resonance in the $\phi=35^o$ valleys, additional resonances extrapolating to a finite energy with $H \to 0$ (Fig.4). The rather low mobility and the small energies involved (around the dielectric anomalies associated with the TO-phonon mode ($\omega_{TO}=21cm^{-1}$ $\hat{=}$2.6meV for x=0.008 [39])) make, however, an unambiguous identification of the existence of a finite splitting of the 0^-, 0^+ levels for $H \to 0$ difficult. Apparently magnetooptical intraband transistions in PbMnTe, PbMnS are influenced by the presence of Mn, but a proper description of the effects is not possible with the molecular field approximation.

ZEROFIELD-"SPIN SPLIT" STATES

It is well known that in a dilute magnetic semiconductor, for the case of a localized carrier interacting with a finite number of magnetic ions a non vanishing "spin-splitting" results even in the absence of an external magnetic field ("bound magnetic polaron"). However, a nonvanishing zero field splitting has been reported in $Pb_{1-x}Mn_xTe$ [17], in $Pb_{1-x}Mn_xS$ ([3, 39, 34]) for <u>delocalized</u> carrier states. Stepniewski [41] has reported such observations also recently for a narrow gap II-VI system $Hg_{1-x}Mn_xTe$ (x=0.09). Also in wide gap materials, Golnik et al. [42] reported systematic deviations of magnetic absorption data from results of mean field models.

Typical results for $Pb_{1-x}Mn_xTe$ are shown in Fig.5. Magnetooptical interband transmission data in Faraday and Voigt geometry (selection rules are shown in the inset) extrapolate for $H \to 0$ to finite splittings. It should be stressed that the zero field energy splittings do not depend on the valley orientation

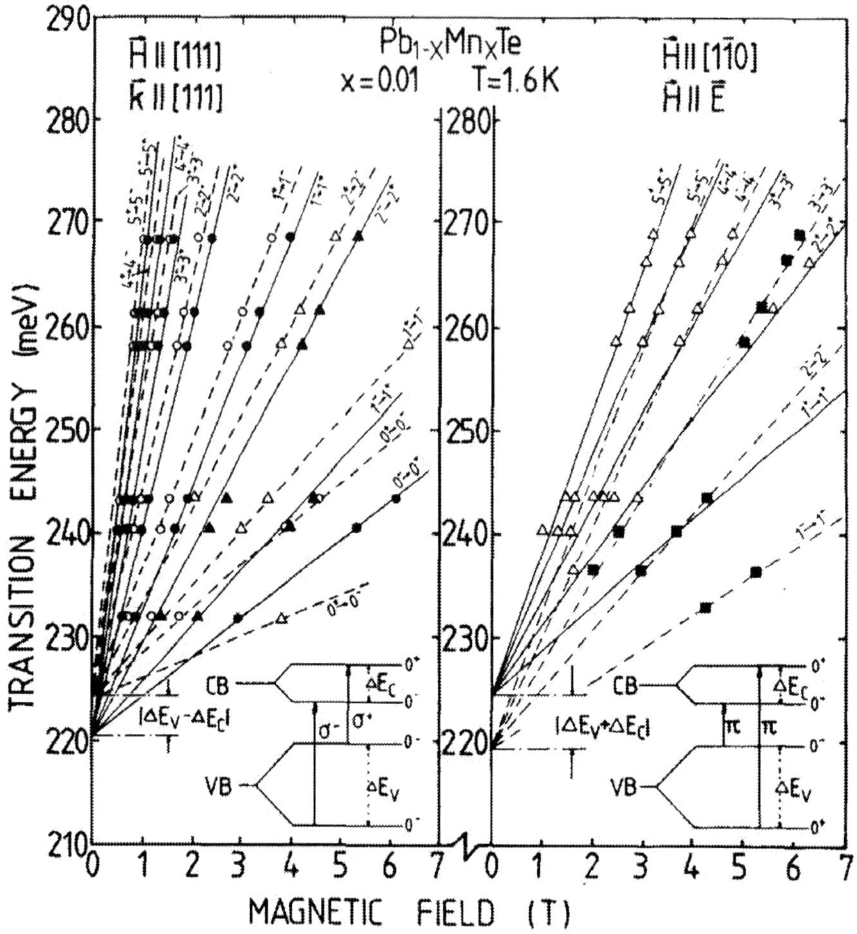

Fig. 5: Interband transition energies vs magnetic field: $Pb_{1-x}Mn_xTe$ lefthandside: Faraday geometry, righthandside: Voigt geometry; ●: σ^+-transitions for $\phi=0^o$, o: σ^- for $\phi=0^o$; ▲,Δ: σ^+,σ^- for $\phi=70,73^o$; ■,Δ: π-polarization. Inserts: selection rules and amount of energy splitting ΔE_c and ΔE_v ([17]).

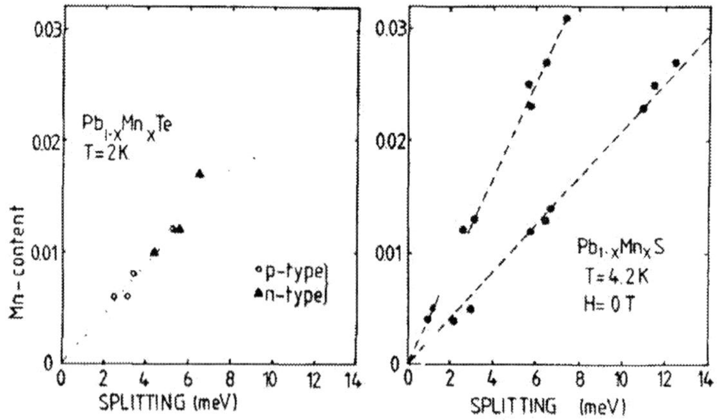

Fig.6: Dependence of the spin-splitting in PbMnTe and PbMnS on composition x as deduced from magnetooptical interband transitions (l.h.s. [35,43]) and laser emission (r.h.s. [3])

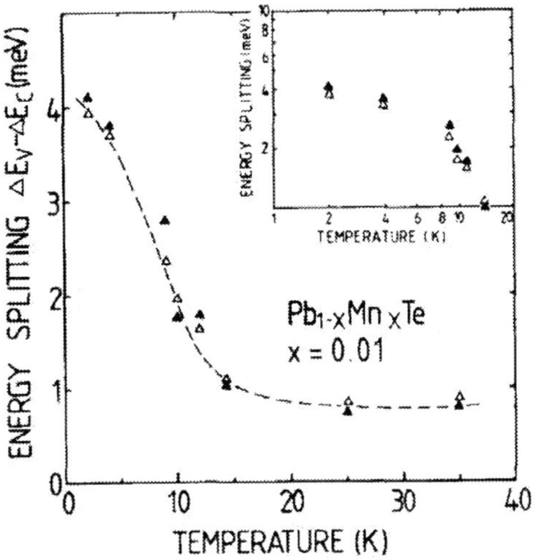

Fig.7: Temperature dependence of ($\Delta E_v-\Delta E_c$) of valence and conduction band spin splitting at $H \cong 2T$ (Δ) and H = 4T (▲) for PbMnTe (x=0.01). After [17].

although the masses have an anisotropy of about 10. The selection rules also imply that in the Faraday configuration with both polarizations σ^+, σ^- the difference of the valence and conduction band splittings $|\Delta E_v-\Delta E_c|$ is measured, whereas in the Voigt configuration one measures the sum $|\Delta E_c+\Delta E_v|$ (see insets). It is immediately evident that $|\Delta E_v| \geq |\Delta E_c|$ but that $\Delta E_c \neq 0$. The data were fit with a model calculation [35] for the Landau levels and this fit yielded values of the valence band splitting ΔE_v about a factor of 10 larger as compared to ΔE_c (this holds for various samples with x up to 1.7%). In Fig.6 (lefthandside), the observed values $|\Delta E_v-\Delta E_c|$ of the extrapolated zero field splitting are shown vs Mn-content [17, 35, 43]. In the righthandside the total conduction and valence band splitting as obtained from p-n-junction laser emission (at H=0) is shown vs Mn-content. There is more than one line due to a total of four observed transitions [3,34]. The values of the splitting are also confirmed by magnetooptical measurements on homogenously doped PbMnS samples [38]. Finally, in Fig.7 the temperature dependence of the splitting $|\Delta E_v-\Delta E_c|$ is shown for $Pb_{1-x}Mn_xTe$. It is apparent that at low temperatures there is a saturation whereas in a narrow temperature region, the splitting decreases quite steeply. As already shown previously [17], the splitting does not at all scale linearly with the magnetization M(T,H). It should be pointed out, that the polarization properties of the transitions (see Fig.5) (selection rules as in PbTe) confirm that free and not bound carrier states are involved (Faraday configuration: $\sigma^+ \to \sigma^-$: $\hbar\omega+(\Delta E_v-\Delta E_c)/2$; $\sigma^- \to \sigma^+$: $\hbar\omega+(\Delta E_c-\Delta E_v)/2$; Voigt configuration: $\sigma^+ \to \sigma^+$: $\hbar\omega+(\Delta E_c+\Delta E_v)/2$; $\sigma^- \to \sigma^-$: $\hbar\omega-(\Delta E_c+\Delta E_v)/2$). Both von Ortenberg [44] as well as Golnik and Spałek [45] tried to explain this behaviour invoking the concept of the free magnetic polaron. According to Spałek [46] these free magnetic polaron (FMP) states incorporate (mobile) effective mass states and the exchange coupling to the localized and fluctuating spins of the magnetic ions. They are characterized by a spacially confined wavefunction with a wave vector k within a region of a finite number of N spins and which decays to zero within the space region of volume proportional to N^{-1}. The electron adjusts its spin adiabatically to the direction of the polarization in this volume but without being localized in it. A readjustment of the electron spin to the varying direction of Mn-spins is possible due to the spin flip part of the exchange interaction. Spałek [45, 46] has shown that a stable solution of this problem is possible with a finite number N. Self trapping only occurs in the extreme case of strong polarization. A full analysis of the concept of the free magnetic polaron is in progress [46].

SUMMARY

The dilute magnetic lead salt alloys differ considerably from the corresponding II-VI compounds. The breakdown of the molecular field approximation for the treatment of the spin-spin exchange interaction is obvious from all experimental data. The different host crystal structure (NaCl- instead of zinc-blende or wurtzite, and the different electronic band structure: carriers being situated close to the L-points of the BZ) is responsible both for the rather weak exchange interaction and peculiarities like spin split electron states in zero external fields and paramagnetic susceptibilities of the materials.

ACKNOWLEDGEMENTS: I would like to acknowledge collaboration with H.Pascher, M.von Ortenberg, L.Palmetshofer and W.Jantsch. It is a pleasure to thank J.Spałek for his suggestions concerning the FMP. Work supported by Fonds zur Förderung der wiss.Forschung Project No. 5002.

REFERENCES

1. J.Niedwodniczanska-Zawadzka and A.Szczerbakow, Solid State Commun. 34, 887 (1980).
2. G.Karczewski, M.Klimkiewicz, I.Glass, and A.Szczerbakow, and R.Behrendt, Appl.Phys. A29, 49 (1982).
3. G.Karczewski and L.Kowalczyk, Solid State Commun. 48,653(1983).
4. A.Krost, B.Harbecke, R.Faymonville, H.Schlegel, E.J.Fantner, K.E.Ambrosch,and G.Bauer, J.Phys.C. 18, 2119 (1985).
5. D.L.Partin,and C.M.Thrush,Appl.Phys.Lett. 45, 193 (1984).
6. M.Averous, B.A.Lombos, C.Fau, E.Ilbnouelghazi, J.C.Tedenac, G.Brun, and M.A.Bartkowski, phys.stat.sol. (b) 131, 759 (1985).
7. J.K.Furdyna, J.Vac.Sci.Technol. A4, 2002 (1986).
8. J.R.Anderson and M.Gorska, Solid State Commun. 52,601 (1984).
9. G.Karczewski, M. on Ortenberg, Z.Wilamowski, W.Dobrowolski and J.Niedwodniczanska-Zawadzka, Solid State Commun. 55,249, (1985).
10. D.G.Andrianov, S.A.Belokon, A.A.Burdakin, V.M.Lakeenov, and S.M.Yakubenya, Sov.Phys.Semicond. 19,770 (1985).
11. M.Escorne, A.Mauger, J.L.Tholence, and R.Triboulet, Phys.Rev. B29, 6306 (1984).
12. T.Hamasaki, Solid State Commun. 32 1069 (1979).
13. R.W.Cochrane, M.Plischke, and J.O.Ström-Olsen, Phys.Rev. B9, 3013 (1974).
14. M.Escorne, A.Ghazali, and P.Leroux-Hugon in Proc.Int.Conf. Phys.Semocond (Teubner, Stuttgart 1974).
15. T.Story, R.R.Galazka, R.B.Frankel, and P.A.Wolff, Phys.Rev. Lett. 56, 777 (1986).
16. G.Bauer, in Physics of Semiconducting Compounds, ed.R.R. Galazka (Ossolineum, Warszawa 1983) p.62.
17. H.Pascher, E.J.Fantner, G.Bauer, W.Zawadzki,and M.v.Ortenberg Solid State Commun. 48, 461 (1983).

18. G.Karczewski, M.v.Ortenberg, and R.R.Galazka, Proc. 8th Int. Conf. Infrared Millimeter Waves, ed. R.L.Tempkin (Miami Beach 1983), p. T5-6.
19. G.Elsinger, L.Palmetshofer, and A.Lopez-Otero, il nuovo cimento 2D, 1869 (1983).
20. Y.Shani, A.Katzir, K.H.Bachem, P.Norton, M.Tacke, and H.M. Preier, Appl.Phys.Lett. 48, 1178 (1986).
21. V.G.Vanyarkho, V.P.Zlomanov, and A.V.Novoselova, Inorg.Mat. 6, 1352 (1970).
22. M.N.Vinogradova, N.V.Kolomoets, and L.M.Sysoeva, Sov.Phys. Semicond. 5, 186 (1971).
23. W.C.Goltsos, A.V.Nurmikko, and D.L.Partin, Solid State Commun. 59, 183 (1986).
24. L.Kowalczyk and A.Szczerbakow, Acta physica polonica A67, 189 (1985).
25. M.Bartkowski, A.H.Reddoch, D.F.Williams, G.Lamarche, and Z.Korczak, Solid State Commun. 57, 185 (1986).
26. D.G.Andrianov, N.M.Pavlov, A.S.Savelev, V.L.Fistul, and G.P.Tsiskarisvili, Sov.Phys.Semicond. 14, 711 (1980).
27. D.G.Andrianov, S.O.Klimonskii, and A.S.Savelev, ibid 19, 765 (1985).
28. T.Suski, J.Igalson, and T.Story, Proc.Int.Conf.Phys.Semicond. Stockholm 1986, in print.
29. D.L.Mitchell and R.F.Wallis, Phys.Rev. 151, 581 (1965).
30. R.Grisar, H.Burkhard, G.Bauer, and W.Zawadzki, in Physics of Narrow Gap Semiconductors, Proc. 3rd Int.Conf., Warsaw 1977 (Elsevier N.Y. 1978) p.115.
31. J.Niedwodniczanska-Zawadzka, G.Elsinger, L.Palmetshofer, A.Lopez-Otero, E.J.Fantner, G.Bauer, and W.Zawadzki, Physica 117B and 188B, 458 (1983).
32. R.L.Bernick and L.Kleinman, Solid State Commun. 8, 569 (1970).
33. J.Niedwodniczanska-Zawadzka, J.Kossut, A.Sandauer, and W. Dobrowolski, Lecture Notes in Physics 133, 245 (1980).
34. G.Karczewski, Ph.D.Thesis, Warsaw University, 1986, unpublished.
35. H.Pascher, G.Bauer, M.von Ortenberg, L.Palmetshofer, and W.Zawadzki, to be published.
36. R.R.Galazka and J.Kossut, Lecture Notes in Physics 133, 245 1980).
37. J.Mycielski, see Ref.16, p.364.
38. G.Karczewski and M.von Ortenberg, in Proc.Int.Conf.Phys. Semicond., ed. J.D.Chadi and W.A.Harrison (Springer N.Y. 1985) p.1435.
39. J.Neuwirth, W.Jantsch, L.Palmetshofer and W.Zulehner, J.Phys.C 19, 2475 (1986).
40. T.Dietl and J.Spałek, Phys.Rev.B28, 1548 (1983).
41. R.Stepniewski, Solid State Commun. 58, 19 (1986).
42. A.Golnik, A.Twardowski, and J.A.Gaj, J.Crystal Growth 72 376 (1985).
43. S.Gerken, Dilomer Thesis 1983, unpublished.
44. M.von Ortenberg, Solid State Commun. 52, 111 (1984).
45. A.Golnik and J.Spałek, J.Mag. and Mag.Mat. 54-57, 1207 (1986).
46. J.Spałek, to be published.

MAGNETIC PROPERTIES OF $Pb_{1-x}Gd_xTe$

M. GORSKA[1] AND J.R. ANDERSON*, AND Z. GOLACKI**
* Department of Physics and Astronomy, University of Maryland, College Park, Maryland 20742
** Institute of Physics, Polish Academy of Sciences, Warsaw, Poland

The magnetization and magnetic susceptibility of Bridgman-grown $Pb_{1-x}Gd_xTe$ have been measured over a temperature range from 2 to 300 K and in magnetic fields from 0.01 to 50 kOe. The x-values of the crystals ranged from 0.03 to 0.07. The magnetic susceptibility followed a Curie-Weiss behavior, $\chi = C/(T + \theta)$, with positive θ implying an antiferromagnetic exchange interaction between Gd ions. The magnetic field dependence of the magnetization was fitted to a modified Brillouin function with parameter values that agreed fairly well with those from Curie-Weiss plots. The magnitude of θ was comparable to the value found for $Pb_{1-x}Mn_xTe$ for similar x values; but since the ion spin is bigger for Gd this suggests that the exchange interaction in Gd-doped PbTe is roughly half the value in Mn-doped PbTe.

Introduction

Magnetization measurements in diluted magnetic semiconductors containing ions such as Mn^{2+} with unfilled 3-d shells have shown that the exchange interaction between neighboring magnetic ions is antiferromagnetic [1]. In addition, the low field results have been described in terms of a Curie-Weiss behavior. Here we are reporting results on a semiconductor containing a rare-earth magnetic impurity Gd for comparison with diluted magnetic semiconductors containing Mn. Since the f-shell of Gd is more localized and shielded in comparison with the d shells of transition metal ions, one would expect a smaller exchange interaction in rare-earth doped diluted magnetic semiconductors.

Some previous measurements have been carried out on $Pb_{1-x}Gd_xTe$ (PGT). From electron paramagnetic resonance in PbTe doped with small amounts of Gd Bartkowski *et al.* [2] found the g-factor to be approximately 2 as expected. Transport and magnetization measurements were carried out by Averous *et al.* on PGT [3].

Estimates of the exchange interaction, based on the Curie-Weiss temperature, have been reported previously for another PbTe-based material, namely $Pb_{1-x}Mn_xTe$ (PMT), by several authors [4-6]. In the present work we report measurements of the magnetization and magnetic susceptibility. Our magnetization measurements were fitted to a modified Brillouin function but, unlike the results reported by Averous *et al.* [3], we do not find evidence for a non-zero internal magnetic field within the errors of our measurement.

Experiment

Magnetization measurements from 0 to 50 kOe have been carried out on single crystals of PGT using a SQUID system. The experimental arrangement has been described previously [7]. The temperature range was from about 2 K to 200 K and in some cases to 300 K. Each value of magnetization was the average of 16 measurements at fixed field and temperature. In the interpre-

[1]On leave from the Institute of Physics, Polish Academy of Sciences.

Mat. Res. Soc. Symp. Proc. Vol. 89. © 1987 Materials Research Society

tation of all measurements we made a constant correction to the susceptibility of -3×10^{-7} emu/g to account for the diamagnetism of PbTe [8].

The crystals of $Pb_{1-x}Gd_xTe$ were cut from larger boules grown by the Bridgman technique. The x-values, determined by microprobe and X-ray fluorescence analyses, ranged from about 0.03 to 0.07 and are given in Table I. We estimate the uncertainty in x-value, which includes the variation throughout a sample, to be about 0.01. The samples were all n type with carrier concentrations of 10^{19} to 10^{20} cm^{-3}.

Results

In Fig. 1 we present the inverse susceptibility χ^{-1} vs. temperature for the five samples indicated in Table I. The measurements were made at fields of 100 Oe or less. It is clear from the figure that all five curves extrapolate to a small negative temperature as χ^{-1} goes to zero, implying a net antiferromagnetic interaction. At low temperatures the curvature of the χ^{-1} vs. T plots is negative for all x-values.

A least squares fit of the susceptibility χ to the Curie-Weiss expression

$$\chi = \frac{P_1}{T + \theta} + \chi_{dia} \qquad (1)$$

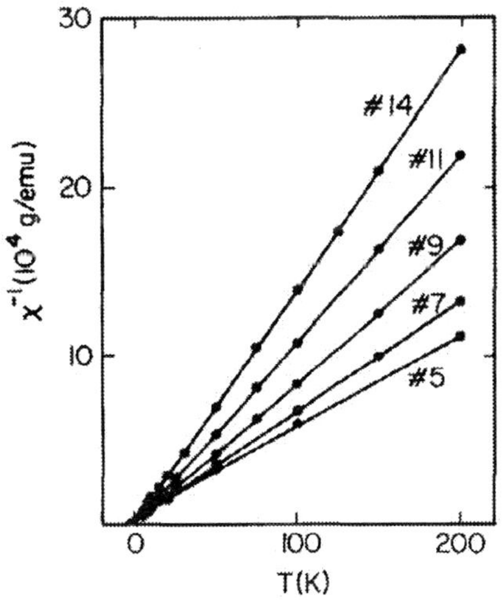

Fig. 1. Inverse susceptibility versus temperature for $Pb_{1-x}Gd_xTe$. The lines are fits to the Curie-Weiss law.

gives the parameters shown in Table I. Here T is the absolute temperature, χ_{dia} is the diamagnetic susceptibility of PbTe, and θ is the Curie-Weiss temperature. Although measurements were made down to 2 K, the fit was carried out over a temperature range from 20 to 200 K. The values of $\bar{x}$, the effective x referring to the net Gd contribution to the susceptibility, are obtained from the expression

$$\bar{x} = (m_1 + m_3) \{S(S+1)(g\mu_B)^2 N_A/3k_B P_1 + m_1 - m_2\}^{-1} , \qquad (2)$$

where m_1, m_2, and m_3, are the atomic masses of Pb, Gd, and Te, respectively, S is the Gd spin, g is the g-factor for Gd, μ_B is the Bohr magneton, and k_B is the Boltzmann constant. The constant P_1 is obtained from the Curie-Weiss fit (Eq. 1). For Gd we take g = 2 and S = 3.5.

We see from the table that, within experimental error, the values of $\bar{x}$ agree with values obtained from x-ray analysis. This is to be expected since the Gd concentrations are low. The values for θ are small and increase with increasing x.

Table I
Curie-Weiss law fitting parameters for $20 < T < 200$ K

Sample #	x	$\bar{x}$	θ(K)	θ_o(K)	$\frac{2J}{k_B}$(K)
5	0.066 ± 0.01[a,b]	0.07 ± 0.01	3.8 ± 0.4	54.3	0.86
7	0.064 ± 0.01[b]	0.066 ± 0.01	2.6 ± 0.2	39.4	0.63
9	0.056 ± 0.01[a,b]	0.053 ± 0.005	1.8 ± 0.2	34.0	0.54
11	0.036 ± 0.005[b]	0.041 ± 0.005	1.5 ± 0.1	36.6	0.58
14	0.033 ± 0.005[a,b]	0.033 ± 0.005	1.7 ± 0.3[c]	51.5	0.82

a. Microprobe measurement

b. X-ray fluorescence measurement

c. Errors are larger because low-field data were taken only at 100 Oe.

Spalek <u>et al</u>. [1] have shown that for values of θ substantially smaller than the measurement temperature both P_1 and θ should vary approximately linearly with x. Therefore one would expect θ_o ($= \theta/\bar{x}$) to be a constant, independent of the sample composition if the distribution of Gd ions on the lattice sites is random. Except for sample #5 with highest Gd concentration, this seems to be the case if one takes the experimental uncertainties into account (see Table I).

From θ_o the nearest neighbor exchange integrals, $2J/k_B$, can be estimated from the relation:

$$\frac{2J}{k_B} = \frac{3\theta_o}{S(S+1)z} \tag{3}$$

where z is the number of nearest neighbors on cation sites. For PGT, which has the NaCℓ structure, $z = 12$. From Table I we see that the value of $2J/k_B$ lies between 0.5 K and 1 K, which is quite small; the average value is 0.69 ± 0.14. Similar measurements for $Pb_{1-x}Mn_xTe$ [5], however, gave values of $2J/k_B$ only about twice as large. Here we should also point out that the exchange integral for $Hg_{1-x}Mn_xTe$ is 14.3 K, a much larger value [1]. Although the HgTe-based materials have a different structure from the PbTe-based materials, we do not believe this difference is important to the magnetic properties because the number of nearest neighbors is the same and the lattice constants of PbTe and HgTe are almost identical. The larger number of carriers in the PbTe-based systems is a more likely source for the small values of these exchange integrals. We should also note, however, that the overlap of the 3-d wavefunctions on the cation sites with the anion wavefunctions is different for the two structures.

The magnetization of PGT as a function of magnetic field is shown for two representative values of x in Figs. 2a and 2b. In all the samples, the magnetization was fitted to a modified Brillouin function [9] of the form

$$M = Sg\mu_B x_B N_o B_{5/2}(\zeta) \,, \tag{4}$$

Table II
Brillouin function fitting parameters

Sample #	5		9		14	
T(K)	x_B	T_B(K)	x_B	T_B(K)	x_B	T_B(K)
2.2	0.064	2.7	0.060	1.8	0.036	1.2
3.5	0.067	3.2	0.061	2.1	0.036	1.3
5.5	0.070	3.7	0.061	2.2	0.036	1.4
7	0.072	4.0	0.061	2.4	0.036	1.6
10					0.037	1.9

where

$$\zeta = g\mu_B H/k_B(T + T_B) \tag{5}$$

and

$$B_s(\zeta) = \frac{2S+1}{2S} \coth\left\{\frac{2S+1}{2S}\zeta\right\} - \frac{1}{2S} \coth\left\{\frac{\zeta}{2S}\right\} \tag{6}$$

is a Brillouin function. Here N_0 is the number of cation sites/gram. There are two fitting parameters, x_B and T_B, which represent an effective occupation probability of a cation site and an effective exchange interaction, respectively. The lines drawn through the data points in Figs. 2a and 2b are given by Eq. 4. The parameters for the Brillouin function fit for several temperatures are given in Table II. The value of x_B and T_B agree fairly well with the values of x and θ as expected, since the x-values are small.

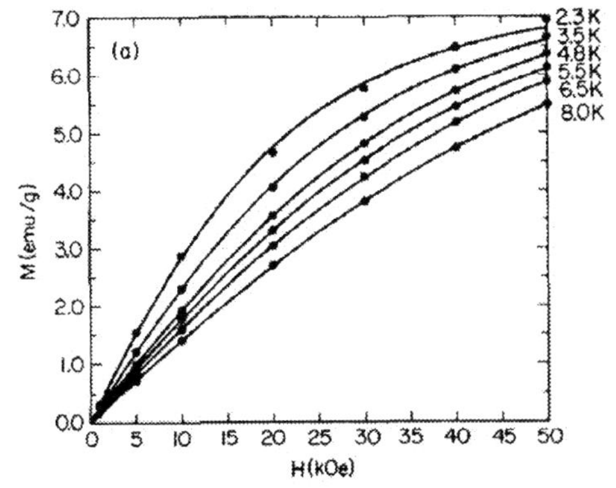

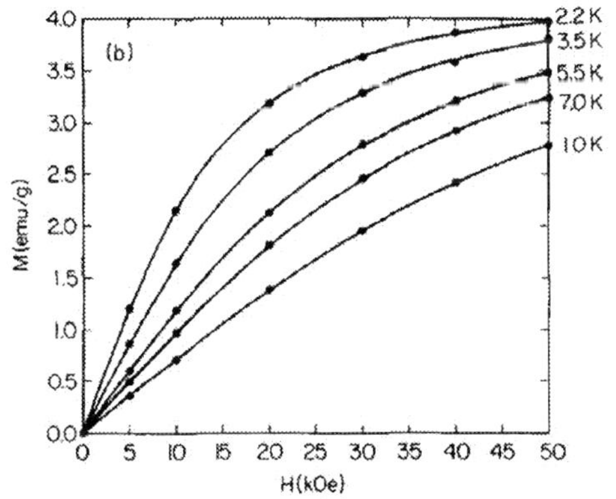

Fig. 2. Magnetization versus magnetic field for $Pb_{-x}Gd_xTe$. The lines are fits to a modified Brillouin function. (a) Sample #5. (b) Sample #14.

Conclusions

Measurements of magnetization and magnetic susceptibility show that the exchange interaction in $Pb_{1-x}Gd_xTe$ is about half the value of $Pb_{1-x}Mn_xTe$ and more than a factor of 10 smaller than the exchange interaction in $Hg_{1-x}Mn_xTe$. In all cases the interaction between magnetic ions is predominantly antiferromagnetic. There is no evidence for a magnetic phase transition in $Pb_{1-x}Gd_xTe$ down to our lowest measurement temperature of 2 K.

Acknowledgments

We wish to acknowledge helpful discussions and the support of J.E. Schirber, E.L. Venturini, and L.J. Azevedo of Sandia Laboratories, at which many of these measurements were carried out. We are also grateful for the help of Richard Frankel of the National Magnet Laboratory at which measurements were also carried out. We also thank M. Meyerhof of Martin Marietta for the microprobe measurements and R. Nielsen for X-ray fluorescence measurements. This work was supported in part by the U.S. Department of Defense Advanced Research Projects Agency (DARPA) and the U.S. Army Research Office under Grant No. DAAG29-85-K-0052.

REFERENCES

1. For references, see J. Spalek, A. Lewicki, Z. Tarnawski, J.K. Furdyna, R.R. Galazka, and Z. Obuszko, Phys. Rev. B33, 3407 (1986).

2. M. Bartkowski, D.J. Northcott, J.M. Park, A.H. Reddoch, and F.T. Hedgcock, Solid State Commun. 56, 659 (1985).

3. M. Averous, B.A. Lombos, C. Fau, E. Ilbnouelghazi, J.C. Tedenac, G. Brun, and M.A. Bartkowski, Phys. Stat. Sol. (b) 131, 759 (1985).

4. M. Escorne, A. Mauger, J.L. Tholence, and R. Triboulet, Phys. Rev. B29, 6306 (1984).

5. J.R. Anderson and M. Gorska, Solid State Commun. 52, 601 (1984).

6. M. Bartkowski, A.H. Reddoch, D.F. Williams, O. Lamarche, and Z. Korczack, Solid State Commun. 57, 185 (1986).

7. J.R. Anderson, M. Gorska, L.J. Azevedo, and E.L. Venturini, Phys. Rev. B33, 4706 (1986).

8. S.N. Lykov and I.A. Chernik, Fiz. Tekh. Poluprovdn. 14, 1861 (1980) [Sov. Phys. Semicond. 14, 1112 (1980)].

9. J.A. Gaj, R. Planel, and G. Fishman, Solid State Commun. 29, 435 (1979).

NMR AND EPR STUDY OF $Sn_{0.98}Gd_{0.02}Te$ AT LOW TEMPERATURES.

B.S. Han, O.G. Symko, D.J. Zheng, Department of Physics, University of Utah, Salt Lake City, Utah 84112 and F.T. Hedgcock, Department of Physics, McGill University, Montreal, PQ H3A2T8

ABSTRACT

Magnetic resonance behavior of $Sn_{0.98}Gd_{0.02}Te$ has been investigated from 2.5 K down to 0.015 K using a SQUID magnetometer and a ^3He-^4He dilution refrigerator. Detection of magnetic resonance with a SQUID allows the detection of NMR and EPR of some of the constituents of the sample. This method is particularly useful for broad lines. We have investigated the NMR for ^{125}Te, ^{119}Sn, and ^{117}Sn, and the EPR for the Gd ion in a field of 265 Oe. The contribution of Gd impurities is maily important in the Knight shift of the Sn isotopes; this shift is 7%. The EPR shows 2 broad lines with structure; one line we attribute to the paramagnetic ions and the second one we attribute to Gd ions which have formed clusters.

INTRODUCTION

We present here an investigation of low field magnetic resonance behavior of a semimagnetic semiconductor as it is cooled from 4 K down to 10 mK. The system is a compound of $Sn_{1-x}Gd_xTe$ where the Gd^{3+} ion forms a very localized moment in the degenerate semiconductor SnTe. There have been studies[1,2,3] of a transition ion impurity Mn^{2+} in SnTe, but few experiments exist for rare-earth impurities. EPR studies[4] on Gd doped PbTe have shown that the impurity is an S-state ion and that near the central line (-1/2,1/2) there is another much broader line present. Although the origin of this line was not clear, it was attributed to "clusters" of Gd ions subject to exchange interactions or to cross relaxation. Such cluster lines have been observed in other systems, such as the compound CdZnMnTe[5] as well as the alloy PtGd[6]. Such measurements have been at relatively high temperatures while the results presented here extend down to temperatures where the internal interactions of clusters become dominant over the thermal energy. Since spin freezing mechanisms and cluster formations are sensitive to the external magnetic field, the measurements presented here are in a relatively low external magnetic field.

Mat. Res. Soc. Symp. Proc. Vol. 89. © 1987 Materials Research Society

As a result, the central line (-1/2,1/2) and the cluster lines are seperated well.

EXPERIMENTAL DETAILS

The sample of $Sn_{1-x}Gd_xTe$ of nominal concentration x of 2% was grown by a Bridgman method. In order to have a well-defined sample geometry the single crystals were ground to a fine powder which was packed into an epoxy resin container, the sample dimensions being 2 mm diameter by 6 mm long. Around this sample was wrapped a small saddle-shaped coil for r.f. excitation of the magnetic resonance. The sample was located inside the mixing chamber of a ^3He-^4He dilution refrigerator and its temperature was measured with a Cerium Magnesium Nitrate (CMN) thermometer. The CMN thermometer also located in the mixing chamber and was calibrated against a commercially calibrated germanium-resistance thermometer. The magnetic resonance was detected by a SQUID magnetometer[7]. An external magnetic field of 265 Oe, was trapped inside a niobium tube surrounding the sample and was determined from the proton NMR in the sample container. It is the steady magnetic field in which all the magnetic resonance was done. This field was trapped near a temperature of 9 K and the sample was cooled in this field down to 10 mK. Magnetic resonance was performed by sweeping the frequency of the r.f. field in this fixed external field. This makes it possible to cover a wide range of frequencies and detect NMR as well as EPR. Such a method of doing magnetic resonance offers extremely high sensitivity for broad lines at low magnetic fields making it very applicable to the present study. The frequency was swept from 100 kHz to 1.3 GHz. However, in principle, it is possible to go to even higher frequencies. The EPR spectrum of Gd^{3+} in a cubic environment[8] consists of a ground state Γ_7 and with Γ_8 and Γ_6 above it. However, in this experiment we have looked at only the Zeeman splitting (-1/2,1/2) in the field of 265 Oe.

RESULTS AND DISCUSSIONS

In the fixed field of 265 Oe we looked at the NMR of ^{117}Sn, ^{119}Sn, and ^{125}Te. Figure 1 shows the spectrum at 15 mK. The corresponding NMR shift was calculated by using the gyromagnetic ratio values[9]. Those are presented in Table I. However, at 100 mK both Sn lines show a Knight shift of 2.5%. This result is similar to that obtained for Mn in Sn-Te, where the Knight shift increases as the temperature is reduced.

Table I: The NMR shift $\Delta K/K$ of ^{117}Sn, ^{119}Sn, and ^{125}Te in $Sn_{0.98}Gd_{0.02}Te$ at T=0.015 K and the gyromagnetic ratio values[9].

Isotope	f_{res} (MHz)	$\Delta K/K$ (%)	$\gamma/2\pi$ ($KHzG^{-1}$)
^{117}Sn	0.372	7.3	1.5168
^{119}Sn	0.388	7.6	1.5869
^{125}Te	0.355	-0.3	1.3454
^{1}H	1.128		4.2576

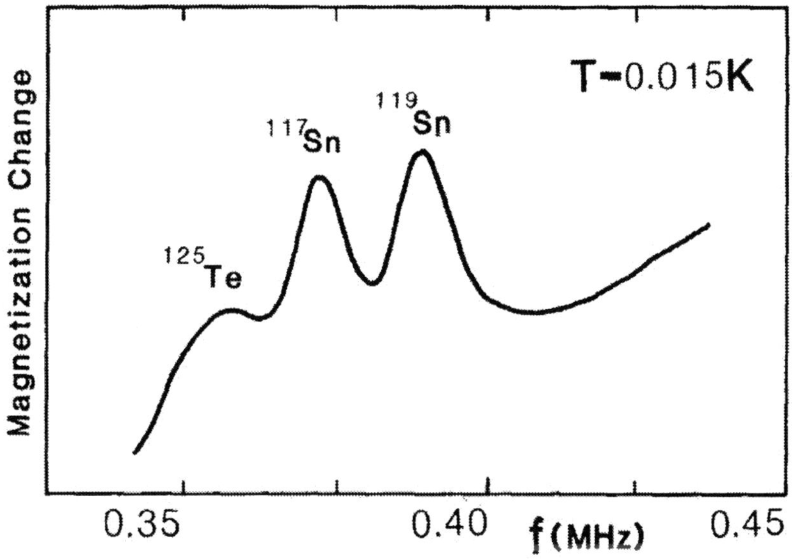

Fig.1 The NMR spectrum of ^{117}Sn, ^{119}Sn and ^{125}Te in $Sn_{0.98}Gd_{0.02}Te$ at T=0.015 K.

The behavior of Te is interesting in that the line has been largely broadened with essentially no Knight shift. Such a result implies that the Te ion is close to the Gd ions which are coupled anti-ferromagnetically. Indeed EPR measurements[10] on other systems conclude that the anion plays a major role in mediating the spin-spin interaction. The Sn ions are further away and they respond to the overall internal magnetic field as evidenced by the magnetic contribution to the Knight shift.

In that same magnetic field and with the same apparatus we looked at the EPR of the Gd^{3+} ions. Figure 2(a) shows the spectrum at 2.5 K. there is a line at 745 MHz corresponding to about zero internal field, which we attribute to the Gd^{3+} ions in the paramagnetic state. Above that frequency

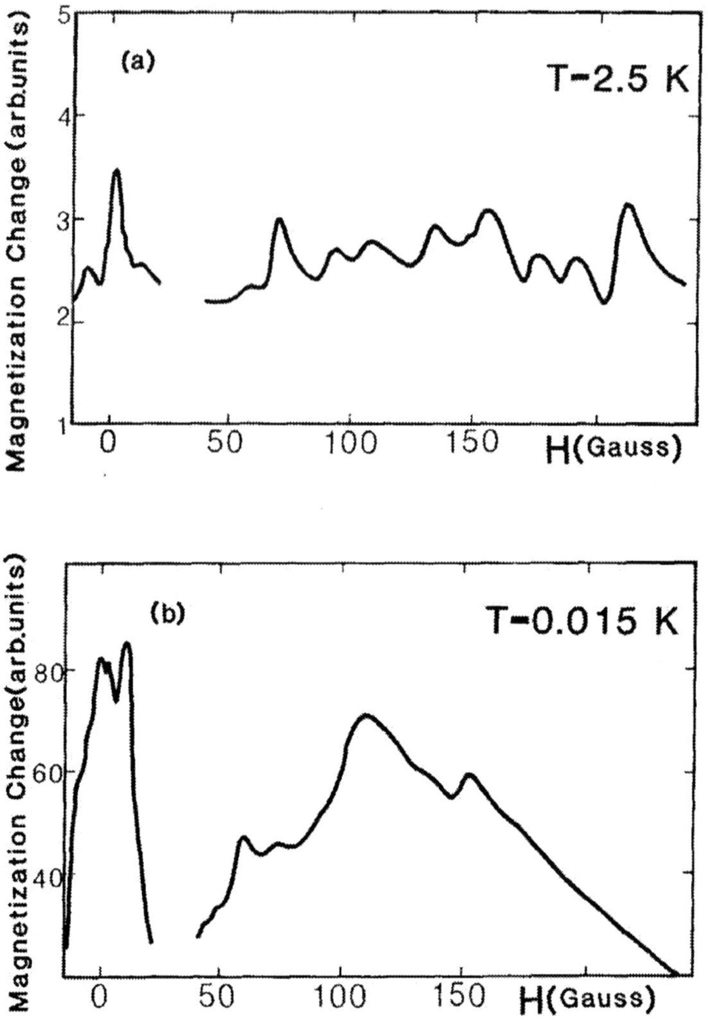

Fig.2 The EPR spectrum of Gd^{3+} in $Sn_{0.98}Gd_{0.02}Te$ (a) T = 2.5 K, (b) T = 0.015 K. H is the internal magnetic field. In both (a) and (b), the magnetization change is in the same arbitrary units.

there are other lines such as at 935 MHz, 1035 MHz, and 1170 MHz corresponding to the internal field of 69 G, 105 G and 155 G respectively, which we attribute to clusters of Gd ions. To investigate this hypothesis, we followed the spectrum to very low temperatures. Figure 2(b) shows the spectrum at 0.015 K. The paramagnetic line has increased in amplitude as the temperature was lowered and the clusters lines also grew merging together to one broad line.

The results that we obtained are consistent with the high temperature data obtained in reference[11] for Gd in PbTe. Their susceptibility data show departures from Curie-Weiss behavior below 2.5 K for a 5% sample. This could indicate the formation of clusters of Gd ions as indicated by our results, forming some sort of super paramagnetic states. Because we did our measurements in low magnetic field, it was easy to see the shifted cluster line and to see it grow as the temperature got reduced. Very likely the growth of such clusters will lead to some sort of magnetic ordering either between the spins or between the clusters at some low temparature[12].

The structure of the paramagnetic line at 0.015 K indicates the dipolar interation between Gd ions. The dipolar energy $3\mu^2/R^3$ corresponding to an average distance R, between two Gd magnetic dipoles μ , usually given by $(4/3)\pi(R/2)^3 = a^3/4x$ for a fcc lattice of lattice constant a, is approximately 25 G, which is in agreement with the experiment.

REFERENCES

(1) R.W. Cochrane, F.T. Hedgcock, and A.W. Lightstone, Can. J. Phys. 56, 68 (1978).

(2) B. Perrin and F.T. Hedgcock, Can. J. Phys. 60, 1783 (1982).

(3) B. Perrin and F.T. Hedgcock, J. Phys. C: Solid St. Phys. 15, 6037 (1982).

(4) M. Bartkowski, D.J. Northcott, J.M. Park, A.H. Reddoch, and F.T. Hedgcock, Solid St. Commun. 56, 659 (1985).

(5) A. Manoogian, B.W. Chan, R. Brun del Re, T. Donofrio, and J.C. Woolley, J. Appl. Phys. 53, 8934 (1982).

(6) M. Hardiman, J. Pellisson, S.E. Barnes, P.E. Bisson, and M. Petter, Phys. Rev. B22, 2175 (1980).

(7) R. Chamberlin and O.G. Symko, J. Low Temp. Phys. 35, 337 (1979).

(8) E. Jaehne and O.G. Symko, Solid St. Commun. 30, 31 (1979).

(9) R.C. Weast Handbook of Chemistry and Physics 53 rd edition (CRC) 1972-1973 P. E-58.

(10) R.E. Kremer and J.K. Furdyna, Phys. Rev. B31, 1 (1985).

(11) F.T. Hedgcock, P.C. Sullivan, J.T. Grembowicz, and M. Bartkowski, to be published in Can. J. Phys.

(12) B.S. Han, O.G. Symko, D.J. Zheng and F.T. Hedgcock, to be published.

SATURATION BEHAVIOUR OF SUPERHYPERFINE STRUCTURE OF Mn^{2+} IN PbTe

A.H. REDDOCH, M. BARTKOWSKI and D.J. NORTHCOTT
Division of Physics, National Research Council of Canada, Ottawa, K1A 0R6, Canada

ABSTRACT

X-band EPR measurements have been made on single crystals and powders of PbTe doped with Mn below 100 ppm. A seven-fold splitting of each Mn hyperfine structure line is shown to arise from tellurium superhyperfine structure. Microwave saturation results are presented which indicate that the spin-lattice relaxation time is different for different superhyperfine lines. This asymmetry is not readily explained by conventional models.

INTRODUCTION

Superhyperfine structure, in the sense of electron paramagnetic resonance hyperfine structure arising from nominally diamagnetic species near a paramagnetic center, has been seen in a number of systems including transition metal ions in diamagnetic insulators [1] and in semiconductors [2,3]. Although a basic understanding is available there appears to be no comprehensive theory of such effects. Thus, in semiconductors shfs is, in some cases, seen from nearest neighbour anions [3] while in other cases it may be seen only from next nearest neighbour cations [2]. There seems to be no explanation of this difference.

Pifer in a study of Mn in PbTe reported a set of spectra in which each Mn hfs line showed a complex lineshape which varied with the concentration and sign of the carriers [4]. No explanation was available.

We report here our work on that same system and show that there is shfs from neighbouring tellurium anions and that these shfs lines have different saturation behaviours. These observations account at least partially for Pifer's spectra.

EXPERIMENTAL

Single crystals of PbTe doped with Mn below 100 ppm were grown by a Bridgman technique. Powder samples were prepared from these. The samples were n-type material. EPR spectra were obtained with a Varian E12 X-band EPR spectrometer. Crystals were mounted in a two-circle goniometer. Cooling was achieved with a gas flow system supplied by a liquid helium transfer line. Temperatures below 4.2K were obtained by pumping. Temperatures were measured with a calibrated carbon-glass thermometer and with a gold-chromel thermocouple. Measurements were made from 2.3 to 100K.

RESULTS

At sufficiently low microwave power each Mn line shows a five-fold structure of equally spaced lines, Fig. 1, with the

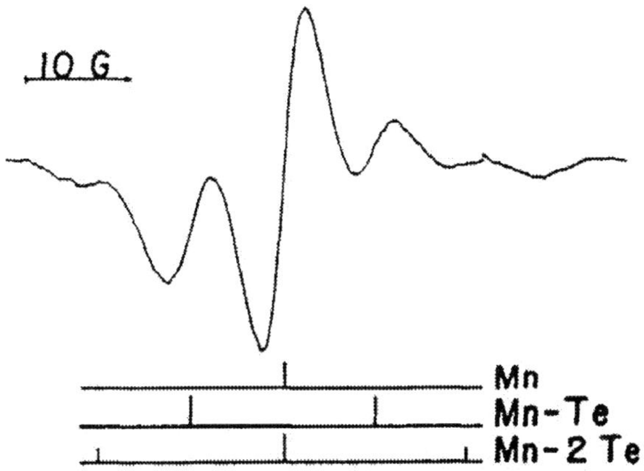

Fig. 1
Superhyperfine structure on lowest field Mn line in powder at 4.2K. Modulation frequency 10kHz, microwave power 50μw.

average intensities of corresponding pairs of lines being 1.0:0.27:0.04 from the center line out. Two additional weak lines can sometimes be detected. The spectrum is isotropic and is readily seen in a powder. The intensity ratios and the isotropy show that the structure does not arise from fine structure which would have intensities 8:5:9:5:8 and would be strongly anisotropic [5].

The observed intensities agree well with what would be expected for hyperfine structures from Te^{123} and Te^{125} in natural abundance [6] on the six nearest neighbour sites. Each nucleus has spin $I = \frac{1}{2}$ and they have similar nuclear moments. Together their abundance is 8%. From this information it can be predicted that 13 lines will occur in the spectrum but that only the central lines will have significant intensity [7]. The relative intensities from the center would be 1:0.25:0.027:0.0016... in good agreement with our observations. The lead isotope Pb^{207}, having an abundance of 21%, would give more lines and would not account for the observations. The coupling constant for the Te is 15.5G or $14.5 \times 10^{-4} cm^{-1}$.

It can be seen in Fig. 2 that at the lowest microwave power the spectrum of essentially three lines is not exactly symmetric in intensity. As the power is increased the asymmetry becomes extreme with the first low field satellite growing until it dominates the spectrum. The second low field satellite also grows although less markedly. The central line decreases in relative amplitude while the high field satellite almost disappears. Note that the behaviour is essentially the same for all six manganese lines.

In Fig. 3 the intensities of the lines as a function of microwave power are presented. Clearly the components respond to the power at different rates.

Since these phenomena arise from increased microwave power, it appears that saturation must be considered. From the Bloch equations [8] this means that the product T_1T_2 must differ for the various components. Since the intensity of the satellites is essentially symmetric at low microwave power, the relaxation time T_2 would not account for the asymmetry by itself. Thus, the effect must arise from the spin-lattice relaxation time T_1.

In particular, T_1 must be an asymmetric function of the spin component m_{Te}. In a time-dependent perturbation approach the spin

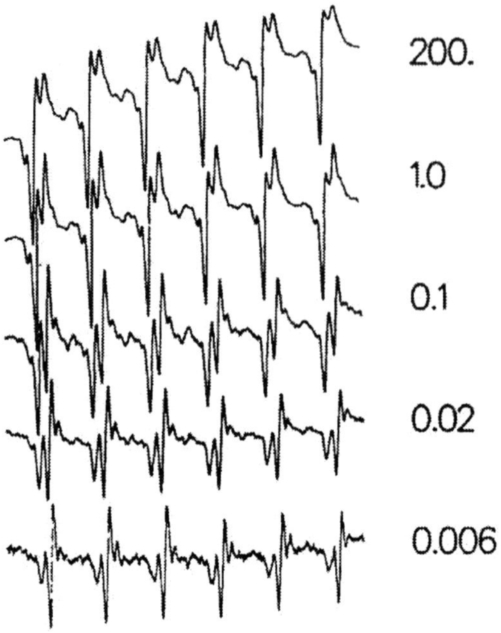

Fig. 2
Spectra at 4.2K at various microwave power levels (in mW).

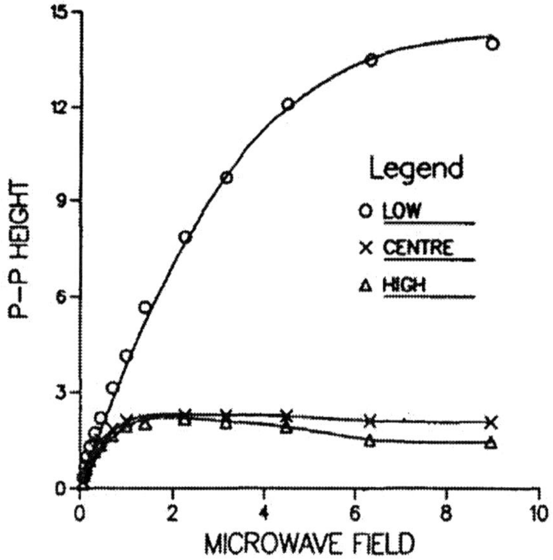

Fig. 3
Saturation behaviour of centre line and low-and high-field tellurium satellites. Heights normalized by species abundance. Temperature 7K. Field units as $mW^{\frac{1}{2}}$.

lattice relaxation time would be proportional to the square of a matrix element. Hence for a simple perturbation any dependence on m_{Te} should be independent of the sign of m_{Te}, i.e. symmetrically placed satellites would have the same saturation behaviour. This approach then requires a simultaneous perturbation of two or more terms in the Hamiltonian. Note that the spin lattice relaxation time does not depend on m_{Mn} although it does depend on m_{Te}. If we do not have a spin bottleneck [9], as the large difference in the g-values of the carriers and the Mn ions [10] suggests, then carrier interactions may provide an effective spin lattice relaxation process. However, direct carrier-nucleus interaction with the tellurium nuclei is not a particularly effective mode of relaxation [11].

The cause of the asymmetric intensities remains unclear. Possibly the electric as well as the spin properties of the carriers must be considered.

The spectra presented here have much in common with some of Pifer's spectra. It appears that superhyperfine structure and asymmetric saturation play a large part in his results.

REFERENCES

[1] T.P.P. Hall, W. Hayes and F.I.B. Williams, Proc. Phys. Soc. 78, 883 (1961).

[2] J. Lambe and C. Kikuchi, Phys. Rev. 119, 1256 (1960).

[3] J.J. Davies, J.E. Nicholls and D. Verity, J. Phys. C. 13, 1291 (1980).

[4] J.H. Pifer, Phys. Rev. 157, 272 (1967).

[5] W. Low, Ann. N.Y. Acad. Sci. 72, 69 (1958).

[6] CRC Handbook of Chemistry and Physics, 66th ed., R.C. Weast, editor, (CRC Press, Boca Raton, Florida, 1986).

[7] M. Bartkowski, D.J. Northcott and A.H. Reddoch, Phys. Rev. B 34, 6506 (1986).

[8] A. Abragam and B. Bleaney, Electron Paramagnetic Resonance of Transition Ions, (Clarendon Press, Oxford, 1970).

[9] S.E. Barnes, Adv. in Phys. 30, 801 (1981).

[10] C.K.N. Patel and R.E. Slusher, Phys. Rev. 177, 1200 (1969).

[11] B. Perrin and F.T. Hedgcock, J. Phys. C. 15, 6037 (1982).

EPR OF Mn^{2+} AND Gd^{3+} IONS IN PbTe AND SnTe SEMICONDUCTORS

MARIAN BARTKOWSKI, D.J. NORTHCOTT, A.H. REDDOCH, D.F. WILLIAMS, F.T. HEDGCOCK* and Z. KORCZAK**
Division of Physics, National Research Council of Canada, Ottawa, K1A 0R6, Canada
*Department of Physics, McGill University, Montreal, H3A 2T8, Canada
**Institute of Physics, M. Curie-Sklodowska University, Lublin 20-031, Poland

ABSTRACT

The EPR spectra of manganese and gadolinium doped PbTe and SnTe semiconductors were measured from 4.2K up to 400K for dopant concentrations ranging from 50 to 50000 ppm.

Resolved fine, hyperfine and superhyperfine structure as well as forbidden and cluster lines were measured. The combination of Lorentzian absorption and dispersion derivatives fits the recorded spectra well. Nonlinear broadening of linewidths with temperature is characteristic in these semiconductors.

A manganese-induced broad maximum in linewidth versus temperature was found in PbTe around 180K for dopant concentrations in excess of 1.5at%.

INTRODUCTION

The rare earth gadolinium and transition metal manganese are the magnetic dopants for narrow gap semiconductors like PbTe and SnTe in our studies of localized-mobile spin interactions. This group of materials offers the opportunity to work with extreme mobilities in electrical transport [1], carrier induced magnetic phase transition [2] and spin freeze-out [3-4] in magnetic properties.

While Mn is known to interact strongly with the carrier spins in various semiconductors to yield the well-known properties of dilute magnetic semiconductors there is little evidence that Gd with its supposedly shielded 4f electrons has similar effects. In this paper we present some recent EPR spectra of manganese and gadolinium and data on temperature and concentration dependence of linewidths. These results indicate that the Gd 4f interactions in semiconductors are not negligible.

EXPERIMENTAL

X-band EPR measurements from 4.2K to 400K were performed on manganese and gadolinium doped crystalline PbTe and SnTe including the angular variation of the Mn^{2+} and Gd^{3+} spectra [5,6]. The magnetic ion content was varied from 50 to 50000 ppm. The crystals were grown by a standard Bridgman method [3], but limiting the free space above the SnTe load by double ampoule sealing. The quenched samples of manganese doped PbTe were prepared as described in [7]. The results were collected for unannealed samples.

Powdered samples were size selected by liquid sedimentation,

Mat. Res. Soc. Symp. Proc. Vol. 89. ©1987 Materials Research Society

imbedded in insulating wax and placed in quartz capillaries with a diameter <3mm. Single crystal samples were shaped into spheres and mounted in a two-circle goniometer [6].

RESULTS

1. Spectra and spin hamiltonian.

The EPR spectra corresponding to the fine structure, hyperfine and superhyperfine terms of the spin hamiltonian as well as forbidden lines and collective phenomena are shown in Fig. 1.

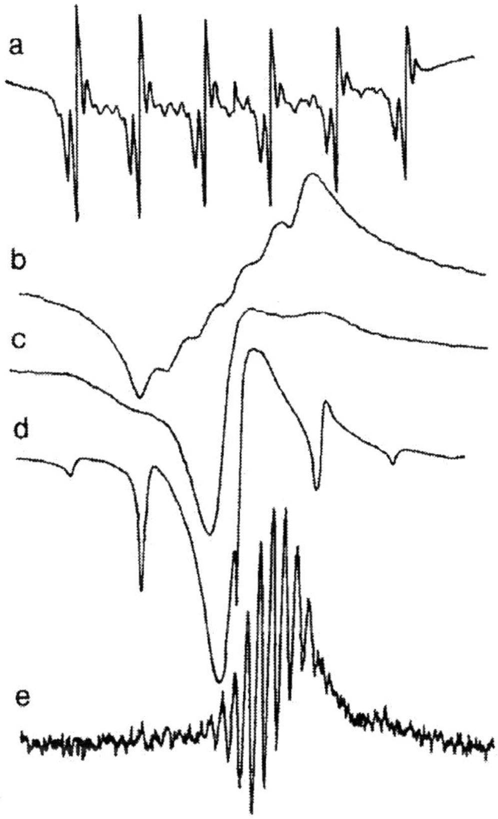

Fig. 1 X-band EPR spectra of manganese in powders of (a)-PbTe, (b)-SnTe and gadolinium in single crystals of (c)-SnTe (d,e)-PbTe, along [111] crystallographic directions.

In Fig. 1a hyperfine, superhyperfine and forbidden lines for manganese doped PbTe are shown. The six strongest lines are those corresponding to EPR transitions involving single manganese ions. The lines on both sides of the hyperfine lines correspond to superhyperfine transitions in which the nearest neighbour Te^{125} spin ½, participates. The pairs of lines centered between the hyperfine lines are forbidden lines corresponding to EPR transitions with $\Delta m_I = \pm 1$. The spectrum is broadened to the limits of resolution for manganese concentrations above 0.12at% and is collapsed into a broad single line for x > 0.9at%.

In Fig. 1b the hyperfine structure of Mn^{2+} ions in SnTe is shown. For samples with a manganese content in excess of 0.1at% structural features disappear creating a single broad line, while for smaller concentrations structure does not follow manganese content indicating the presence of a line broadening mechanism unrelated to Mn-Mn distance.

Fig. 1c shows the fine structure spectrum of Gd^{3+} ions in SnTe crystals along the [111] direction. In this case, fine structure lines are broader and inseparable from the cluster line.

Fig. 1d consisting of five lines superimposed on a broad cluster line clearly depicts the resolved fine structure of Gd^{3+} ions in the cubic environment of the PbTe crystal along the [111] crystallographic direction.

For PbTe crystals with a gadolinium content of <0.1at% the cluster line is not detected, while for samples with a Gd content >1at% the fine structure lines are only just resolved.

Fig. 1e shows the superhyperfine structure of Gd^{3+} ions attributed to second nearest neighbour Pb^{207} nuclei with spin $\frac{1}{2}$.

The Gd^{3+} fine structure originates from the splitting of the ground state in the presence of a cubic crystal field. At present we are able to provide precise values of b_4 and b_6 for PbTe. At 4.2K they are -110MHz and 2.46MHz respectively [6]. For SnTe b_4 is approximately 100 MHz.

Fine structure lines in the Mn^{2+} spectrum in PbTe and SnTe are absent.

The set of six hyperfine lines of Mn^{2+} in PbTe and SnTe originates from the splitting due to the spin 5/2 of the Mn^{55} nucleus. The hyperfine constants a_{Mn} of 65.7G ± 0.1G in PbTe and 62.5G ± 1G in SnTe are close to those reported [8,9]. However, our value for a_{Mn} in SnTe is larger by about 3G. The reported difference [8,9] between hyperfine constants in PbTe and SnTe, twice as much as found in these experiments, could be due to the sample heat treatment.

The superhyperfine spectra of (Mn-Te) and (Gd-Pb) are identified on the basis of the intensity ratio between lines in the particular set. In the case of tellurium superhyperfine spectra, a total of 13 lines is expected for each hyperfine transition, while for the lead superhyperfine structure a total of 25 lines [10] is expected assuming all 12 nuclei are equivalent.

Note that the manganese superhyperfine structure is isotropic, while that of gadolinium is resolved only on the $[-\frac{1}{2},\frac{1}{2}]$ fine structure line within $\pm 2^0$ of the [100] or [111] crystallographic directions. The superhyperfine constants are a_{Te} = 15.5G[5] and a_{Pb} = 2.0G for manganese and gadolinium doped PbTe respectively.

The forbidden lines are the strongest 45^0 from [100] in the $(1\bar{1}0)$ crystallographic plane, pointing to axial symmetry at the manganese site as their source. However, precise values for the D term have not been established.

The cluster line was found to be angularly dependent increasing in intensity at the collapsing angle of $\sim 30^0$ from [100] in the (110) plane. Similar behaviour of the gadolinium cluster line in Pt [11] was interpreted in terms of a cross relaxation mechanism, which presumably is partly responsible for cluster line appearance.

2. Lineshapes and Linewidths

In all studied samples EPR lines have shown a characteristic distortion around the base line due to the presence of a skin effect (see Fig. 1).

Such lines can be described as a linear combination of the derivative of absorption and dispersion parts of the magnetic suceptibility [12]. In the reported fittings the Lorentzian line shape of absorption and dispersion components was used as in the work of Poole [13]. The ratio of absorption to dispersion for a sample depends on its shape and the ratio of skin layer depth to the sample dimensions. To achieve sample homogeneity, powders were selected by liquid sedimentation. For fine powders, with grain size of a few microns, the dispersion component becomes negligible and this limit was used to study linewidths of superhyperfine and hyperfine lines of Mn^{2+} ions in PbTe and SnTe.

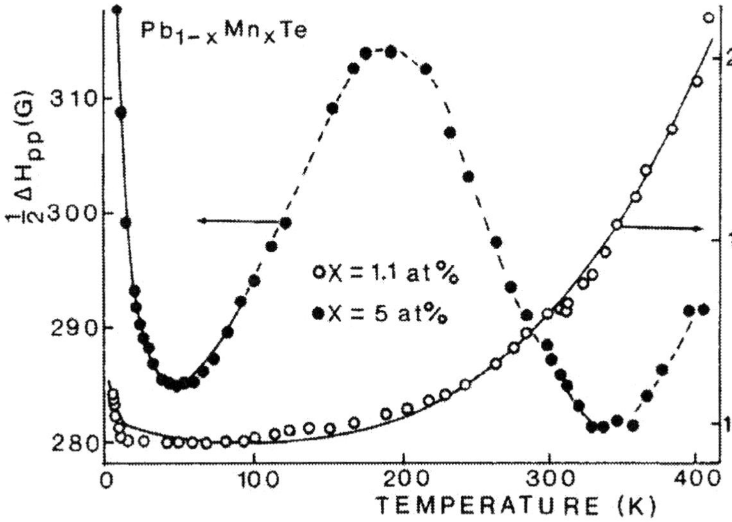

Fig. 2 Linewidth versus temperature for manganese doped PbTe.

In Fig. 2 the linewidth of the single broad resonance for manganese doped PbTe versus temperature is shown. For the PbTe samples with 1.1at% manganese, the linewidth broadens at low and high temperatures. Similar broadening was observed for gadolinium fine structure linewidths.

The temperature variation of linewidth is clearly nonlinear. We find that the following function fits the experimental data:

$$\Delta Hpp = \Delta H_r + C_1/T + C_2 \times T^2$$

for Gd^{3+} fine structure lines. For Mn in Fig. 2 a C_3T^3 term is added to achieve the fit shown. The linear term in the above formula is omitted because its uncertainty is comparable to its value. Note that for metals the line broadening contains a linear term (Korringa term) reflecting the free carrier contribution to line broadening.

For manganese doped PbTe in excess of 1.5at%, we find a broad maximum in linewidth versus temperature, as shown in Fig. 2 for a sample with 5at% of manganese content.

No crystallographic changes are detected for these samples at 300K and 77K which indicates a manganese-induced broadening mechanism. Note that goodness of fit for samples with $x < 1.1$ varies similarly to the linewidth of the 5at% manganese sample. This suggests that changes in lineshape around 180K are reflected strongly in linewidth for samples with a manganese content in excess of 1.5at%.

The residual ΔH_r and C_1/T broadening terms are common features of our manganese and gadolinium doped PbTe.

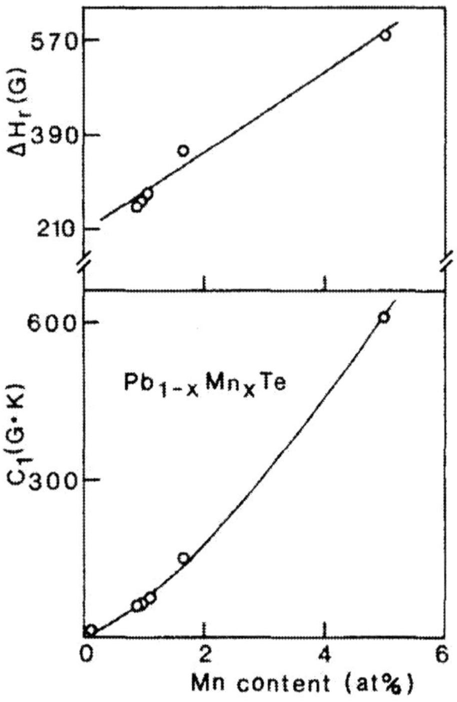

Fig. 3 Residual linewidth ΔH_r and low temperature broadening rate C_1 versus Mn content in $Pb_{1-x}Mn_xTe$.

In Fig. 3 we show the residual linewidth ΔH_r and low temperature broadening rate C_1. The residual linewidths vary linearly with manganese concentration in PbTe, with a rate of 70G/at%, suggesting dipolar broadening.

For the low temperature broadening term we find quadratic functions of x give the best fit. The parameters for the curve shown are 66.7 (GK/at%) for the linear term and 11.2 GK/(at%)^2 for quadratic one. It has to be stressed that the point at $x \approx 0$ corresponds to a broadening rate of a single manganese line in the superhyperfine set.

For the manganese doped SnTe and PbSnTe alloys, low temperature linewidth narrowing is a common feature. However, the Korringa term as applied in [14] is inadequate to describe linewidth changes in a wider temperature range, due to a clear nonlinearity.

CONCLUSION

The fine structure of Gd^{3+} ions in PbTe and SnTe semiconductors have been resolved. Superhyperfine structure due to Te^{125} and Pb^{207} have been identified in the EPR spectra of manganese and gadolinium doped PbTe. Linewidth broadening reflected by a peak in linewidth around 180K induced by manganese in excess of 1.5at% in $Pb_{1-x}Mn_xTe$ semiconductor was found. For $Pb_{1-x}Mn_xTe$ residual linewidth and low temperature broadening rate were found to vary linearly and quadratically with total manganese content respectively.

Our results show that at least in some cases it is possible to obtain detailed information about both Mn and Gd centres in IV-VI compounds through superhyperfine structure. Linewidth variation in such systems appears to be dominated by a residual width probably of dipolar origin, magnetization effects at low temperature and Korringa-like carrier interactions at high temperatures. Since the Gd spectra show cluster lines and broadening from carriers as well as superhyperfine structure from next-nearest neighbours it is clear that the 4f electrons have significant interactions with their environment. In this respect these systems have analogies to Gd in metals [11]. Work is in progress to examine the interactions in more detail.

REFERENCES

[1] M. Averous, B.A. Lombos, C. Fau, E. Ilbnouelghazi, J.C. Tedenac, G. Brun and M. Bartkowski, Phys. Stat. Sol. (b)131, 759 (1985).

[2] T. Story, R.R. Galazka, R.B. Frankel and P.A. Wolff, Phys. Rev. Letters, 56, 777 (1986).

[3] F.T. Hedgcock, P.C. Sullivan, J.T. Grembowicz and M. Bartkowski, Can. J. Phys. 64, 1345 (1986).

[4] M. Escorne, A. Mauger, J.L. Tholence, R. Triboulet, Phys. Rev. B 29, 6306 (1984).

[5] M. Bartkowski, D.J. Northcott and A.H. Reddoch, Phys. Rev. B 34, 6506 (1986).

[6] M. Bartkowski, D.J. Northcott, J.M. Park, A.H. Reddoch and F.T. Hedgcock, Solid State Commun. 56, 659 (1985).

[7] Z. Korczak and M. Subotowicz, Phys. Stat. Sol. (a)77, 497 (1983).

[8] M. Inoue, H. Yagi, T. Muratani and T. Tatsukawa, J. Phys. Soc. Jpn. 40, 458 (1976).

[9] T. Hejwowski and M. Subotowicz, Phys. Stat. Sol. (b)106, 373 (1981).

[10] A. Abragam and B. Bleaney, Electron Paramagnetic Resonance of Transition Ions, (Clarendon Press, Oxford 1970).

[11] M. Hardiman, J. Pellisson, S.E. Barnes, P.E. Bisson and M. Peter, Phys. Rev. B 22, 2175 (1980).

[12] A.M. Kahn, Phys. Rev. B 16, 64 (1977).

[13] C.P. Poole, Jr., Electron Spin Resonance, 2nd ed., (John Wiley & Sons, New York 1983).

[14] R.W. Cochrane, F.T. Hedgcock and A.W. Lightstone, Can. J. Phys. 56, 68 (1978).

VERY HIGH MOBILITY IN SEMIMAGNETIC SEMICONDUCTORS WITH RARE EARTH

AVEROUS M., LOMBOS* B.A., BRUNO A., LASCARAY J.P., FAU C. and LAWRENCE* M.F.
Groupe d'Etude des Semiconducteurs, U.S.T.L., Place E. Bataillon,
34060 - MONTPELLIER-Cédex, FRANCE.
* Concordia University, 1455 Maisonneuve Bd - Montréal, Québec, H3G IM8, CANADA.

ABSTRACT

A semimagnetic semiconductor with rare earth like magnetic ion is studied : $Pb_{1-x}Gd_xTe$. Very high Hall mobilities were found up to 10^7 $cm^2/V.s$. The magnetization is well fitted at 4.2K by a Curie-Weisslaw, with a Gaj's parameter of 1K. The value of J_{NN}, the exchange in a Gd pair deduced from the susceptibility measurements is small, as expected, due to the specific character of Gadolinium (S state and 4f shell) : $J_{NN}/k = -0.3°K$ for x = 0.025.

INTRODUCTION

Due to the exchange interactions induced by the magnetic moments localized at the magnetic ions, the SMSC exhibit new and interesting physical properties |1|, |2|, |3|. The Mn as magnetic ion was the most extensively studied and in the case of small gap material leads to a non linear splitting of the Landau levels |2| with the magnetic field in $Hg_{1-x}Mn_xTe$ for example. Due to the phase diagram restrictions the introduction of Fe is more difficult, however some investigations were devoted to this material as well; the results obtained in the case of HgFeTe are rather different from HgMnTe due to the peculiar magnetism of the Fe ions |4|, |5|. This could be explained in terms of Van Vleck type paramagnetism.

In this work, the magnetic ions chosen are gadolinium and the material is $Pb_{1-x}Gd_xTe$. There are some main differences between the two SMSC families (transition metals or rare earths).

The interaction between the crystal and the carrier of a magnetic moment depends on the depth of the half filled shell. It is stronger when the shell responsible of the magnetic moment is an external one : so this interaction is large in the case of iron ion group and small in the case of rare earth-group ($4f^7$ shell).

In the rare earth, the spin orbit coupling is higher than the crystal field coupling, so the later introduces only a decomposition of the fundamental multiplet and lift partially the degeneracy of the order (2J+1).

EXPERIMENTAL RESULTS AND DISCUSSION

Two method were used to prepare samples : a modified Bridgman method, and a heap pipe controlled zone melter. The nominal Gd content was 0.01, 0.02 and 0.05 (in the crucible). The x in $Pb_{1-x}Gd_xTe$ was determined by electron micropobe analysis and electron-spin resonance. The results are for sample A (below detection), sample B :0.01, sample C:0.025.The band gap was determined

Mat. Res. Soc. Symp. Proc. Vol. 89. ©1987 Materials Research Society

by optical absorption measurements in the temperature range 77°K - 300°K. Its temperature dependence is : dE_g/dT = 3.6 meV/K rather near to the one reported for PbTe. |6|

Transport measurements between 4.2 K - 300K were performed : the Fig. 1 and Fig. 2 represent the resistivity and the mobility deduced from Hall constant as a function of T.

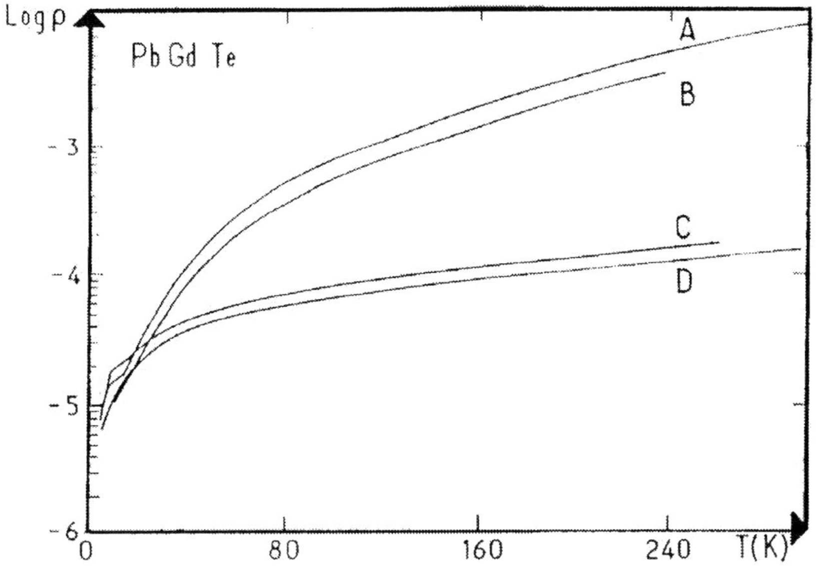

Fig. 1 : Resistivities as a function of temperature of $Pb_{1-x}Gd_xTe$ (x = sample A trace ; sample B 10^{-2} ; sample C 2.5 10^{-2}).

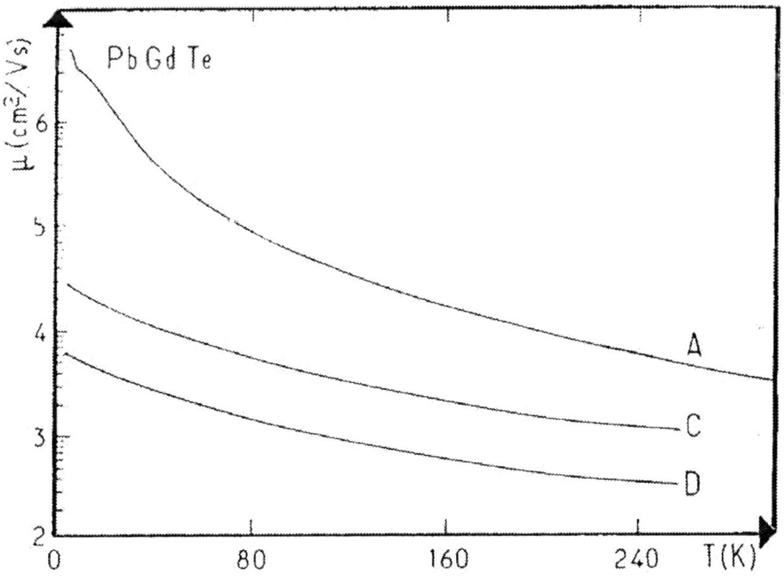

Fig. 2 : Hall mobilities as a function of temperatures.

The low temperature values of μ are very high. For sample A, after 3 days annealing under vacuum in the presence of PbTe powder at 450°C, with the sample at 400°C, this value increases up to 10^7 cm^2/Vs. It is in our knowledge the higest mobility ever measured on material with this gap.

Probably it is mainly due to the high value of the dielectric constant, so we have determined from the reflectivity, the static dielectric constant ε_0, related to ε_∞ by the Lyddane - Saches-Teller equation and found :

$$\varepsilon_\infty \,(300\ K) = 30 \qquad \varepsilon_0 \,(300\ K) = 380$$

We can understand that this large value of ε_0 screens this scattering by ionized impurities at 4.2K.

On Fig.3 we have plot the theoretical curve obtained for long range potential scattering (screened Coulomb potential) and the short range one, following Ravitch et al |7| with :

$$\tau_{coul} = 3\pi\, \varepsilon_0^{\,2}\, \hbar^3/2\, e^4\, N\, m^*_d\, (1 + 2\, \varepsilon_F/\varepsilon_g)\ \phi\,(\delta_0) \qquad (1)$$

with $\phi\,(\delta_0) = \ln\,(1 + \delta_0^{\,-1}) - (1 + \delta_0)^{-1}$

where $\delta_0 = (2\, k_F\, r_0)^{-2}$

r_0 screening radius for a medium with dielectric constant ε_0.

$k_F = \left(\frac{3\,\pi^2\, n}{N}\right)^{1/3}$ n the carrier concentration for one valley

and :

$$\tau^{\,i}_{core} = \tau^{0}_{core}\left\{\left[1 - \frac{\varepsilon_F}{\varepsilon_g + 2\varepsilon_F}\left(1 - \frac{u_v^{(i)}}{u_c^{(i)}}\right)\right]^2 - \frac{8}{3}\,\frac{\varepsilon_F(\varepsilon_g + \varepsilon_F)}{(\varepsilon_g + 2\varepsilon_F)^2}\,\frac{u_v^{\,i}}{u_c^{\,i}}\right\} \quad (2)$$

$\tau^{(o)}_{core}$ is the relaxation time in the case of parabolic band. The subscripts "v" and "c" refer to conduction and valence band.

We have plotted the experimental points (crosses) on Fig.3.

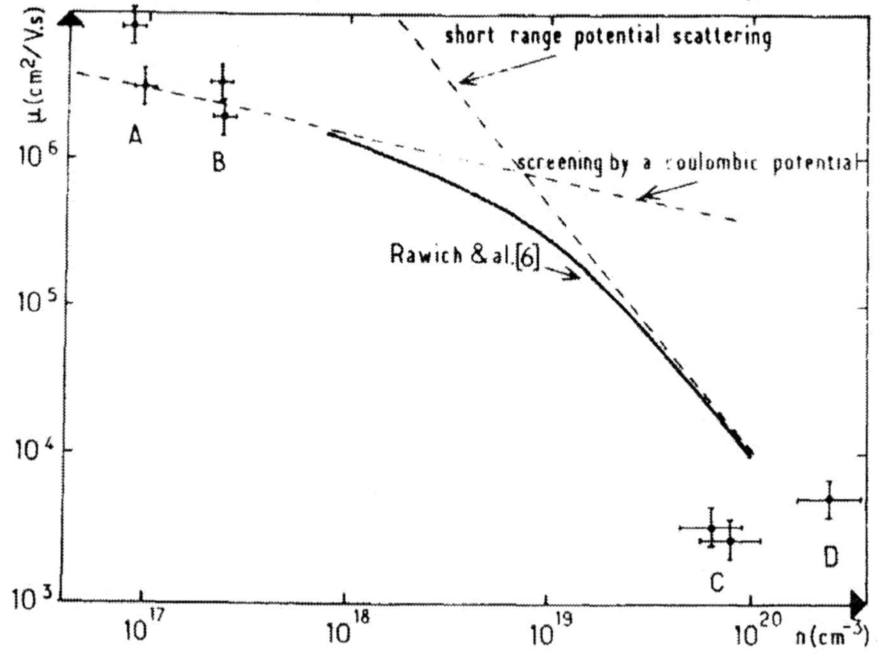

Fig. 3 : Mobility versus the carrier concentration at 4.2K. crosses : experimental data ; lines : theoretical calculation.

It could be noted, that the main effect is due to the large value of ε_0. However at low carrier concentration the experimental values are of the order of magnitude of the theoretical ones, even slightly higher. This fact is certainly due to specific properties of Gd ions : for example the presence of magnetic moments decreases the scattering due to the spins.

Fig. 4 shows the magnetization as a function of magnetic field at 4.2K. The experimental data (crosses) are very well fitted by a Brillouin function, will a Gaj parameter |9|. T_o = 1K. T_o is a fitting parameter which represents the phenomenological antiferromagnetic interaction of the relatively isolated Gd^{3+} with distant neighbors.

One can conclude that at 4.2K the contributions of the Gd-Gd pair is negligible, since no step could be detected |8|. This is due to the small value of J_{NN} in that case. This result was anticipated because the Gd ion has a spherical symmetry (S state). Then we try to obtain J_{NN} from the susceptibility measurements.

Assuming a random distribution and exchange interaction between NN's only, θ is given by |10|:

$$\theta = 2\ z\ x\ J_{NN} S(S+1)/3\ k$$

where J_{NN} is the exchange constant between a central Gd spin and another Gd spin situated on the first coordination sphere. z is the number of cations in this sphere : for NaCl structure, z = 12.

From Fig.5 one deduce θ ≅ - 1K that give J_{NN} = - 0.3 K. One remarks that J_{NN}/k is smaller compare to the value obtained for example in the case of $Cd_{1-x}Mn_xTe$ (J_{NN} = - 7.8°K). |11|

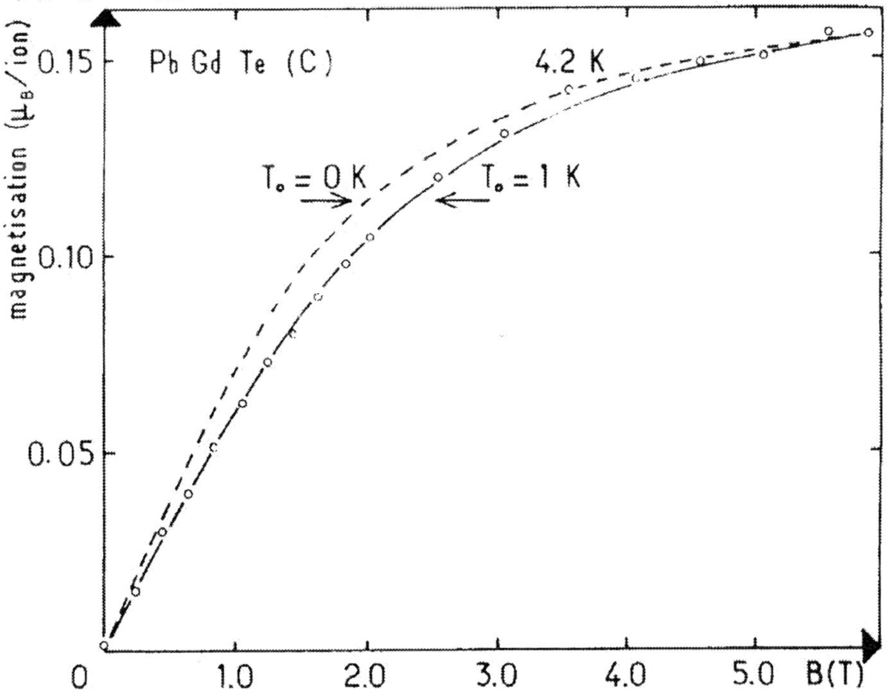

Fig. 4 : Magnetization versus magnetic field intensity ;
circles : experimental data
lines : fitting using the Brillouin function.

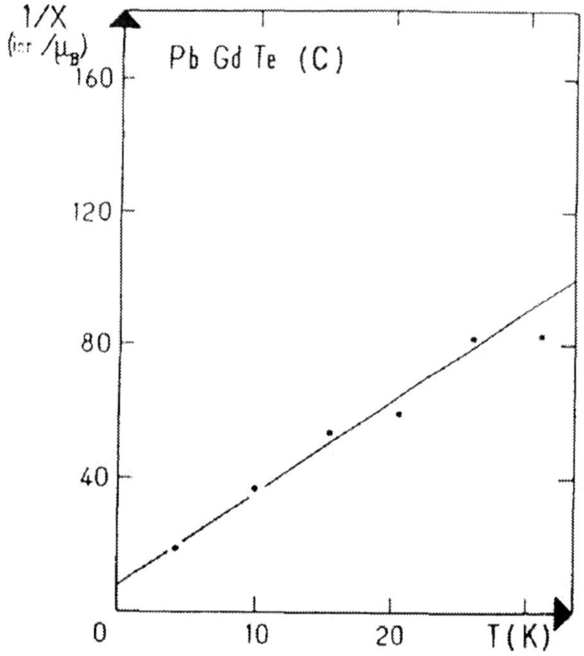

In a forth coming paper we will give a study of J_{NN} as a function of Gd concentration up to 9%, deduced from susceptibility and magnetization measurements. The magnetization is performed at smaller T.

Fig. 5 : The reciprocal magnetic susceptibility as a function of temperature (Line : Curie-Weiss law Point : experimental data).

CONCLUSION

We have described new and interesting results obtained on $Pb_{1-x}Gd_xTe$ namely a very high value of the mobility which seems to be the highest one obtained on material with same gap, and a small value of J_{NN}, which represents the exchange in a pair of Gadolinium. This small J_{NN} value was expected due to the specific of Gd ions (spherical symmetry and $4f^7$ shell).

REFERENCES

|1| R.R. GALAZKA, Proc. Internat. Conf. Phys. Narrow Gap. Semicond. Linz (Austria) Ed. E. Gornik, H. Heinrich and L. Palmestshofer, Springer Verlag, 1981, p. 294.

|2| G. BASTARD, Proc. Internat. Conf. Phys. Narrow. Gap Semicond. Ed. J. Rauluszkiewicz, M. Gorska and E. Kaczmarch, P.W.N. Polish Sci. Publ. Warsaw, 1978, p. 63

|3| G. BASTARD, J.A. GAJ, R. PLANEL and C. RIGAUX, J. Phys. C5, 247, 1980.

|4| Y. GULDNER, C. RIGAUX, M. MENANT, D.P. MULIN and J.K. FURDYNA Solid State Commun. 33, 133, 1980.

|5| S. ABDEL-MAKSOUD, C. FAU, J. CALAS, M. AVEROUS, B.A. LOMBOS, G. BRUN and J.C. TEDENAC, Solid State Commun. 54, 811, 1985.

|6| J. NIEWODNICZANSKA-ZAWADZKA, G. ELSINGER, L. PALMETSHOFER, E.J. FANTNEV, G. BAUER and W. ZAWADZKI, Proc. of the 16th Intern. Conf. on the Phys. of Semicond. Ed. M. Averous, North Holland, Amsterdam, 1983, p. 458.

|7| Y.I. RAVICH, B.A. EFIMORA and V.I. TAMARCHENKO, Phys. Stat. Sol.b,1971,453

|8| B.E. LARSEN, K.C. HASS and R.L. AGGARWAL, to be published in Phys.Review.

|9| J.A. GAJ, R. PLANEL and G. FISHMAN, Solid State Commun. 29, 435, (1979).

|10| J.H. VAN VLECK, J. Chem. Phys. 9, 85, (1941).

|11| R.L. AGGARWAL, S.N. JASPERSON, P. BECLA and R.R. GALAZKA, Phys. Rev. B.32, 5132, (1985).

Magnetic Properties of Gd-Substituted Yttrium Nitride

R. B. van Dover, L. F. Schneemeyer, and E. M. Gyorgy, AT&T Bell Laboratories Murray Hill, NJ 07974

Abstract

$Gd_xY_{1-x}N$ is an instantiation of the broad series of diluted magnetic materials composed of lanthanide nitrides in a Sc, Y or La nitride host. These lanthanoid nitrides are interstitial compounds which all form in the B1 structure and with similar lattice constants, so solid solutions are readily obtained. We have prepared single-crystal thin films by epitaxial growth on sapphire substrates, and obtained material with moderate carrier densities ($n_{Hall}\sim10^{21}$ cm^{-3}) and mobilities ($\mu_{Hall}\sim10$ $cm^2V^{-1}sec^{-1}$). Below the ordering temperature $T_c\sim15$ K the $Gd_xY_{1-x}N$ films ($0.25<x<0.40$) exhibit a ferromagnetic-like M-H loop. The saturation magnetization measured at 15 kOe and 4.2 K is anomalously low, representing an effective moment of only 4.2 μ_B per Gd compared to the expected value of at least 7.2 μ_B. The piecewise-linear M-H curve is inconsistent with conventional models for a simple anisotropic ferromagnet. These features illustrate some of the unusual properties we are beginning to identify in the behavior of these new materials.

Introduction

The pseudobinary compounds of the form $Gd_xY_{1-x}N$ are examples of a broad series of magnetic materials we have described recently: the rare-earth substituted Sc, Y, and La pnictides.[1] In the range $0.2<x<0.45$ the Gd-doped compounds exhibit semiconductor-like conductivities and mobilities, and are also readily magnetized at low temperatures. Thus they may prove suitable for some of the applications proposed for Diluted Magnetic Semiconductors (DMS).[2]

The lanthanoid pnictides are all refractory materials, and the melting points of the nitrides are especially high. YN is reported[3] to melt at >2670°C while GdN melts[4] at >2500°C. In order to obtain high-quality material we have resorted to reactive cosputtering at a slow rate and high substrate temperature in a UHV environment. As we will show, we have obtained epitaxial growth on sapphire substrates, yielding single-crystal specimens.

Gadolinium was chosen for this initial investigation because of all of the lanthanide candidates which could be substituted on Y sites, Gd should result in the simplest magnetic behavior. This is because the Gd^{3+} ion is in the spherically symmetric $^8S_{7/2}$ state. In a cubic lattice this implies the absence to first order of crystal anisotropy—though growth-induced anisotropy is certainly not precluded in cosputtered thin films.

GdN has been extensively investigated in the past,[5–8] though the interpretation of the properties of this material is complicated by the fact that it is evidently a semimetal with a minimum carrier density of about 0.06 carriers per Gd.[7] The magnetic properties are strongly dependent on the carrier density: Gambino, *et al.*,[5] have claimed that GdN is ferromagnetic, while Wachter and Kaldis[7] have argued that undoped GdN is a metamagnet and that only GdN doped with O to increase the carrier density ($GdN_{0.96}O_{0.04}$) is ferromagnetic. Narita and Kasuya[8] have presented theoretical arguments showing that under the conditions present in GdN the energies of the ferromagnetic and antiferromagnetic states can be very similar, due to indirect exchange interactions. It is likely that many of these considerations may apply as well to the diluted system $Gd_xY_{1-x}N$, in which the carrier density can be much lower, and in which the Gd density is itself a variable. This could be manifested, for example, in unusual spin configurations or heightened sensitivity to small anisotropies. Thus the $Gd_xY_{1-x}N$ system may prove to be an interesting system from the point of view of new physics as well.

In this paper we begin by discussing the technique used to synthesize single-crystal thin films, and then summarize the results of structural and electrical measurements we have used to characterize them. The magnetic properties of these films are rather complex and are not yet fully understood, so we will succinctly present the important features, describe our current interpretation, and indicate the significant questions which remain outstanding.

Mat. Res. Soc. Symp. Proc. Vol. 89. © 1987 Materials Research Society

Synthesis

The films were prepared by reactive magnetron cosputtering in a diffusion-pumped high-vacuum system. Two separate magnetron guns were employed to sputter from Gd and Y targets with independent rates. The guns are aligned to deposit onto a row of substrates, thus allowing a phase spread of varying composition (typically 10-40 at. % Gd) to be prepared simultaneously, facilitating comparison between samples with different compositions but otherwise identical. The substrates were $(1\bar{1}02)$-cut sapphire, cleaned and mounted on a molybdenum block which could be heated to 900 °C. The guns and substrates were enclosed by a liquid-nitrogen-cooled copper Meissner trap, which efficiently getters reactive background gasses, yielding an effective background pressure of *ca.* 5×10^{-10} Torr during sputtering, as estimated by other experiments.

Research grade Ar and N_2 (99.999% purity) were introduced through independent flow controllers to give typically 10-20 mole % N_2 with a total flow rate of 50 sccm. The diffusion pump was throttled during sputtering so that this flow resulted in a chamber pressure of 5×10^{-3} Torr as monitored by a capacitance manometer. The sputtering rates were monitored by a crystal microbalance and were typically 0.05-0.1 nm-sec^{-1}; the total film thickness was 100-200 nm. When the desired thickness was obtained the substrates were cooled as quickly as possible to minimise any loss of N from the surface of the films. Upon reaching 200 °C a third magnetron gun was used to sputter a 10 nm AlN layer on the films to protect them from oxidation and hydrolysis in moist air, a problem which evidently discouraged previous investigations.[9] This overlayer is sufficiently thin to allow good electrical contact yet is thick enough to protect the films during short exposures to laboratory air and indefinite storage in a dessicator.

The thickness of the films was determined by weight gain, rate calculations, profilometry using a Dektak II[10], and Rutherford Backscattering Spectrometry (RBS). The agreement between these various measurements was within approximately 10%, which is gratifying for such small samples. The compositions were determined by RBS and rate calculations, which agreed within 2 at. %.

Structure

Deposition onto (1102)-oriented sapphire substrates held at 650 °C or higher yields single-crystal epitaxial films of $Gd_xY_{1-x}N$. This is demonstrated in Figure 1, which is a composite representation of two Read camera x-ray photographs taken with Cu Kα radiation. The Read camera is essentially a wide-film Debye-Scherrer instrument, which allows a large section of reciprocal space to be examined efficiently. Polycrystalline films yield powder-like patterns in this camera; textured films give spotty patterns depending on the degree of orientation. The fact that both the (200) and (222) reflections (obtained by adjusting the incident angle to half the Bragg angle 2θ for each reflection) are delta-function-like indicates that the film is a single crystal. Quantitative analysis of the structure using a four-circle goniometer diffractometer is currently being pursued.

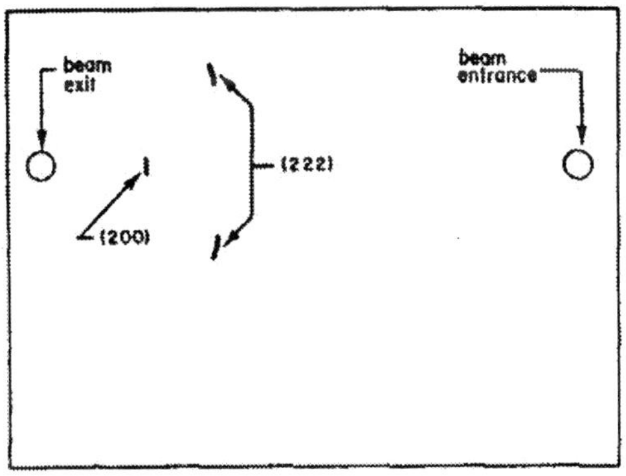

Figure 1. Composite representation of two Read camera photographs, showing that the sample (D) is a single crystal. If the sample had been an untextured polycrystal, the reflections would take the form of arcs, and all of the allowed reflections would be present.

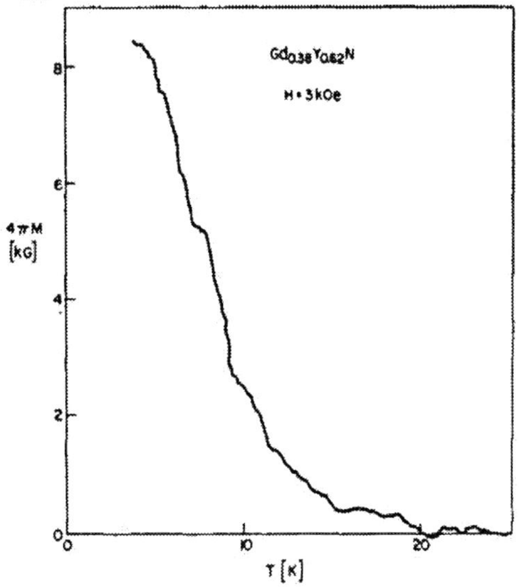

Figure 2. Magnetization *versus* temperature for sample D. Data taken on a VSM in an applied field of 3 kOe.

Electrical Properties

The properties of YN films synthesized in the same manner as these $Gd_xY_{1-x}N$ films are discussed in detail elsewhere.[11] The Hall effect and resistivity of the Gd-substituted films were measured at room temperature on the van der Pauw configuration[12] using tungsten point contacts. The Hall constant R_H could only be measured at room temperature in films with Gd because at low temperatures anomalous scattering (the extraordinary Hall effect) complicates interpretation of the data. Table I lists the carrier density $n_{Hall}=(R_H ec)^{-1}$ and Hall mobility R_H/ρ for various films. Deposition at 850 °C generally yields films with superior properties—lower carrier density (presumably because there are fewer N vacancies) as well a higher mobilities (because there are fewer defects in the crystalline lattice, as well as less scattering by charged impurities).

The properties of the GdN film are interesting *per se*. GdN is consistently described in the literature as a semimetal due to overlap between the metal d-band and hybridized s-p states in the valence band, an interpretation which is consistent with the optical properties we have measured.[1] The carrier density of sample E is smaller than the minimum reported in the literature[7] and we have obtained even lower values in other GdN films. We believe that the excess carrier density arises not from inadvertent O incorporation, but more likely from N deficiencies due to the difficulty of obtaining stoichiometric material, as well as impurities in the Gd target.

YN, on the other hand, is found to have a relatively low carrier density. There is some controversy in the literature[13] over whether YN is a semimetal or semiconductor, which has not been definitively resolved. Our preliminary investigations indicate that YN can be doped with acceptors to reduce the carrier density. In any case even the carrier densities we have achieved may be sufficiently small to admit some modulation, *e. g.*, by photoexcitation or field-effect. We note parenthetically that if the doping level in this film were due entirely to O incorporation, it would represent a composition of $YN_{0.996}O_{0.003}$, *i. e.*, that O is at most a very minor constituent of these films.

The intermediate $Gd_xY_{1-x}N$ samples are consistent with these extremes. The carrier densities are generally comparable to or smaller than those found in the GdN samples. This may be due to the lower purity of the Gd target (99.9% spectroscopic purity) compared to the Y target[14] (30 ppm impurity, maximum). In any case the carrier densities of the Gd-substituted samples are fairly low, while the mobilities are relatively high, and we expect that the magnetic properties we measure are not dominated by impurities except as they influence the measured electron concentrations.

Table I. Properties of selected $Gd_xY_{1-x}N$ samples. T_D is the deposition temperature, T_c is the Curie temperature, H_K is the anisotropy field measured with the applied field perpendicular to the substrate, and M_S is the saturation magnetization. H_k and M_S were measured at 4.2 K, while n_{Hall} and μ_{Hall} were measured at room temperature.

$Gd_xY_{1-x}N$							
sample	x	T_D [°C]	T_C [K]	H_K [kOe]	$4\pi M_S$ [kG]	n_{Hall} [$10^{21}cm^{-3}$]	μ_{Hall} [$cm^2V^{-1}sec^{-1}$]
A	0	850	-	-	-	0.077	10.4
B	0.22	850	12	11	3.0	0.64	19.2
C	0.36	850	17	10.5	8.4	0.83	15.2
D	0.38	650	13	7	8.0	2.1	0.33
E	1.00	850	40	28	23	1.1	0.44

Magnetic Properties

All of the samples listed in Table I with $x>0$ had ordering temperatures above 4.2 K. A sample with 15 at. % Gd also ordered at T>4.2 K, but was less extensively investigated. In general, samples with less Gd have lower ordering temperatures, but we have not yet addressed the interesting question of what sort of order, if any, occurs for very dilute Gd solutions.

Figure 2 presents a typical M-T curve, specifically for $Gd_{0.38}Y_{0.62}N$ (sample D), taken with a vibrating sample magnetometer (VSM) with a fixed 3 kOe field parallel to the plane of the substrate. This figure reveals the difficulty of identifying the ordering temperature. As listed in Table I, we have assigned the value T_c=17 K, which is based on a rough extrapolation of the linear portion of the M-T curve. Since we lack a detailed model of the magnetic state near T_c, we are reluctant to attempt curve-fitting using conventional clustering theories[14] to obtain a more accurate value for T_c nor have we applied the conventional experimental technique described by Arrott[15]. Nevertheless while the absolute accuracy of the T_c assignments is no better than ±10%, relative changes can be judged more accurately, and the difference between samples B and D, for example, is easily measured.

A comparison of the ordering temperatures of samples A, B, C, and E reveals the main trend, that T_c increases monotonically with x for films prepared under otherwise identical conditions. But the pair C, D shows that the Gd fraction is not the only consideration, and indeed suggests that the increased carrier density in sample D has decreased T_c. This is an unanticipated result. However, large discrepancy in mobilities between samples C and D, which is due to the difference in deposition temperature renders this conclusion tentative. Further work is contemplated in which the carrier density is manipulated directly by doping rather than by varying the deposition temperature.

The M-H curve of sample D, measured at 4.2 K with a VSM, is shown in Figure 3. The solid and dotted lines correspond to measurements with the field parallel and perpendicular to the substrate, respectively. The diamagnetic contribution from the substrate was determined by measurement of a second substrate cut from the same sapphire wafer, and has been subtracted from the data. Curves measured on samples cooled in nominally zero field and cooled in a field of 3 kOe showed no significant differences. This sample had a coercive force of 10 Oe and a remanence of 2 kG (measured with the field parallel), neither of which is resolved in the figure. This curve differs markedly from previously published data for GdN and $GdN_{1-x}O_x$.[7]

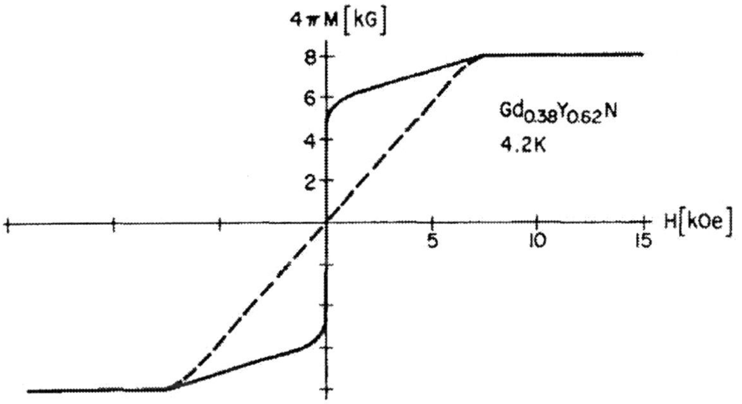

Figure 3. M-H curve of sample D. The solid line represents VSM data taken with the applied field parallel to the plane of the sample, while the dashed line was taken with the field perpendicular.

The overall form of this curve is similar to the curve that would be obtained from a thin film specimen with an easy axis (axes) not oriented in either of the measurement directions. However, measurement of the in-plane loop at a bias field of 500 Oe as a function of ϕ, the angle between the field and a fixed direction in the plane, shows only a small (13%) modulation in M, which is neither large enough nor of the right periodicity to support that simple interpretation. A small modulation was also found in the more sensitive torque measurement of this sample as a function of ϕ, but again the magnitude and periodicity are inconsistent with the simple anisotropic thin film model.

Furthermore the amplitude of the saturation magnetization is much smaller than expected. The net moment per Gd is calculated to be 5.0 μ_B, while the value expected for a temperature of approximately $T_c/3$ would be 7.2 μ_B or greater (depending on the degree of induced electron polarization). This discrepancy implies that for some reason a large fraction of the Gd in the sample is not aligned with the field even at 15 kOe. It is not known whether this is a surface effect associated with one or both surfaces of the film, or a bulk effect. A similar discrepancy is found in the other $Gd_xY_{1-x}N$ samples, as well as the GdN sample (E).

Torque curves were measured at 4.2 K as a function of θ, the angle between the applied field and the normal to the film, in a digitally controlled torque magnetometer. Measurements taken at 45° are plotted in Figure 4 on axes which linearize the data for films with a uniaxial anisotropy.

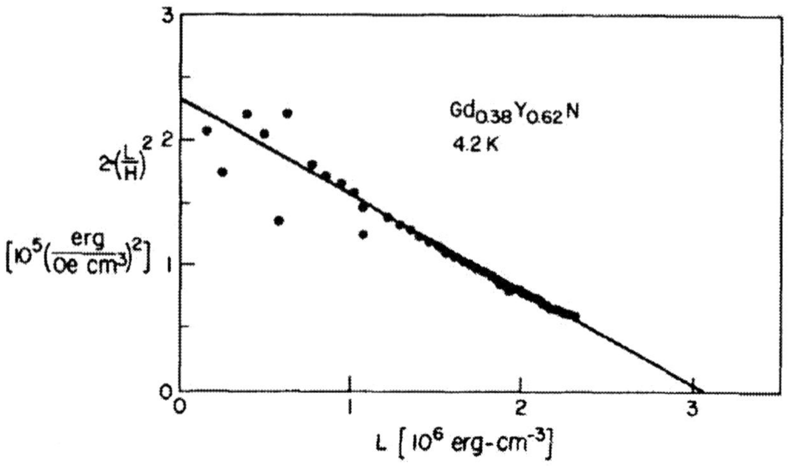

Figure 4. Torque data for sample D. The axes (torque/applied field)^2 *versus* (torque) linearize the data for the case of a sample with a well-defined magnetization and uxiaxial anisotropy. See text.

Following the analysis of Miyajima *et al.*,[16] we identify the saturation magnetization $4\pi M_s$= 6.0 kG and K_{eff}=3.1×10^{6}erg–cm^{-3}, from which we expect H_k=$2K_{eff}/M_s$= 12.8 kOe. These values of $4\pi M_s$ and H_k are completely inconsistent with the VSM data. We conclude that this film cannot be described in terms of a well-defined magnetic moment, M_s, and a uniaxial anisotropy, so that the analysis of the torque data using the method of Miyajima does not apply, even though the data is quite linear and the axis-intercepts can be identified unambiguously. Thus at present the appropriate interpretation of the VSM and torque data, which measure closely related aspects of the magnetic energy, must await further experiments.

Conclusions

We have investigated the properties of $Gd_xY_{1-x}N$, a new diluted magnetic system with interesting properties. Thin films prepared by reactive magnetron cosputtering have low impurity levels and exhibit semiconductor-like electrical behavior. The films order magnetically at temperatures in the vicinity of 15 K, depending on the Gd concentration and other factors. The magnetic state below T_c is ferromagnetic-like, but with features which are not as yet understood.

Acknowledgements

The authors would like to thank J. F. Dillon, Jr., for useful discussions regarding magnetic anisotropy in thin films and J. M. Graybeal for facilitating the RBS measurements.

References

1. L. F. Schneemeyer, R. B. van Dover and E. M. Gyorgy (to be published in Proceeding, 1986 Conf. on Magnetism and Magn. Matls.).
2. R. R. Galazka, Proc. 14th Intnl. Conf. on the Physics of Semiconductors, Edinburgh, 1978 (Inst. Phys., London, 1979), p. 135.
3. C. Kempter, N. H. Krikorian, and J. C. McGuire, J. Phys. Chem. **61**, 1237 (1957).
4. G. Busch, P. Junod, O. Vogt and F. Hulliger, Phys. Lett. **6**, 79 (1963).
5. R. J. Gambino, T. R. McGuire, H. A. Alperin and S. J. Pickart, J. Appl. Phys. **41**, 933 (1970).
6. R. A. Cutler and A. W. Lawson, J. Appl. Phys. **46**, 2739 (1975).
7. P. Wachter and E. Kaldis, Solid State Comm. **34**, 241 (1980).
8. F. Hulliger, "Rare Earth Pnictides," in *Handbook on the Physics and Chemistry of Rare Earths*, K. A. Gschneider and L. Eyring, Eds., (North Holland, Amsterdam, 1979).
9. J. P. Dismukes, W. M. Yim, J. J. Tietjen, and R. E. Novak, RCA Review **31**, 680 (1970).
10. Sloan Technology Corp., Santa Barbara, California 93103.
11. R. B. van Dover and L. F. Schneemeyer (to be published).
12. L. J. van der Pauw, Phillips Res. Reports **13**, 1 (1958).
13. obtained from Ames Material Preparations Center, Ames Laboratory, Iowa State Univ, Ames, Iowa.
14. see, *e. g.*, S. Chikazumi and S. H. Charap, *Physics of Magnetism*, (Krieger, Malabar, 1978), p. 70*ff*.
15. A. Arrott, Phys. Rev. **108**, 1394 (1957).
16. H. Miyajima, K. Sato and T. Mizoguchi, J. Appl. Phys. **47**, 4669 (1976).

THE RELEVANCE OF LONG-RANGE INTERACTIONS IN DILUTED MAGNETIC SEMICONDUCTORS

DE JONGE W.J.M.*, TWARDOWSKI A.** AND DENISSEN C.J.M.*
*Department of Physics, Eindhoven University of Technology, 5600 MB Eindhoven, The Netherlands
**Institute of Experimental Physics, University of Warsaw, Warsaw, Poland

ABSTRACT

Data are presented yielding evidence on the long-range character of magnetic interactions in Diluted Magnetic Semiconductors. The radial dependence of these interactions is deduced from the concentration dependence of the freezing temperature. It is shown that calculations of the thermodynamic properties, which include this long-range interaction, are in fair agreement with the experimental data.

INTRODUCTION

During the past years extensive investigations have been performed on the magnetic behavior of Diluted Magnetic Semiconductors (DMS) [1]. The efforts have been devoted almost exclusively to DMS of the type $A_yMn_xB_z$, such as: (CdMn)Te, (HgMn)Te, and corresponding Selenides, and $(CdMn)_3As_2$, $(ZnMn)_3As_2$, to which we also will restrict ourselves in this paper.

The essential features which are commonly observed in all these DMS, at low Mn concentrations x, are [1]:

1. Curie-Weiss behavior of the magnetic susceptibility χ at high temperatures, indicating AF interactions;
2. A cusp or kink in the low temperature χ, indicating a spin-glass like transition at a temperature depending on x;
3. A continuous magnetic contribution to the specific heat with a broad maximum in the ^{4}He temperature region, shifting to higher T with x;
4. A field dependence of the magnetization M indicating AF Mn-Mn interactions in some cases accompanied with steps.

Originally this magnetic behavior has been interpreted as arising from interactions between Mn-ions, situated at nearest neighbor sites in the host lattice [2]. The use of this model was strongly supported by the original observation that the spin-glass transition was restricted to DMS with a magnetic ion concentration x <u>above</u> the percolation limit, x_c, of the host lattice. From this observation it was concluded that the transition was induced by short range AF (nearest neighbor) interactions causing topological frustration effects due to the high symmetry of the host lattice. Subsequent calculations on the basis of this conjectured short-range interaction only, however, gave rise to a wide spread of exchange parameters deduced from various sets of data and the questionable need to adjust the random distribution of the magnetic ions [2]. To our knowledge, no consistent set of interaction parameters, explaining all the data simultaneously, has been obtained on this basis.

In this paper we would like to emphasize that, as we have shown recently [3-6], this is mainly due to the underestimation of the long-range character of the exchange interaction between distant Mn-ions.

THE SPIN-GLASS TRANSITION

In a first report on the spin-glass behavior of $(CdMn)_3As_2$ [7] we noted that in this DMS a freezing transition, as monitored by a cusp in the χ, exists for Mn concentrations far below the percolation limit. This fact has recently been documented more extensively [3], whereas also in other DMS like $(ZnMn)_3As_2$ [4], (ZnMn)Se [8], (ZnMn)Te [8], (CdMn)Te [9], (CdMn)Se [10], (HgMn)Te [11] and (HgMn)Se [12] the same features have been observed. Fig. 1 shows a typical example of $\chi(T,x)$ as observed in $(ZnMn)_3As_2$.

This extension of T_F down to zero in the dilute limit, cannot be understood on the basis of nearest neighbor interactions, but strongly indicates the existence of a long-ranged interaction as the driving mechanism behind this transition. If we accept that the nature of this transition is a spin-glass freezing (and a number of experiments have been performed to corroborate this conjecture [13]), then a scaling analysis should be applicable [14] and a radial dependence of the exchange between distant Mn-ions can be deduced from $T_F(x)$.

If R denotes the distance (in units of nearest neighbor distance) between two Mn-ions one may then derive:

$$\ln T_F \sim \alpha\, x^{-1/3}, \text{ for } J(R) \propto e^{-\alpha R} \text{ or} \qquad (1)$$

$$\ln T_F \sim n/3 \ln x, \text{ for } J(R) \propto R^{-n}. \qquad (2)$$

In the present analysis we will mainly employ the power-law behavior (2) since it gives a better description of the majority of the data. In Fig. 2 we plotted most of the available data on a double logarithmic scale. From this figure one may conclude that a simple power-law dependence of $J(R) \sim R^{-n}$, as reflected in $T_F(x)$, seems to describe the data of all these DMS in the whole concentration range. At the percolation limit, which depends on the structure of the host lattice and varies from 33% for tetragonal Cd_3As_2 to 17% for fcc CdTe, no dramatic effects are observed. The exponent n deduced from Fig. 2 for the various DMS, ranges from 6.8 to 3 and is tabulated in Table I.

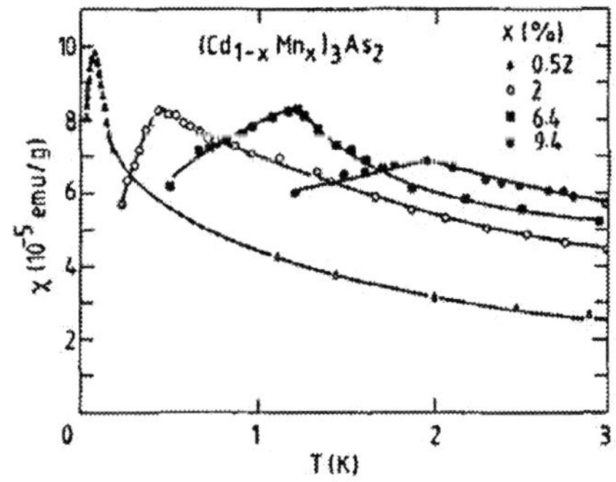

Fig. 1. Low-temperature susceptibility of $(CdMn)_3As_2$.

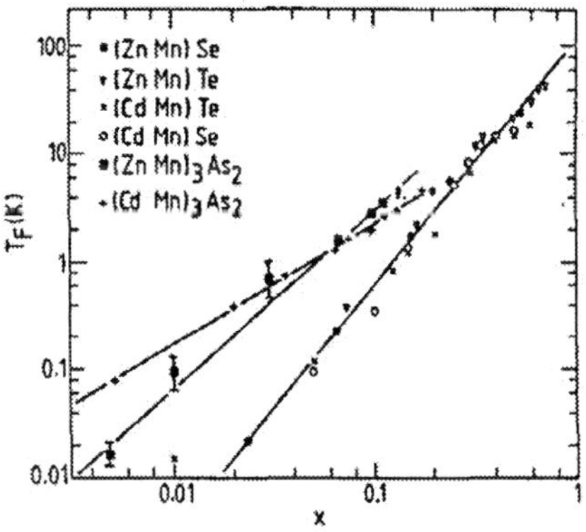

Fig. 2. Freezing temperature T_F as function of the Mn concentration x for various DMS on logarithmic scale. References on the origin of the data are given in the text. The drawn lines are fit to the data yielding the power dependence $J(R) \sim R^{-n}$ as tabulated in Table I.

Table I

Material	type	x range	Eg (eV)	n	Ref.
$Zn_{1-x}Mn_xS$	II-VI	0.3 - 0.4	~3.8	~6.8	15
$Zn_{1-x}Mn_xSe$	..	0.02 - 0.5	2.8 - 3	6.8	8
$Zn_{1-x}Mn_xTe$	..	0.07 - 0.6	2.4 - 2.8	6.8	8
$Cd_{1-x}Mn_xSe$	..	0.05 - 0.5	1.8 - 2.6	~6.8	10
$Cd_{1-x}Mn_xTe$	..	0.01 - 0.6	1.6 - 2.5	~6.8	9
$Hg_{1-x}Mn_xTe$	..	0.02 - 0.5	~0 - 1.1	~5	11
$(Zn_{1-x}Mn_x)_3As_2$	II-V	0.005 - 0.1	~1	4.5	4
$(Cd_{1-x}Mn_x)_3As_2$	II-V	0.005 - 0.2	0 - 0.2	3.5	3
$Pb_{1-x}Mn_xTe$	IV-VI	0.03 - 0.1	0.2 - 0.4	3	16

It is obvious from these results that a considerable difference exists in the range (proportional to 1/n) of interaction J(R) between the various compounds. If superexchange would be the driving physical mechanism behind the interaction between Mn ions, an increasing range is predicted to exist in the case of an increasing covalent bonding [17]. Roughly speaking, such an increasing covalency may be conjectured comparing the II-VI compounds with II-V or IV-VI materials. However, definite statements about this await further study; specificially spectroscopic investigations on the location of the empty Mn d-orbitals with respect to the valence band would be very fruitfull.

The increasing range of interaction when going from wide gap materials to smaller gap materials is also qualitatively not inconsistent with the Bloembergen-Rowland exchange interaction mechanism [18], although in that case an exponential decay of J(R) would have been expected.

In that respect we feel a comment should be made. Although it is clear that $J(R) \propto R^{-n}$ (as monitored by $T_F(x)$) yields a better fit to the experimental data than an exponential decay in the concentration range from far below to far above the percolation limit, one should not exclude the principal possibility that the x dependence of the freezing temperature T_F could be different above and below the percolation limit. Therefore one has to be careful to draw pertinent conclusions from these observations.

MAGNETIC PROPERTIES

As quoted in the introduction, no consistent description of specific heat C_m, magnetisation M and susceptibility χ has been obtained so far on the basis of a nearest neighbor interaction only. It is our aim to show that, in principle, such a consistent and simultaneous description of these data can be obtained provided that the long-range interaction is included. We therefore calculated the C_m, M and χ of a number of DMS with a model including the R^{-n} dependence of the interaction as observed from $T_F(x)$.

The calculations were performed with the so-called Extended Nearest Neighbour Pair Approximation. The essential ingredient of this approximation, first suggested by Matho [19], is the assumption that the partition function of the ensemble of random spins can be factorized in the contribution of pairs. In this case a pair consists of a Mn-spin coupled to its nearest neighbouring Mn-spin, which can be located anywhere as prescribed by the statistics in a random array.

Since the format of this paper is rather limited we are not able to report on the model parameter, and results for each DMS system in detail. For more extensive information we refer in that respect to [3-6].

In general the exponent n in the power law dependence of the exchange was obtained from $T_F(x)$, and is tabulated in Table I. For the magnitude of the interaction we inserted (if available) the average value reported for the nearest neighbor exchange for the specific system (see also next section).

For the present purpose we have chosen to examplify the results by the figures 3-5, which all refer to (CdMn)Te. Results for other systems show basically the same features. We would like to emphasize that no efforts have been undertaken yet to adjust the parameters in order to obtain a better overall fit to the data.

From the comparison shown in figures 3-5 we would like to conclude that, on the whole, the agreement between calculated results, using reported values for the interaction, and the data is fair, in particular, since one has to recall that the pair approximation is only valid for low concentrations.

Nevertheless it is obvious from the comparison between the calculations and the data that in some cases significant deviations exist, indicating that the present model and/or the parameters are not completely correct. At least part of these deviations can be substantially reduced by fitting of one or more parameters in the theoretical model. We did however not perform such a detailed analyses since it is outside the scope of this paper in which, we only aimed to investigate whether a model which includes the long-range nature of the interaction is in principle capable of explaining the overall magnetic behavior in DMS. As more detailed comparison and will be published elsewhere.

CONCLUDING REMARKS

A first comment concerns the relation between the long-range interaction and the nearest neighbor exchange parameters which can be derived from direct measurements such as the "steps" in the magnetization [20,21]. In principal the nearest neighbor interaction J_{nn}, the next nearest neighbor interaction J_{nnn} and so on, are included in the analytic expression for the long-range interaction J(R). (R = 1 yields J_{nn}, R = $\sqrt{2}$ yields J_{nnn}). However, one has to realize that $T_F(x)$ probes mainly the "tail" of the interaction range. We feel that it is quite conceivable that the mechanism governing the interaction between distant spins (mediated by carriers) is supplemented with other, more direct, interactions when the magnetic ions are on neighbouring sites, yielding a deviating value for J_{nn}. This is illustrated in Fig. 6 where we have plotted the long-range interactions in (CdMn)Te as indicated by the present experiments as function of the Mn-Mn distance. It is interesting to note that indeed the actual data for the interaction between non-neighbouring Mn-ions, matches the radial dependence $J(R) \propto R^{-6.8}$, as deduced from $T_F(x)$, fairly well (dotted curve), whereas J_{nn} strongly deviates in accordance with the conjecture made above.

Finally we would like to comment on the existence of a spin-glass transition in semimagnetic semiconductors. To start with, we would like to emphasize that all evidence gathered so far indicates dominant antiferromagnetic interactions between the magnetic moments in all of the semimagnetic semiconductors which have been studied. According to the present understanding of spin-glasses, randomness combined with either competition or frustration seem to be essential ingredients for the existence of a spin-glass. It has been shown theoretically [22] that a diluted magnetic array on a fcc host lattice coupled by nearest-neighbor antiferromagnetic interactions results in a spin-glass only when the concentration of magnetic ions x exceeds the percolation limit x_c. For long range interactions competing AF and F interactions seem essential.

For DMS, it has been shown that the spin-glass transition also exists for concentrations below x_c and extends in fact down to x = 0. At the same time the long-range interactions are AF. It seems to us that these observations of spin-glass behavior for $x<x_c$ cannot be understood on the basis of the above-mentioned theoretical predictions. The situation for

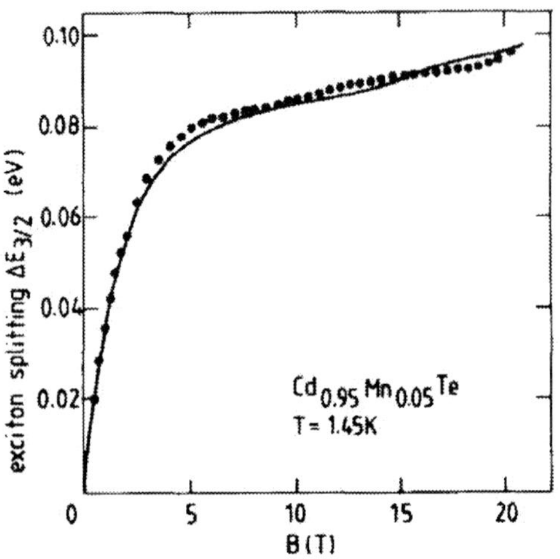

Fig. 3. Magnetization of $Cd_{0.95}Mn_{0.05}Te$ as monitored by the exciton splitting $\Delta E_{3/2}$ (Aggerwal et. al [20]. Solid line represents pair correlation calculations using $J(R) = -10\ R^{-6.8}$. The "steps" in the calculation are shifted to higher field values because J_{nn} in this calculation is too high (see fig. 6).

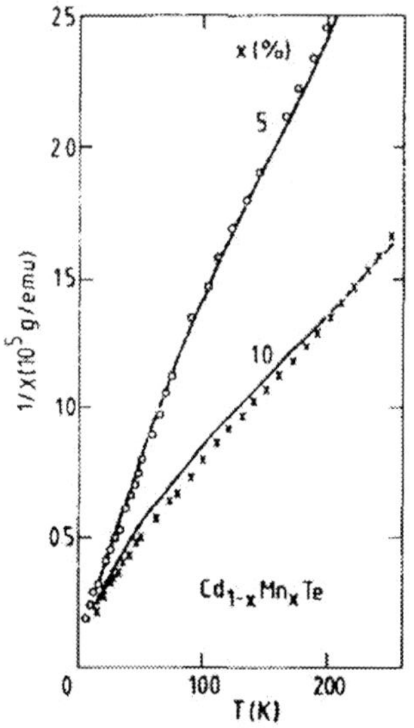

Fig. 4. Inverse high temperature susceptibility of (CdMn)Te (Galazka [2]). Solid lines are calculations with $J(R) = -10\ R^{-6.8}$.

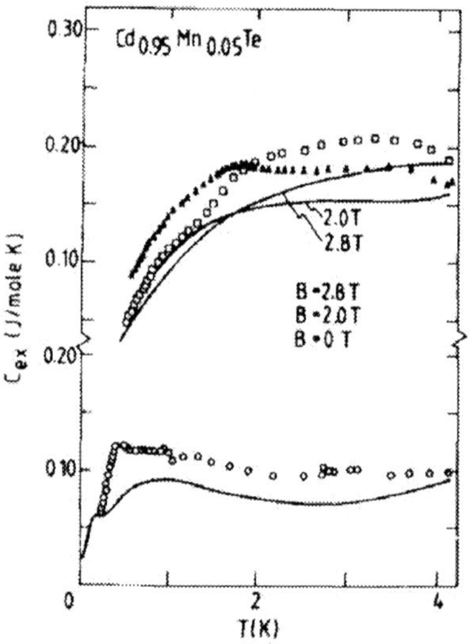

Fig. 5. Magnetic specific heat of $Cd_{0.95}Mn_{0.05}Te$ according to Galazka et al. [2]. Solid lines represent pair correlation calculations using $J(R) = -10\ R^{-6.8}$.

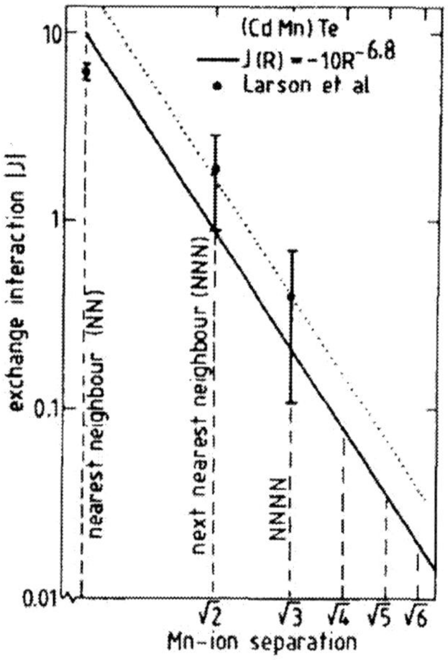

Fig. 6. Logarithmic plot of J(R) in (CdMn)Te. Dotted line represents $J(R) = -17\ R^{-6.8}$. Data are obtained from analysis of optical data [21].

$(Cd_{1-x}Mn_x)_3As_2$ is even more pronounced since in this case, for $x>x_c$, the nearest-neighbor frustration mechanism is also excluded due to the simple cubic symmetry of the host lattice.

In conclusion, we feel that the spin-glass behavior of DMS cannot be understood on the basis of the present ideas about the essential characteristics of the spin-glass formation.

REFERENCES

[1] N.B. Brandt and V.V. Moshchalkov, Adv. Phys. 33, 193 (1984); J.K. Furdyna, J. Appl. Phys. 53, 7637 (1982); R.R. Galazka, J. Cryst. Growth 72, 364 (1985).

[2] R.R. Galazka, S. Nagata and P.H. Keesom, Phys.Rev. B. 22, 3344 (1980).

[3] C.J.M. Denissen, H. Nishihara, J.C. van Gool and W.J.M. de Jonge, Phys. Rev. B. 33, 7637 (1986).

[4] C.J.M. Denissen, Sun Dakun, K. Kopinga, W.J.M. de Jonge, H. Nishihara, T. Sakakibara and T. Goto; to be published.

[5] C.J.M. Denissen, W.J.M. de Jonge, Solid State Commun.59, 503 (1986).

[6] C.J.M. Denissen, Ph.D.Thesis, Eindhoven Univ. of Technology, (1986).

[7] W.J.M. de Jonge, M. Otto, C.J.M. Denissen, F.A.P. Blom, C. van der Steen and K. Kopinga, J. Magn. Magn. Mat. 31-34, 1373 (1983).

[8] A. Twardowski, C.J.M. Denissen, W.J.M. de Jonge, A.T.A.M. de Waele, M. Demianuk and R. Triboulet, Solid State Commun. 59, 199 (1986); S.P. McAllister, J.K. Furdyna, W. Giriat, Phys.Rev.B29, 1310 (1984).

[9] M.A. Novak, O.G. Symko, D.J. Zheng, S. Oseroff, Phys.126B, 469 (1984).

[10] M.A. Novak, O.G. Symko, d.J. Zheng and S. Oseroff, J. Appl. Phys. 57 (1), 3418 (1985).

[11] A. Mycielski, C. Rigaux, M. Menaut, T. Dietl and M. Otto, Solid State Commun. 50, 257 (1984).

[12] R.R. Galazka, private communication.

[13] S. Oseroff, Phys. Rev. B25, 6584 (1982).

[14] J. Souletie, J. de Phys. C2, 3 (1978).

[15] Y.Q. Yang, P.H. Keesom, J.K. Furdyna and W. Giriat, J. Solid State Chem. 49, 20 (1983).

[16] M. Escorne, A. Mauger, J.L. Tholence and R. Triboulet, Phys. Rev. B29, 6306 (1984).

[17] W. Geertsma, C. Haas, G.A. Sawatzky and C. Vertogen, Physica 86-88A, 1039, (1977).

[18] N. Bloembergen and T.J. Rowland, Phys. Rev. 97, 1679 (1955).

[19] K. Matho, J. Low. Temp. Phys. 35, 165 (1979).

[20] R.L. Aggarwal, S.N. Jasperson, P. Becla and R.R. Galazka, Phys. Rev. B32, 5132 (1985).

[21] B.E. Larson, K.C. Hass and R.L. Aggarwal, Phys. Rev. B33, 1789 (1986).

[22] L. De Seze, J. Phys. C10, L353 (1977); J. Villain, Z. Phys. B33, 31 (1979); G.S. Grest and E.F. Gabl. Phys. Rev. Lett., 43, 1182 (1979).

II-VI COMPOUNDS WITH Fe - NEW FAMILY OF SEMIMAGNETIC SEMICONDUCTORS

ANDRZEJ MYCIELSKI
Institute of Physics, Polish Academy of Sciences, Al. Lotników 32/46, 02-668 Warsaw, Poland

ABSTRACT

Several experimental methods: absorption, photoemission and transport measurements were used to determine the energy position of substitutional $Fe^{2+}(3d^6)$ donor state in the band structure of the semimagnetic semiconductor $Hg_{1-v-x}Cd_vFe_xSe$ for $0 \leq v \leq 0.7$ and $v+x=1$, and $x \leq 0.15$. For $v \leq 0.40$, $Fe^{2+}(3d^6)$ level is a resonant donor located in the conduction band. For v=0 (HgSe) we obtain 230 meV for the position of $Fe^{2+}(3d^6)$ level with respect to the bottom of the conduction band which coincides with the position of the Fermi level for electron concentration $N \cong 5 \times 10^{18}$ cm^{-3}. Surprisingly, the mobility of free electrons (T~4.2K) is abnormally high and the Dingle temperature measured in quantum magnetoresistivity oscillations (SdH effect) and magnetooptical measurements is abnormally low. Because of the Coulomb interaction between the ionized donors, at low T, there will appear some correlation of their positions. This may lead to a kind of "liquefying" of the system of ions and to its "crystallisation" (i.e. formation of a superlattice or hyperlattice of ionized donors) at even lower T. The space-ordering of ionized donors influences dramatically the free-carrier scattering and correspondingly explains the high mobility and low Dingle temperature. Finally, we shall also present some magnetic properties of these new semimagnetic materials.

INTRODUCTION

In recent years the already broad family of diluted magnetic (semimagnetic) semiconductors based on manganese like $Hg_{1-x}Mn_xTe$, $Cd_{1-x}Mn_xTe$ was extended to include iron in selenium and tellurium , II-VI compounds. It was found that atoms (Fe^{2+}) enter substitutionally for Zn, Cd or Hg sites. In tellurium II-VI compounds the solubility of iron is about 2-5 at%[1-5]. However, in sellenium II-VI compounds the solubility of iron reaches about 15 at% [6-9]. It was experimentaly demonstrated that the predominant part of the physical properties of these alloys depends on location of the Fe^{2+} $(3d^6)$ states relative to the top of the valence band or bottom of the conduction band. According to the general tendency due to Hund's rule the $Fe^{2+}(3d^6)$ level is located much higher in the semiconductor band structure [10-12, and references therein] than the $Mn^{2+}(3d^5)$ level which is well below the top of the valence band,- see for example [13,14] for CdMnTe, [15,16] for CdMnSe, and [17] for HgMnSe. To describe the influence of the presence of the iron atoms and location of the $Fe^{2+}(3d^6)$ state in the band structure on the physical properties of II-VI semiconductor compounds we will present the experimental data which were obtained for $Hg_{1-v-x}Cd_vFe_xSe$ mixed crystals with $0 \leq v \leq 0.7$ and v=1, and with iron concentration $0 < x \leq 0.15$.

In this paper we report on the experimental results which enabled the location of the iron level in the band structure of the alloy HgCdSe. We shall also present the results of quantum transport and magnetooptical measurements on free carriers as well as the magnetic properties of these new semimagnetic materials. The theoretical model and experimental evidence for the space ordering of the ionized resonant iron donors $Fe^{2+}(3d^6)$ and for the formation of a superlattice (or hyperlattice) of these donors in HgFeSe will also be discussed.

Mat. Res. Soc. Symp. Proc. Vol. 89. © 1987 Materials Research Society

ABSORPTION MEASUREMENTS

The schematic band structure of the $Hg_{1-v}Cd_vSe$ solid solution ranging from HgSe to CdSe is shown for T=4.2K in figure 1.

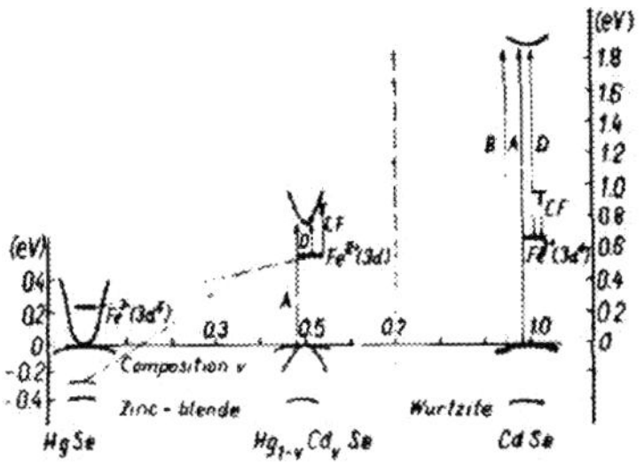

Figure 1. The schematic band structure of solid solution $Hg_{1-v}Cd_vSe$ ranging from HgSe to CdSe for T=4.2K. The possible optical transitions are indicated by arrows. The position of the iron level Fe^{2+} ($3d^6$) in crystals of different composition is shown according to the results of the paper [8].

The position of the iron level in the crystals with different compositions is shown according to the results of the paper [8]. It is known that for the composition $v \leq 0.7$ the alloy has a zincblende structure, while for $v \geq 0.8$ it has a wurtzite structure. The change in structure from zincblende to wurtzite takes place at values of v between 0.7 and 0.8. Measurements of the optical absorption in $Hg_{1-v}Cd_vSe$ alloys have been reported for samples with $0.15 < v \leq 0.68$ [18]. X-ray diffraction measurements confirmed that the samples used in their studies had the zincblende structure and were free of the inclusions containing the wurtzite phase.

In our studies we have also confirmed by X-ray analysis that the samples have a pure zincblende structure up to v=0.7. The composition and the homogeneity of the samples were determined by density measurements and electron microprobe analysis.

In the zincblende structure solid solution the dependence of the energy gap on v (at T=4.2K) is indicated in figure 1 by a full line (E =-0.27 eV for HgSe up to E=1.03 eV for $Hg_{1-v}Cd_vSe$, v=0.68) in agreement with previous experimental results([19] and [18] respectively). In figure 1 the possible optical transitions are indicated:

(i) from the split valence band in the wurtzite structure to the conduction band (exciton lines A and B);

(ii) from the donor $Fe^{2+}(3d^6)$ level to the conduction band (transition D);

(iii) crystal field transitions $^5E(^5D)\text{-}^5T_2(^5D)$ within the $3d^6$ configuration of Fe^{2+} ions on substitutional cation sites.

In CdSe iron acts as a deep donor. The position of the $Fe^{2+}(3d^6)$ donor level in doped CdSe (with concentration $N_{Fe} = 1.5\times10^{19}$ cm^{-3}) has been established at 0.63 ± 0.02 eV above the top of the valence band, and found to be independent of temperature in the range 80-300K by optical absorption measurements [20].

In our experiments the photo-ionization and exciton absorption of $Cd_{1-x}Fe_xSe$ has been obtained for iron concentrations up to $x \leq 0.15$ ($\sim3\times10^{21}$ cm^{-3}). An example of the photo-ionization and exciton absorption for x=0.03 is shown in figure 2.

The tail of the absorption edge was measured on the sample 150μm thick. This absorption we have attributed to the photo-ionization D transition (see figure 1). To interpret the tail absorption we have to subtract the linear background absorption from the total. According to [20-22], the remaining absorption α is then plotted in the $(\alpha/\hbar\omega)^{2/3}$ against $\hbar\omega$ coordinates (see figure 2). The threshold of the photo-ionization absorption gives the position of the $Fe^{2+}(3d^6)$ donor level with respect to the bottom of the conduction band.

The tail of the absorption edge is due to photo-ionization from the $Fe^{2+}(3d^6)$ states to the conduction band. For two samples of $Cd_{1-x}Fe_xSe$ with the concentration 3 at.% ($\sim6\times10^{20}$ cm^{-3}) and 0.4 at.% ($\sim8\times10^{19}$ cm^{-3}) the absorption coefficients for $\hbar\omega$ =1.5 eV are, 173cm^{-1} and 23cm^{-1} respectively and, thus, are proportional to the concentration of iron atoms. The exciton absorption (A and B) indicated in figure 2 was measured on the same sample etched to a thickness of less than 1 μm (the thickness of the sample after etching was not measured). At T=2K two sharp exciton lines were observed. With increasing temperature the B line disappears and the A line decreases but remains (at T=300K) in the form of the knee on the absorption edge.

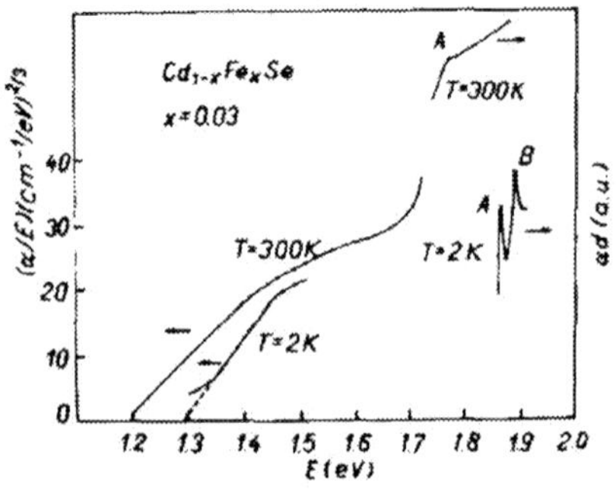

Figure 2. Photo-ionization (D transitions) and exciton absorption (T=2 and 300K) for $Cd_{1-x}Fe_xSe$, x=0.03. The exciton absorption was measured on a sample etched to less than 1 μm thick (the thickness was not measured)[8].

The presence of sharp exciton lines at T=2K found for all iron concentrations (up to 15 at.%) gives evidence of the good quality of the measured crystals.

Subtracting the photo-ionization threshold energies from the energy gaps it was found that the position of the $Fe^{2+}(3d^6)$ donor level above the top of the valence band in CdSe is, within the experimental error, independent of temperature and iron concentration (up to 15 at.%) and equals 0.58 ± 0.03 eV.

The energy position of the excition A and B for $Cd_{1-x}Fe_xSe$ against iron concentration (at T=2K) is shown in figure 3. It is found that the change in the energy gap is: dE_g/dx=14.8 meV/at.% (for comparison for $Cd_{1-x}Mn_xSe$ dE_g/dx=15.4 meV/at.% [23]).

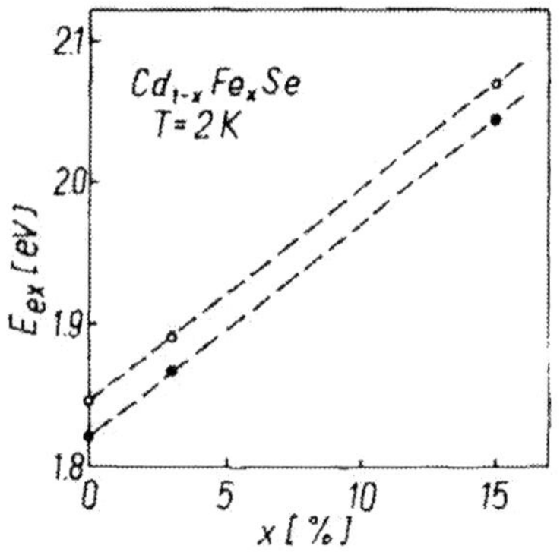

Figure 3. The energy position of excitons A and B against iron concentration for $Cd_{1-x}Fe_xSe$. T=2K[8].

The solid solution $Hg_{1-v-x}Cd_vFe_xSe$ has been obtained for the whole range of composition corresponding to the zincblende structure (v< 0.7), and for an iron concentration with x⩽ 0.15. Absorption results for $Hg_{1-v-x}Cd_vFe_xSe$ for v=0.55 and x=0.05, taken at 4.2K are shown in figure 4. At the energy 170-220 meV a clear threshold has been observed, connected with the transition from the $Fe^{2+}(3d^6)$ level to the conduction band (D transition). Between 300 and 450 meV the characteristic for $Fe^{2+}(3d^6)$ in all II-VI compounds [24,25] is crystal field absorption due to $^5E(^4D)$-$^5T_2(^5D)$ transitions and this was observed. At the energy of 770-800 meV the transition from the valence band to the conduction band (A transition) appears. The absorption coefficient for D transitions in this sample equals approximately 300-350 cm^{-1} which is in qualitative agreement with data for $Cd_{1-x}Fe_xSe$

x=0.03 (see figure 2). The average for the absorption coefficient for the $^5E-^5T_2$ crystal field transitions is about 1750 cm^{-1}.

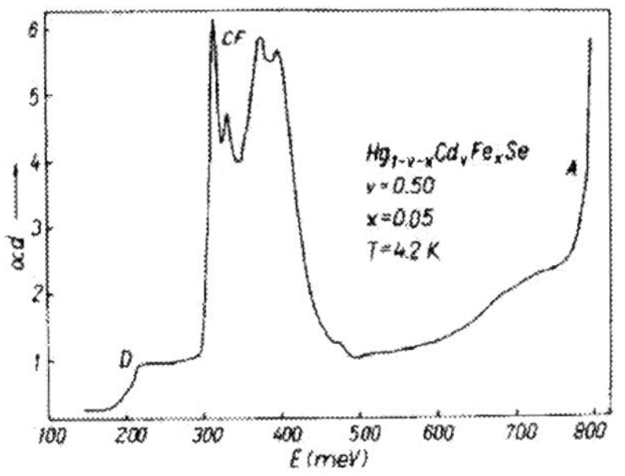

Figure 4. Absorption results for $Hg_{1-v-x}Cd_vFe_xSe$ for v=0.55 and x=0.05, taken at 4.2K. The thresholds connected with D transitions, crystal field absorption (CF) and absorption edge (transition A) are indicated[8].

This high value for the absorption coefficient is connected with the high concentration of iron (5 at.%). These results are in agreement with the calibration of the absorption coefficient against iron concentration done, for example, on ZnS:Fe [24]. The relative sharpness of the zero-phonon line of the crystal field transition suggests a good homogeneity in the samples investigated.

PHOTO-EMISSION MEASUREMENTS

In $Hg_{1-x}Fe_xSe$ with zero band gap measurements of the optical absorption similar to those described above cannot be performed because of very strong interband absorption. However, we have measured the photo-emission yield for pure HgSe and $Hg_{1-x}Fe_xSe$ with x up to 0.12, and in the photon energy range up to 6 eV. The crystals have been cleaved in UHV (10^{-9} Torr) and were investigated in situ at room temperature. Two poto-emission thresholds have been observed. The experimental results are shown in figure 5.

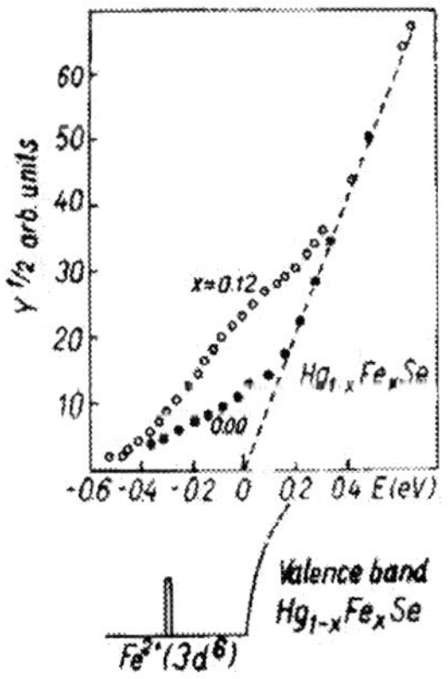

Figure 5. Experimental curves of the photo-emission yield for HgSe and $Hg_{1-x}Fe_xSe$, x=0.12 (T=300K). The lower part shows schematically the position of the additional density of states given by $Fe^{2+}(3d^6)$[8].

The higher energy threshold corresponds to the photo-emission of electrons from the valence band edge of HgSe. The energy in figure 5 is measured from the top of the valence band, that is, for each sample the experimental curve is shifted to overlap at high yields with the curves for other samples. The question remains whether at high yields the curves follow the $(E-E_T)^2$ (broken line in the figure 5 or $(E-E_T)^{5/2}$ law [26].

For the sample $Hg_{1-x}Fe_xSe$, x=0.12, the lower energy threshold is about 0.4±0.15 eV below the higher one. We have observed that height of the step corresponding to the lower energy threshold increases with iron concentration. The lower part of figure 5 shows schematically the position of the additional density of states given by $Fe^{2+}(3d^6)$ which is responsible for the lower energy photo-emission threshold. These experiments do not determine the localization of the $Fe^{2+}(3d^6)$ states but indicate the appearance of the additional density of states which is localized higher than top of the valence band. In our opinion this result agrees well with ref. [27] where additional density of states near Fermi level for HgFeSe in comparison to HgSe was observed.

TRANSPORT MEASUREMENTS

Important information on the position of the $Fe^{2+}(3d^6)$ level in the band structure of HgSe may be obtained from the measurements of the Hall coefficient as a function of iron concentration. In figure 6 the experimental dependence of the free-electron concentration N_e on the iron concentration N_{Fe} for T=4.2K is shown for $Hg_{1-x}Fe_xSe$ with N_{Fe} ranging from 10^{18} up to approximately 10^{21} cm^{-3} (x=0.05). The curve is an average of the results for eight samples. Up to an iron concentration of $N_{Fe} \sim 5 \times 10^{18}$ cm^{-3}, N_e is roughly equal to N_{Fe}. For $N_{Fe} > 5 \times 10^{18}$ cm^{-3}, N_e is independent of N_{Fe}.

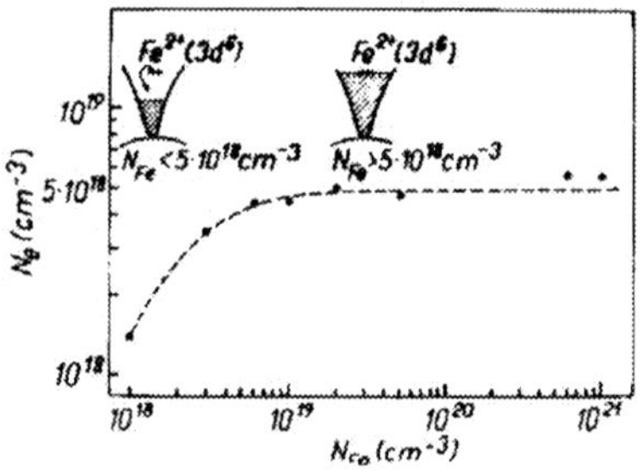

Figure 6. Experimental dependence of the free-electron concentration N_e on the iron concentration N_{Fe} for $Hg_{1-x}Fe_xSe$ (T=4.2K). The curve is an average of the results for eight samples [8].

These experimental facts are consistent with the assumption that the $Fe^{2+}(3d^6)$ level is a resonant donor state located in the conduction band (see the upper part of figure 6). For low iron concentration all these states are ionized, that is, are in the $Fe^{3+}(3d^5)$ state. For high N_{Fe}, the Fermi level in the conduction band reaches and stabilizes at the resonant donor level and any further increase in N_{Fe} will not influence N_e.

From the helium temperature measurements of Shubnikov-de Haas oscillations and of the reflectivity in the range of the plasma edge (in and without an external magnetic field) which will be discussed in next chapter we have determined the electron effective mass at the Fermi level for samples with $N_{Fe} > 5 \times 10^{18}$ cm^{-3} up to x=0.05. The result is that this effective mass is equal to that for HgSe doped to the concentration $N_e = 5 \times 10^{18}$ cm^{-3} with, for example, indium. This only suggests that the influence of iron on the conduction band shape and energy gap E_0 is not very essential. Therefore, for an iron concentration $N_{Fe} \cong 5 \times 10^{18}$ cm^{-3}, that is for x~0.0002, the influence of iron on the conduction-band shape is surely negligible. Thus, we can use the HgSe band structure and the value $N = 5 \times 10^{18}$ cm^{-3} to determine the position of the Fermi level, that is the position of the $Fe^{2+}(3d^6)$ level with respect to the bottom of the conduction band (or to the top of the valence band for HgSe). We obtain 230 ± 10 meV, which does contradict the photo-emission result, in the rather broad error bars of the latter.

Finally, our experimental results enable us to located the $Fe^{2+}(3d^6)$ level in $Hg_{1-v-x}Cd_vFe_xSe$ above the top of the valence band by energy: 230 ± 10 meV (v=0, $X \leq 0.15$), 580 ± 30 meV (v=0.55, $x \leq 0.15$), and 580 ± 30 meV (v+x=1, $x \leq 0.15$). Assuming a linear dependence of the level position on v in the range of zincblende structure (see the broken line in figure 1) we can expect that the donor $Fe^{2+}(3d^6)$ is a resonant one up to v~0.40.
Substitutional iron ion acts in $Hg_{1-x}Fe_xSe$ as a neutral resonant donor. For low iron concentrations $N_{Fe} < 5 \times 10^{18}$ cm^{-3} all resonant donors are ionized to the $Fe^{3+}(3d^5)$ states (see figure 6). This was confirmed by the studies on oriented HgFeSe samples of angular dependence of the EPR spectra characte-

ristic for the $3d^5$ shell (see for example the results for ZnTe:Mn [28]. In the range $N_{Fe} = 1\times10^{18}$ cm^{-3} to $N_{Fe} = 5\times10^{18}$ cm^{-3} the intensities of the EPR lines are proportional to the N_{Fe}.

MAGNETOOPTICAL AND MAGNETO TRANSPORT MEASUREMENTS AND SPACE-ORDERING OF IONIZED DONORS IN HgFeSe

As a consequence of ionization of the Fe^{2+} state (for $N_{Fe} < 5\times10^{18}$ cm^{-3}) the free electron concentration N_e roughly equals the iron concentration. For $N_{Fe} \simeq 5\times10^{18}$ cm^{-3} the Fermi level reaches the resonant donor level $Fe^{2+}(3d^6)$ and becomes pinned to it. The further increase of N_{Fe} does not influence either N_e nor the concentration of ionized donors. The pinning will occur also in the presence of an external quantizing magnetic field. In an increasing magnetic field this should result in oscillatory changes of N_e, instead of the usual small oscillatory changes of the Fermi level which keep N_e constant. Such a phenomenon (which in fact was observed earlier for the resonant acceptor states located in the conduction band in $Hg_{1-x}Cd_xTe$ [29]) is a direct proof of the pinning of the Fermi level to the resonant state. The oscillatory changes of the free electron concentration should lead to oscillatory changes in the plasma frequency ($\omega_p \simeq \sqrt{N_e}$). To check these expectations the reflectivity measurements in the magnetic fields up to 7T ($T \simeq 10$K) have been performed. The parallel Voigt (E∥H) configuration was used because the position of the plasma edge is then independent of the magnetic field, and depends only on the plasma frequency. Fig. 7 presents a sample of the experimental recordings [35].

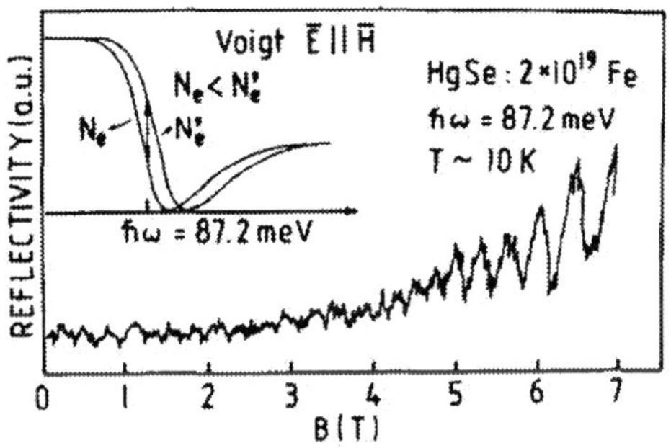

Fig. 7. Magnetoreflectivity oscillations in $Hg_{1-x}Fe_xSe$. The insert shows schematically the origin of the reflectivity changes.

The oscillations of the plasma frequency in the increasing magnetic field lead to the oscillations of the measured reflectivity (see Fig. 7). The observed oscillations have the relative amplitude up to about 1% and are periodic in 1/B. This gives N_e consistent with the Hall value. The accompanying measurements of the σ_{xy} component of the conductivity tensor also show an oscillatory behaviour in the magnetic field.

When the Fermi level is pinned to the resonant state (i.e. for $N_{Fe} > 5\times10^{18}$ cm^{-3}) only a fraction of the resonant donors $Fe^{2+}(3d^6)$ are ionized (to the $Fe^{3+}(3d^5)$ state). Due to the screened Coulomb interaction at low temperatures these donors become ionized not in the random way but keeping the distances between $Fe^{3+}(3d^5)$ ions as large as possible (to minimize interaction energy). This may lead (for $N_{Fe} \gg 5\times10^{18}$ cm^{-3}) to a kind of "liquefying" of the system of ions and even to its "crystallization" (i.e. formation of a superlattice (or hyperlattice) of ionized donors) at still lower temperatures. The above effect was theoretically analyzed by J. Mycielski [30]. The basic assumption of the above model is that the neutral donor state Fe^{2+} is very narrow (long living). This assumption justifies the omission of the resonance scattering of the free electrons. It is also assumed that the resonant state is sufficiently localized to avoid formation of an impurity band from the resonant states. According to this model if the ionized donor system becomes ordered in the form of a superlattice the low momentum transfer scattering processes usually very important for the Coulomb scattering, are excluded. As a result the free electron mobility μ should increase and the Dingle temperature T_D (the parameter describing the broadening of the Landau levels) should decrease very strongly. These effects were observed by us in magnetoresistivity (Shubnikov - de Haas effect) measurements on $Hg_{1-x}Fe_xSe$, $0.00005 < x < 0.05$ (i.e. $1x10^{18} < N_{Fe} < 1x10^{21}$ cm^{-3}). The experimental recordings for samples with $N_{Fe} = 3x10^{18}$ and $6x10^{18}$ cm^{-3} are shown in Fig. 8 [35].

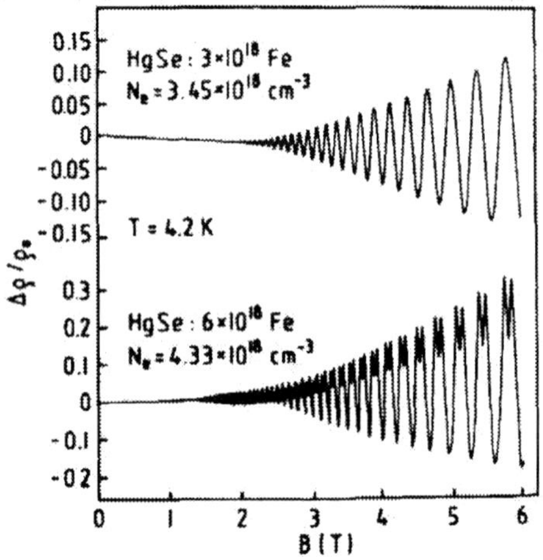

Fig. 8. Shubnikov - de Haas oscillations in $Hg_{1-x}Fe_xSe$ for two different values of iron concentration.

The most significant feature of these recordings is that for the sample with the higher iron concentration (and higher electron concentration!) the oscillations are more pronounced and spin splitting is observed. The results of an analysis of the experimental data (together with result obtained by Vaziri et al [31,32]) are presented in Fig. 9 [35].

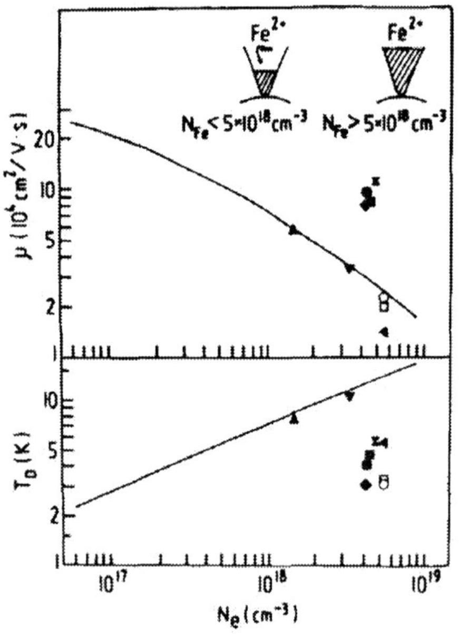

Fig. 9. Dingle temperature T_D and mobility μ vs electron concentration in $Hg_{1-x}Fe_xSe$. Solid lines are theoretical for singly ionized donor scattering in HgSe (after Dietl [34]. The different symbols refer to the samples with following iron concentrations N_{Fe}:
our data: ▲ - $1x10^{18}$, ▼ - $3x10^{18}$,
(in cm^{-3}) ◆ - $6x10^{18}$, ● - $1x10^{19}$,
⧗ - $2x10^{19}$, ■ - $5x10^{19}$,
and ◭ - x=0.045
after Vaziri et al [31,32]:
○ - x=0.03, □ - x=0.05.

It can be seen that for the samples doped with iron to N_{Fe} above $5x10^{18}$ cm^{-3} the Dingle temperature drops down and is abnormally low, i.e. is about 5 times smaller than for HgSe with comparable electron concentration. The mobility of the electrons also exhibits a change and becomes about twice as high as for HgSe (doped with e.g. indium) with similar N_e (these observations are also in qualitative agreement with μ values obtained in [33]. The observed drop of the electron mobility for the samples with N_{Fe} greater than a few percent could be partially attributed to the alloy scattering.

MAGNETIC PROPERTIES

Magnetic susceptibility of monocrystalline samples of HgFeSe, HgCdFeSe and CdFeSe has been studied both in high- and in low-temperature regions [36-38]. In the former case it obeys the Curie-Weiss low. The exchange interactions are antiferromagnetic for all concentrations studied. At low temperatures a Van Vleck paramagnetism has been observed for the CdFeSe and also for ZnFeSe, ZnFeTe [3,7], and HgFeTe [2]. The interpretation of the above results is based on the $Fe^{2+}(3d^6)$ electronic configuration. The ground state of the Fe^{2+} single ion ^5D is split by a tetrahedral crystal field into a ^5E orbital doublet and ^5T orbital triplet. Spin-orbit interaction splits the ^5E term into singlet A_1, triplet T_1, doublet E, triplet T_2 and singlet A_2 [39,40]. The energy separation between these states is approximately equal ≈ 1.5 meV. Thus the ground state is a magnetically inactive singlet A_1 which leads to Van Vleck - type paramagnetism at $T \lesssim 20K$.

The suprising result for the $Hg_{1-v-x}Cd_vFe_xSe$ compounds with $v \leqslant 0.5$ and low x is the absence of Van Vleck paramagnetism at low temperatures [37,38]. On the contrary, it is observed in CdFeSe systems [40,38] where the Fe^{2+} state is located well below the conductions band edge. Hence, the absence of the Van Vleck paramagnetism may be connected with a resonant nature of Fe^{2+} $(3d^6)$ state in the corresponding compounds.

Finally, like Mn compounds also for Fe compounds the freezing temperatures of the transition to the spin-glass state have been determined. For $Hg_{1-x}Fe_xSe$ with x=0.10 and 0.105 we observe a transition to spin-glass state with freezing temperatures T =2.4K and 2.8K, respectively. For $Hg_{1-x}Mn_xSe$ this transition is observed for larger x, i.e. for x=0.145 and x=0.176, with T =2.12K and 3.35K, respectively.

ACKNOWLEDGEMENTS

I am grateful to all my colleagues and coworkers in Warsaw, particularly to W. Dobrowolski, K. Dybko, M. Arciszewska, M. Dobrowolska, J.M. Baranowski, J. Wróbel, A. Lewicki, J. Spałek, B. Witkowska and U. Blinowska. Without them nothing could have been made possible.

REFERENCES

1. Y. Guldner, C. Rigaux, M. Menant, D.P. Mullin and J.K. Furdyna, Solid State Commun. 33, 133 (1980).
2. H. Serre, G. Bastard, C. Rigaux, J. Mycielski and J.K. Furdyna, Proc. 4th Int. Conf. Physics of Narrow Gap Semiconductors, edited by E. Gornik, H. Heinrich and L. Polmetshofer (Springer, Berlin, 1982), p. 321.
3. A. Twardowski, A.M. Hennel, M. von Ortenberg and M. Demianiuk, Proc. 17th Int. Conf. Physics of Semiconductors, San Francisco, edited by J.D. Chadi and W.A. Harrison (Springer, New York, 1985) p.1439.
4. N.V. Joshi and L. Mogollon, Prog. Cryst. Growth Charact. 10, 65 (1985).
5. S. Abdel-Maksoud, C. Fau, J. Calas, M. Averous, B. Lombos, G. Brun and J.C. Tedenac, Solid State Commun. 54, 811 (1985).
6. A. Lewicki, Z. Tarnawski and A. Mycielski, Acta Physica Polonica A67, 357 (1985).
7. A. Twardowski, M. von Ortenberg and M. Demianiuk, J. Cryst. Growth 72, 401 (1985).
8. A. Mycielski, P. Dzwonkowski, B. Kowalski, B.A. Orłowski, M. Dobrowolska, M. Arciszewska, W. Dobrowolski and J.M. Baranowski, J. Phys. C: Solid State Phys. 19, 3605 (1986).
9. M. Vaziri, U. Dębska and R. Reifenberger, Appl. Phys. Lett. 47 (4), 407 (1985).

10. P. Vogl and J.M. Baranowski, Acta Physica Polonica,A67, 133 (1985).
11. P. Vogl and J.M. Baranowski, Proc. 17th Int. Conf. Physics of Semiconductors, San Francisco, edited by J.D. Chadi and W.A. Harrison (Springer, New York 1985), p. 623.
12. J.M. Langer and Heinrich, Phys. Rev. Lett. 55, 1414 (1985).
13. B.A. Orłowski, Phys. Status Solidi b95, K31 (1979).
14. M. Taniguchi, L. Ley, R.L. Johnson and M. Cardona, Phys. Rev. B33, 1206 (1986).
15. B.A. Orłowski, K. Kopalko and W. Chab, Solid State Commun. 50, 749 (1984).
16. A. Franciosi, Shu Chang, R. Reifenberger, U. Dębska and R. Riedel, Phys. Rev. B32, 6682 (1985).
17. A. Franciosi, C. Caprile and R. Reinfenberger, Phys. Rev. B31, 8061 (1985).
18. C.J. Summers and J.G. Broerman, Phys. Rev. B21, 559 (1980).
19. A. Mycielski, J. Kossut, M. Dobrowolska and W. Dobrowolski, J. Phys. C: Solid State Phys. 15, 3293 (1982).
20. J.M. Baranowski and J.M. Langer, Phys. Status Solidi (b) 48, 863 (1971).
21. J.W. Allen, J. Phys. C: Solid State Phys. 2, 1077 (1969).
22. J.M. Langer, Phys. Status Solidi (b) 47, 443 (1971).
23. P. Wiśniewski and M. Nawrocki, Phys. Status Solidi (b) 117, K43 (1983).
24. G.A. Slack, F.S. Ham and R.M. Chrenko, Phys. Rev. 152, 376 (1966).
25. J.M. Baranowski, J.W. Allen and G.L. Pearson, Phys. Rev. 160, 627 (1967).
26. O.E. Kane, Phys. Rev. 127, 131 (1962).
27. A. Wall, C. Caprile, A. Franciosi, M. Vaziri, R. Reinfenberger and J.K. Furdyna, J. Vac. Sci. Technolog (in press); A. Franciosi, A. Wall, S. Chang, P. Philip, N. Troullier, A. Raisanem, R. Reifenberger, F. Pool, J.K. Furdyna presented at the 18th Int. Conf. Physics of Semiconductors, Stockholm, 1986.
28. S. Kunii and E. Hirahara, J. Phys. Soc. Japan 19, 1258 (1964).
29. A. Kozacki, S. Otmezguine, G. Weill and C. Verie, Proc. 13th Int. Conf. Physics of Semiconductors, edited by F.G. Fumi, (Roma 1976), p. 476.
30. J. Mycielski, Solid State Commun. (in press).
31. M. Vaziri, D.A. Schwarzkopf and R. Reinfenberger, Phys. Rev. B31, 3811 (1985).
32. M. Vaziri and R. Reifenberger, Phys. Rev. B32, 3921 (1985).
33. N.G. Gluzman, L.D. Sabirzyanova, I.M. Tsidilkovsky, L.D. Paranchich and S.U. Paranchich, Fiz. Tekh. Poluprovodn. 20, 94 (1986).
34. T. Dietl, J. de Phys. (Paris) 39, C6 - 1081 (1978).
35. W. Dobrowolski, K. Dybko, A. Mycielski, J. Mycielski, J. Wróbel, S. S. Piechota, M. Palczewska, H. Szymczak and Z. Wilamowski, Proc. 18th Int. Conf. Physics of Semiconductors, (Stockholm 1986), in press.
36. A. Lewicki, Z. Tarnawski and A. Mycielski, Acta Physica Polonica A67, 357 (1985).
37. A. Lewicki, A. Mycielski and J. Spałek, Acta Physica Polonica (in press) (1986).
38. A. Lewicki, J. Spałek and A. Mycielski, J. Phys. C: Solid State Phys. (in press).
39. W. Low and M. Weger, Phys. Rev. 118, 1119 (1960).
40. J. Mahoney, C. Lin, W. Brumage and F. Dorman, J. Chem. Phys. 53, 4286 (1970).

ELECTRONIC TRANSPORT PROPERTIES OF $Hg_{1-x}Fe_xSe$

F. POOL, J. KOSSUT*, U. DEBSKA, R. REIFENBERGER AND J.K. FURDYNA
Department of Physics, Purdue University, W. Lafayette, IN 47907, USA

ABSTRACT

The electrical resistivity, electron concentration, and mobility of $Hg_{1-x}Fe_xSe$ are reported for $4.2K < T < 300K$ and for $0.0001 < x < 0.12$. The data are interpreted within an electronic band structure model that assumes the existence of resonant donors (due to the presence of Fe ions) whose ground state energy coincides with the conduction band continuum. The electron concentration data enable determination of the value of the donor energy as a function of the temperature and the crystal composition. The low temperature electron mobility for $\sim 0.0003 \le x \le 0.01$ is considerably higher than expected and indicates a reduction of the charged impurity scattering effects at low temperatures.

INTRODUCTION

$Hg_{1-x}Fe_xSe$ belongs to a relatively new group of semiconducting compounds known as diluted magnetic semiconductors (DMS), i.e., semiconducting mixed crystals whose lattice contains substitutional magnetic ions [1-3]. These materials have attracted considerable attention due to their many interesting and novel properties, as well as their potential for device applications. In general, magnetic ions modify strongly (via sp-d coupling) the electronic band structure of a DMS in the presence of the magnetic field and thus affect various physical properties of these materials. The majority of DMS studied so far incorporated Mn ions in various II-VI semiconducting hosts. The growth of $Hg_{1-x}Fe_xSe$ extends this group of semiconducting alloys to Fe compounds. Recent investigations of $Hg_{1-x}Fe_xSe$ [4-10] have revealed that this system shows striking differences from the behavior found for HgSe and $Hg_{1-x}Mn_xSe$. In particular, Fe is believed to act as a donor (or gives rise to an iron-related donor complex) whose energy is resonant with the conduction band. The presence of a donor state possessing magnetic properties and also resonant with the conduction band is new to DMS materials.

The motivation for the experiments reported below was to complete a systematic study of the transport properties of $Hg_{1-x}Fe_xSe$ to gain greater insight into the electronic band structure of the material. In particular, an analysis of the data was performed to further investigate the plausibility of the hypothesis involving charged donor ordering [11-12] which has recently been suggested as an explanation of the observed enhancement of the low temperature mobility.

RESULTS AND ANALYSIS

In this paper we report measurements of the resistivity, electron concentration and Hall mobility for crystalline $Hg_{1-x}Fe_xSe$ samples with $0.0001 < x < 0.12$ over a temperature range $4.2K < T < 300K$. The data acquisition system used in this study has been described elsewhere [13]. Here we concentrate only on the interpretation of our experimental results.

Mat. Res. Soc. Symp. Proc. Vol. 89. © 1987 Materials Research Society

Electron Concentration

A quantitative analysis of our electron concentration data was made by solving numerically the neutrality equation

$$n = N_{Fe} - N^{*}_{Fe} \tag{1}$$

for the Fermi energy E_F, where

$$n = \int_0^{\infty} \rho_c(\varepsilon) f_{F-D} d\varepsilon \tag{2}$$

with n being the conduction electron concentration, N_{Fe} the total concentration of Fe ions, $N_{Fe} = 4x/a^3$, a is the lattice constant, N^{*}_{Fe} is the concentration of the occupied (neutral) donors, f_{F-D} is the Fermi-Dirac distribution function, and ρ_c denotes the density of states in the conduction band, which can easily be derived from $\rho_c^{-1} = \pi^2/k^2(d\varepsilon/dk)$, where ε is taken in the form appropriate for a narrow-gap semiconductor [14]. The values of the material parameters and their temperature dependences were assumed to be equal to those in HgSe [15,16].

To calculate the concentration of occupied donors, N^{*}_{Fe}, one has to know also the related density of states ρ_{Fe} since

$$N^{*}_{Fe} = \int_0^{\infty} \rho_{Fe}(\varepsilon) f_{F-D} d\varepsilon \ . \tag{3}$$

For this study, $\rho_{Fe}(E)$ has been assumed to have a very simple, delta-like form

$$\rho_{Fe}(\varepsilon) = N_{Fe}\delta(\varepsilon - E_{Fe}) \tag{4}$$

corresponding to a sharply defined energy of the ground state of Fe donors.

The system of Eqs. (1)-(4) contains a free parameter E_{Fe}, which was chosen to fit the measurement of n at lowest temperatures. The possibility of a slight temperature dependence in E_{Fe} was taken into account using

$$E_{Fe}(T) = E_{Fe}(0) + \alpha[E_g(T) - E_g(0)] \tag{5}$$

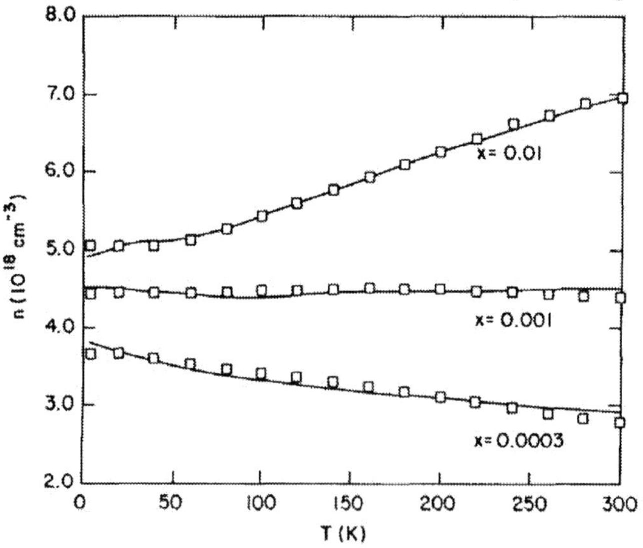

Fig. 1
Electron concentration as a function of temperature in $Hg_{1-x}Fe_xSe$ for x=0.0003, x=0.001 and x=0.01. Solid lines are calculated by utilizing a delta function density of states for the ground state of the Fe donors.

where α was treated as an adjustable parameter. Examples of the fits are depicted in Fig. 1, where it can be seen that even though the samples display a broad variety of temperature dependences, the model is indeed capable to render them correctly. The resulting values of $E_{Fe}(x)$ at 4.2K are:

$$E_{Fe}(x) = 0.21 + 0.52x \quad [eV] \tag{6}$$

with the coefficient α in Eq. 5 equal approximately to 0.2-0.3.

Mobilities

We also calculated the electron mobilities for our samples using the procedure described in detail by Szymanska and Dietl in Ref. 17 and successfully applied to HgSe and $Hg_{1-x}Cd_xSe$, [18,19]. The material parameters used in our analysis were taken from Ref. 18. Calcualtions of the mobilities resulting from the scattering by polar optical phonons, nonpolar optical phonons and acoustic phonons, and charge center scattering were carried out as discussed in Ref. 17. A plot of the data taken at 4.2K as a function of x and the results of the calculation are shown in Fig. 2. As can be seen from

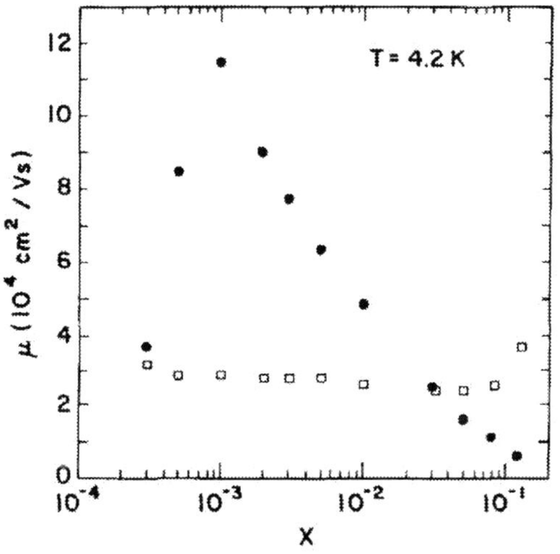

Fig. 2
Semilog plot of the experimental mobility of $Hg_{1-x}Fe_xSe$ as a function of x at 4.2K. The solid dots are experimental data, the open squares are the results of calculations based on Ref. 18.

this figure, the mobility measured at 4.2K can be a factor of ~ 4 greater than the calculated values. In contrast to this discrepancy found at low temperatures, mobilities calculated at room temperature were found to coincide with the data. Since in the low temperature region the most efficient mobility limiting scattering mechanism is due to ionized impurities, it is probable that its modification is responsible for the observed enhancement.

Quite recently an interesting hypothesis was put forward [11,12] where it was proposed that the anomalous behavior of electron mobility in $Hg_{1-x}Fe_xSe$ is due to spatial redistribution (due to Coulomb repulsion) of the electrons in the system of incompletely ionized donors, resulting in the formation of an ordered super-structure of these scattering centers. The hypothetical ordering of the system of ionized Fe donors does qualitatively account for many of the anomalous features observed by us.

In order to further investigate this superlattice model more quantitatively, we found it convenient to characterize the degree of ordering by a parameter σ which describes the randomness of the superlattice. More precisely, σ is defined as the standard deviation of the charged Fe donor lattice site positions from perfect periodicity. Using this parameter, it is now straightforward to write an expression for the momentum relaxation time in the case of scattering by screened Coulomb potentials forming a "superlattice" with a Gaussian disorder:

$$\frac{1}{\tau_{\vec{k}}} = \frac{4me^4}{3\pi\kappa^2\hbar^3} \int_0^{2k_F} \left(1-e^{-q^2\sigma^2}\right) \frac{q^3dq}{(q^2+\lambda^{-2})^2} \qquad (7)$$

where λ is the appropriate screening length and κ is the semiconductor dielectric constant. Eq. (7) does not account for the nonparabolicity and mixing of the electron wave functions in narrow-gap semiconductors, however such generalization of Eq. (7) is not difficult to do and has been included.

Calculations of the mobilities resulting from the scattering by polar optical phonons, nonpolar optical phonons and acoustic phonons, and modified charge center scattering as a function of temperature were carried out. By adjusting σ at each temperature, it is possible to fit mobility data which is higher than predicted for a totally random array of scatters ($\sigma = \infty$). By adjusting σ in the appropriate generalization of Eq. (7) to fit our data, detailed estimates of the temperature variation of σ were obtained. These results for σ are shown in Fig. 3 and can be compared to the mean distance between Fe ions, $\langle r_{Fe}\rangle = a(4/x)^{1/3}$, which are indicated by the dashed lines in Fig. 3. It is evident that the ratio $\sigma/\langle r_{Fe}\rangle$ is smaller than unity only in the lowest temperature region.

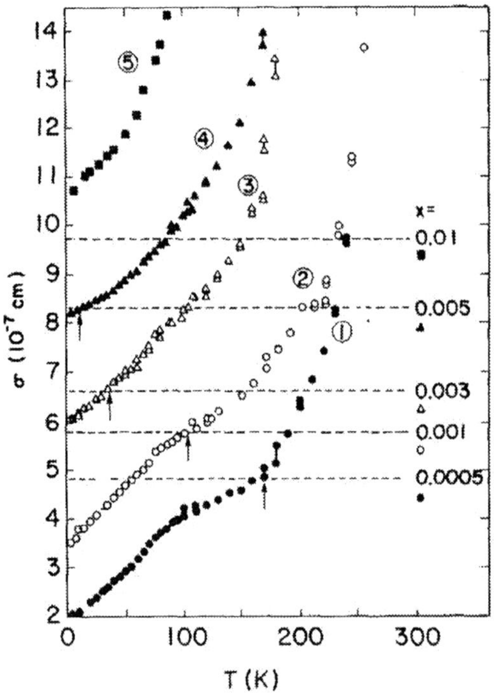

Fig. 3
Fitted values of σ as a function of temperature for samples which display an enhanced mobility. The horizontal dashed lines represent $\langle r_{Fe}\rangle$. Arrows mark temperatures, T^*, for which $\sigma(T)$ intersects $\langle r_{Fe}\rangle$. Note that only values of σ for curve 1 (x=0.0005) represent the true value of this parameter, all others have been shifted by $2(n-1) \times 10^{-7}$ cm (n=2,3,4,5) for clarity.

The increasing trend with increasing x shown by the values of $\sigma/\langle r_{Fe}\rangle$ determined by us at a given temperature is in agreement with expectations [11,12] and stems from the fact that for increasing x the ratio of unoccupied donors to occupied donors decreases, making the charge distribution process less likely. It has been shown [11,12] that the temperature of "liquification" of the charged impurity system, T^*, should be roughly proportional to $[\ln(N_{Fe}/n)]^{-1}$. Fig. 4 shows a plot T^*, the temperature at which σ becomes equal to $\langle r_{Fe}\rangle$ as a function of this quantity. In view of the simplified nature of the theory and the roughness of the above estimate

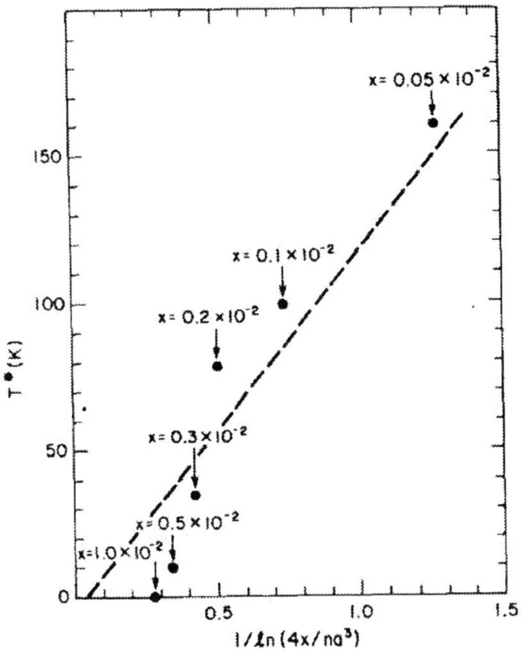

Fig. 4
Temperature at which σ becomes equal to $\langle r_{Fe} \rangle$ as a function of $[\ln(N_{Fe}/n)]^{-1}$. This "liquification temperature", T^*, is approximately proportional to $[\ln(N_{Fe}/n)]^{-1}$ as predicted by the theory. The broken line was found by a linear regression analysis using the data points in this figure.

of the "liquification" temperature, the dependence exhibited in Fig. 4 seems to approximate the strict proportionality predicted by the theory.

CONCLUSIONS

The study of transport properties of $Hg_{1-x}Fe_xSe$ with $0.0001 \le x \le 0.12$ revealed anomalously large values of the electron mobilities at low temperatures for samples in the range $\sim 0.0003 \le x \le 0.01$. The high mobilities are interpreted in terms of a decreased effectiveness of the ionized impurity scattering which is a consequence of an ordering (at least partial) of spatial position of the charges within the donor system. The degree of ordering is a decreasing function of the Fe molar fraction x. The degree of the ordering can be parameterized by the variable σ and indications are that the ordering is far from perfect. The values of σ obtained by analysing electron mobilities are in agreement with similar analysis of the Dingle temperature determined in Shubnikov-de Haas and de Haas-van Alphen experiments [20]. The composition corresponding to the greatest mobility enhancement, $x = \sim 0.001$, also marks the boundary between entirely different types of temperature dependence in the electron concentration: below this value of x the concentration decreases with increasing temperature, while a reverse trend is observed for higher values of x. The analysis of electron concentration data as a function of temperature in the x = 0.12 sample indicates that the absolute value of the energy gap E_g does increase as x increases. A rough estimate shows that E_g is near zero for $x \sim 0.3 - 0.4$.

ACKNOWLEDGMENTS

This work has been supported by National Science Foundation Grant DMR 84-13739. The assistance of M. Miller and M. Vaziri in preparation of this experiment is gratefully acknowledged.

REFERENCES

* On leave from the Institute of Physics, Polish Academy of Sciences, Warsaw, Poland.

1. J.K. Furdyna, J. Appl. Phys., 53, 7637 (1982).
2. J. Mycielski, Prog. Cryst. Growth Char., 10, 101 (1985).
3. J.K. Furdyna, J. Vac. Sci. Technol., A4, 2002 (1986).
4. M. Vaziri, U. Debska and R. Reifenberger, Appl. Phys. Lett., 47 407 (1985).
5. J. Jurewicz, I. Pilecka, K. Dybko, W. Dobrowolski, A. Mycielski and J. Wrobel, Acta Phys. Polon., A69, (1986).
6. A. Mycielski, P. Dzwonkowski, B. Kowalski, B. Orlowski, M. Dobrowolska, M. Arciszewska, W. Dobrowolski and J.M. Baranowski, J. Phys. C, 19, 3605 (1986).
7. A. Wall, C. Caprile, A. Franciosi, M. Vaziri, R. Reifenberger and J.K. Furdyna, J. Vac. Sci. Technol., A4, 2010 (1986).
8. M. Vaziri and R. Reifenberger, Phys. Rev., B32, 3291 (1985).
9. M. Vaziri and R. Reifenberger, Phys. Rev., B33, 5585 (1986).
10. H. Serre, G. Bastard, C. Rigaux, J. Mycielski and J.K. Furdyna, Proc. Int. Conf. Phys. of Narrow Gap Semicond., Linz 1981, Lecture Notes in Physics, 152, (Springer Verlag, Berlin 1982) p. 321.
11. J. Mycielski, Acta Phys. Polon, A69, (1986).
12. J. Mycielski, Solid State Commun. (in press).
13. F. Pool, J. Kossut, U. Debska, R. Reifenberger, submitted to Phys. Rev. B.
14. E.O. Kane, J. Phys. Chem. Solids, 1, 249 (1957).
15. A. Mycielski, J. Kossut, M. Dobrowolska and W. Dobrowolski, J. Phys. C, 15, 3293 (1982).
16. M. Dobrowolska, W. Dobrowolski and A. Mycielski, Solid State Commun. 34, 441 (1980).
17. W. Szymanska and T. Dietl, J. Phys. Chem. Solids, 39, 1025 (1978).
18. T. Dietl and W. Szymanska, J. Phys. Chem. Solids, 39, 1041 (1978).
19. R.J. Iwanowski, T. Dietl and W. Szymanska, J. Phys. Chem. Solids, 39, 1059 (1978).
20. M.M. Miller and R. Reifenberger, Bull. Amer. Phys. Soc., 31, 254 (1986) and to be published.

SYNCHROTRON RADIATION STUDIES OF TERNARY SEMIMAGNETIC SEMICONDUCTORS

A. FRANCIOSI
Department of Chemical Engineering and Materials Science
University of Minnesota, Minneapolis, MN 55455

ABSTRACT

A number of spectroscopic techniques that exploit synchrotron radiation are providing new insight into the properties of ternary semimagnetic semiconducting alloys. Photon energy-dependent photoemission methods are being used to characterize the electronic structure and the elemental and orbital contributions to the valence states. Extended X-ray Absorption Fine Structure Spectroscopy (EXAFS) is providing a detailed understanding of the atomic structure and of the local phenomena that make the virtual crystal approximation inadequate. The new experimental information is stimulating novel theoretical work on the electronic structure, lattice stability and thermodynamics of these materials. We will review the results of a number of recent experimental studies and emphasize systematic trends and promising new areas of research.

INTRODUCTION

Renewed interest in the properties of ternary semimagnetic semiconductor alloys has been stimulated by a number of proposed optoelectronic device applications [1]-[3]. The successful synthesis of high quality epitaxial layers by Molecular Beam Epitaxy [4]-[5], together with recent theoretical developments [6]-[7], suggest that interest will grow in the future with the development of semimagnetic superlattices and Multiple Quantum Well Structures.

In general, the semimagnetic semiconductors studied to date are obtained by replacing some of the cations in the II-VI semiconductor lattice with magnetic Mn or Fe atoms. Random ternary alloys of this kind have composition-dependent transport and optical properties and can be grown as single-phase single-crystals over a wide range of composition. These compounds exhibit new magnetotransport and magneto-optical properties because of the spin-spin exchange interaction between the localized 3d magnetic moments and the band and impurity states [1]-[3]. The resulting large g-factors, giant magnetoresistance, and large values of the Faraday rotation bring intriguing magnetic phenomena into the domain of semiconductor physics [1]-[7].

To understand and optimize the properties of these materials one has to assess the nature of the bonding of the magnetic impurity in the compound semiconductor matrix, localization and hybridization of the 3d orbitals and, conversely, the influence of the magnetic impurity on the stability of the host. However, theoretical investigation of the electronic structure of the alloy or estimates of the alloy stability via total energy calculations are complicated by the double localized/itinerant nature of the 3d states and by the random nature of the alloy.

Recently, a number of experimental breakthroughs have been obtained by exploiting synchrotron radiation [8]-[17]. Resonant photoemission spectroscopy has allowed, for example, isolation of the 3d contribution [8]-[14] to the electronic density of states (DOS) below the Fermi level E_F. Extended X-ray Absorption Fine Structure measurements have shown [15]-[17] that the nearest-neighbor distances between anions and cations exhibit a bimodal distribution and do not follow Vegard's law but stay almost unchanged when the alloy composition is varied. We are now at an exciting time when experimental investigation is providing input parameters indispensable for

Mat. Res. Soc. Symp. Proc. Vol. 89. © 1987 Materials Research Society

theoretical developments, and defining a more complete picture to compare with the theoretical results. In fact, a number of theoretical studies [18]-[26] are shedding light on the structural, magnetic and thermodynamic properties of these materials.

This paper is organized in four sections. Section 2 reviews recent synchrotron radiation photoemission studies. Section 3 summarizes EXAFS work, while section 4 includes a general discussion of the experimental outlook and examines new areas in which synchrotron radiation spectroscopies could be exploited.

SYNCHROTRON RADIATION PHOTOEMISSION STUDIES

Synchrotron radiation emitted by charged particles circulating in a storage ring exhibits a continuum spectrum extending from the infrared to the X-ray region and is linearly polarized in the plane of the orbit. In a typical photoemission experiment a monochromatic synchrotron radiation beam impinges on the sample surface at the focus of an electron energy analyzer. Angle integrated photoelectron energy distribution curves (EDC's) provide information on the energy distribution of the electron states below the Fermi level. In an ideal one-electron picture the EDC's reproduce the initial electron DOS distorted by the matrix element of the optical excitation and superimposed on a smooth secondary background due to inelastically scattered electrons.[27]-[28] However, since the final state of the excitation involves a valence or core hole and electron screening, the one-electron picture may not apply in full, especially in the case of highly localized initial states. In such a situation photoemission determined binding energies may differ from the ground state energy of the system.

The use of synchrotron radiation allows one to tune the photon energy and exploit the different energy dependence of the photoexcitation probability of electron states with different elemental and orbital character. Constant Initial State (CIS) and Constant Final State (CFS) spectroscopies are also possible. In the CIS mode, one scans the photon energy and the analyzer pass energy at the same time in order to keep the initial state constant while varying the photon energy. This allows direct monitoring of the partial photoionization cross section of a given electron state. In the CFS mode one scans the photon energy while keeping the analyzer pass energy fixed, and this yields the overall optical absorption coefficient for an appropriate choice of the kinetic energy (partial yield). [27]-[28]

Photon energy-dependent photoemission, including CIS and partial yield techniques, have been applied to a number of ternary semimagnetic semiconductor single crystals cleaved in situ, including $Hg_{1-x}Mn_xSe$ [8]-[9], $Cd_{1-x}Mn_xSe$ [9]-[10], $Hg_{1-x}Mn_xTe$ [11], $Hg_{1-x}Fe_xSe$ [12], $Cd_{1-x}Mn_xTe$ [13]-[14], and $Cd_{1-x}Mn_xS$ [29]. Typical EDC's for the valence band emission from $Cd_{1-x}Mn_xS$ (x=0.1) and CdS single crystals [29] are shown in fig. 1, with binding energy referred to the Fermi level E_F, both samples being n-type. Visible are the shallow Cd 4d core levels, 9-11eV below the valence band maximum E_V, and several DOS features in the 0-6eV binding energy range. In the binary CdS we observe two major DOS features (1.5 and 4.3eV below the top of the valence band) associated primarily with S p states and with hybrid Cd s-S p bonding combinations respectively [29]. In the ternary alloy we observe increased emission in the 2-7eV region and a sharp new DOS feature emerging 3.5eV below the valence band maximum.

The photon energy dependence of the 3.5eV emission feature is a fingerprint of Mn 3d character, as shown in fig. 2. In the top section of fig. 2 we show the optical absorption coefficient of Mn metal [30] in the photon energy range of the 3p-3d excitation. In the midsection of fig. 2 we plot the relative partial photoionization cross section of the states at 2.5eV. This was obtained by measuring CIS's at a constant initial energy of 3.5eV (relative to E_V) for the ternary and for the binary semiconductors, and plotting the ratio of the two in fig. 2. The characteristic three-fold

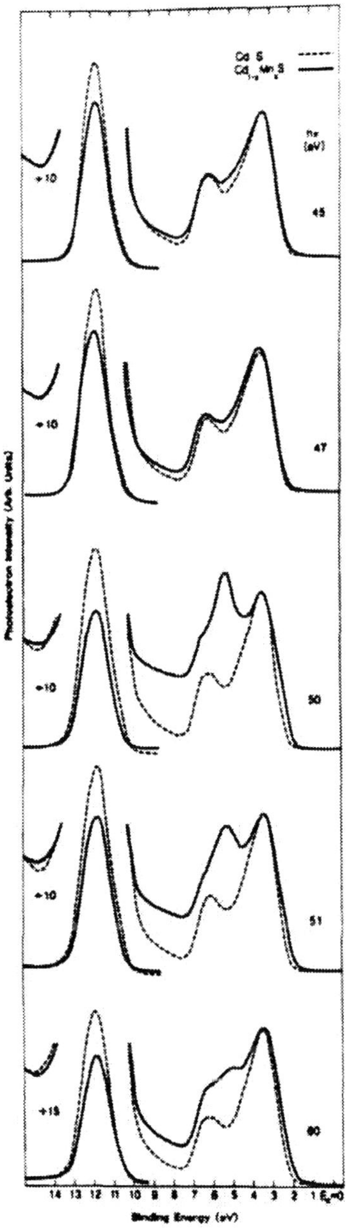

Fig. 1 EDC's for the valence band and Cd 4d core emission from CdS (dashed line) and $Cd_{1-x}Mn_xS$ (x=0.1, solid line) [29].

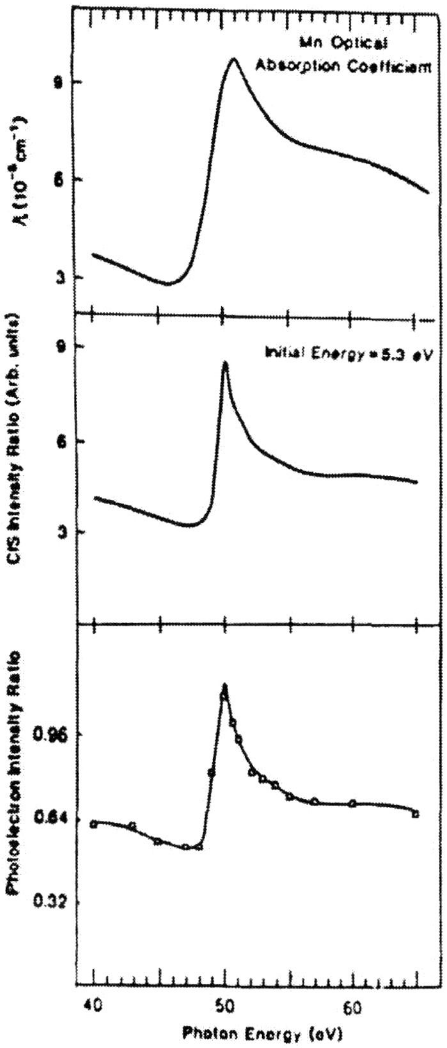

Fig. 2 Top: Optical absorption coefficient for Mn metal in the region of the 3p-3d excitation [30]. Mid-section: Relative partial photoionization cross section of the Mn-derived DOS feature 3.5eV below the valence band maximum, from CIS spectra for $Cd_{1-x}Mn_xS$ and CdS. Bottom: Relative cross section of states at 3.5eV vs. states at 1.5eV, where S p states dominate [29].

enhancement of the cross section at resonance (50eV) versus antiresonance (47eV) is observed in both atomic and metallic Mn, and reflects the quantum-mechanical equivalence of different processes leading from the ground state to the same final state [10]. One process is the direct excitation:

$$3p^63d^54s^2+h\nu\rightarrow 3p^63d^44s^2\varepsilon f$$

The other involves a 3p core excitation and a super Coster-Kronig decay:

$$3p^63d^54s^2+h\nu\rightarrow 3p^53d^64s^2\rightarrow 3p^63d^44s^2\varepsilon f$$

The interference of the two processes yields a characteristic Fano lineshape in the overall excitation cross section. In compounds, the strength of the resonant enhancement is related to the overlap of the p core hole and the d valence states. If the d states become hybridized the overlap is reduced and enhancement diminishes. Correspondingly, some resonant enhancement appears in the cross section of the extra-atomic states that gained d character through hybridization. In the bottom-most section of fig. 2 we show the relative photoionization cross section of the Mn states at 3.5eV versus the S p states 1.5eV below E_V. This was obtained by taking CIS's for $Cd_{1-x}Mn_xS$ at initial energies of 3.5 and 1.5eV, and plotting the ratio of the two in fig. 2. The observed three-fold enhancement at resonance is similar to that observed in the midsection of fig. 2, and argues for a reduced admixture of Mn 3d character at 1.5eV. The result of our analysis for $Cd_{1-x}Mn_xS$ shows similarities with that for the selenide series [8]-[10] shown in fig. 3, the main difference being a somewhat larger modification of the valence states just below E_V in the sulphide. In fig. 3 we show EDC's for $Hg_{0.85}Mn_{0.15}Se$, $Cd_{0.79}Mn_{0.21}Se$ and $Zn_{0.80}Mn_{0.2}Se$ (solid lines) and for the corresponding binaries (dashed lines) after subtraction of a smooth background interpolated between 9 and 0eV. The emergence of a Mn-derived 3d DOS feature 3.4-3.5eV below E_V is emphasized in the ternary-binary difference curves (dot-dashed line) in fig. 3.

For the intermediate-band-gap semiconductor $Cd_{1-x}MnTe$, conventional photoemission studies led to some controversies about the character of the Mn 3d level. Earlier work on sputter-cleaned samples identified a Mn 3d feature 6.8eV below E_V [31]. More recent work by Webb et al. [32] showed a non-dispersive emission feature 3.5eV below E_V, while the deeper feature reported earlier was interpreted as a shake-up satellite of the main 3d feature. The results of refs. [31] and [32] are in contrast with conclusions of Oelhafen et al. [33] who reported measurements on cleaved single crystals and concluded that there is no evidence in their data of localized 3d levels. Strong hybridization of the Mn d levels with Te p states would yield a

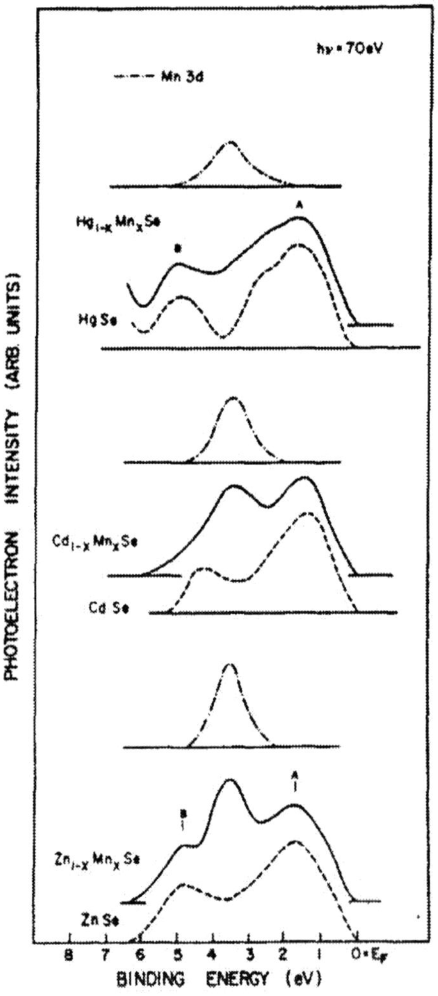

Fig. 3. EDC's for $Hg_{0.85}Mn_{0.15}Se$, $Cd_{0.79}Mn_{0.21}Se$ and $Zn_{0.80}Mn_{0.20}Se$ (solid line) and the corresponding II-VI semiconductors (dashed line) after subtraction of an interpolated background. The ternary-binary difference curves (dot-dashed line) emphasize the Mn contribution to the Density of States (DOS) [9].

valence band in which there is no "energetically sharp" 3d level [33]. A solution to the controversy was given by the first synchrotron radiation study of $Cd_{1-x}Mn_xTe$ single crystals cleaved in situ, recently performed by Taniguchi et al. [13]-[14]. Their results are in good agreement with ours [34]. In fig. 4 we show angular integrated EDC's for $Cd_{0.68}Mn_{0.32}Te$ in the valence band region as a function of photon energy [34]. Fig. 4 unambiguously shows a Mn 3d emission feature emerging at resonance some 3.5eV below the top of the valence band. Taniguchi et al. [13]-[14] investigated the hybridization of Mn 3d with Te p states by comparing the *photoionization cross section of* valence states *for* the ternary versus binary semiconductor (away from the 3p-3d resonance). This is shown in fig. 5 (from ref. [13] where the intensity of the first DOS feature 1.5eV below E_v is plotted versus photon energy using the amplitude of the Cd 4d core level as a reference. This is done for $Cd_{0.35}Mn_{0.65}Te$ and CdTe in the top section of fig. 5. In $Cd_{0.35}Mn_{0.65}Te$ the relative intensity exceeds that of CdTe of a factor of about five at 140eV [13]. This change in cross section reflects the admixture of Mn 3d character. The Mn 3d states show a tenfold increase in the relative photoemission cross section between 90 and 140 eV, as shown in the bottom-most section of fig. 5, and Taniguchi et al. [13]-[14] conclude that strong Mn 3d admixture to the Te p states may involve an average of two d states per Mn atom.

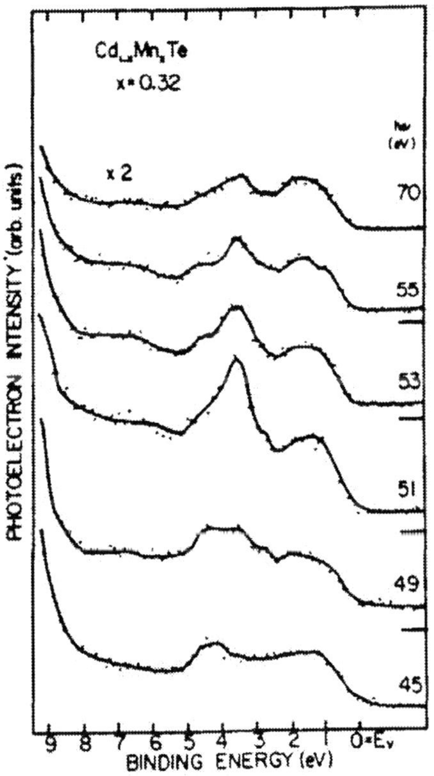

Fig. 4 Detail of the valence band emission for $Cd_{0.68}Mn_{0.32}Te$ showing the emergence of Mn 3d features at resonance [34].

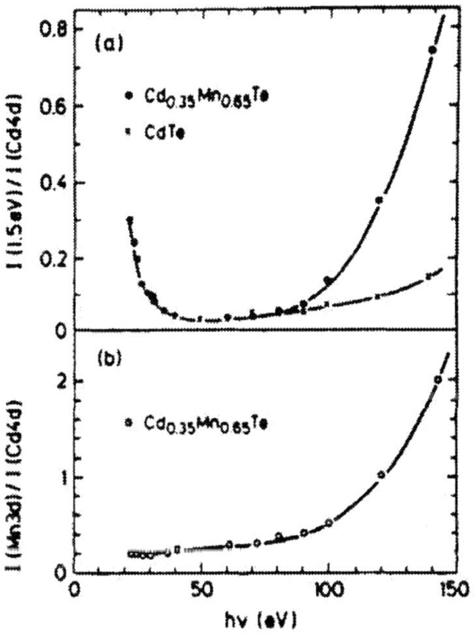

Fig. 5 Normalized amplitudes of the valence band at 1.5eV (a) normalized to the Cd 4d core emission, and the 3.5eV Mn feature (b) with the same normalization. The data shown have been corrected for the reduced Cd emission in the ternary. The change in cross section indicate 3d admixture at 1.5eV. From ref. [13].

EXTENDED X-RAY ABSORPTION FINE STRUCTURE SPECTROSCOPY

EXAFS spectroscopy involves the measurement of an X-ray absorption coefficient as a function of photon energy in the energy region 40-1000eV, above the threshold of a core excitation [37]-[38]. The final state of the photoexcitation involves the spherical wave of an ejected photoelectron interfering with the backscattered waves originating on the neighboring atoms. The interference gives rise to a sinusoidal variation of the absorption coefficient known as EXAFS. In the short-range, single-electron single-scattering theory, the EXAFS signal is represented in k space by a sum of individual waves due to the different types of neighboring atoms or different bonding distances [37]-[38]. Through a Fourier transform technique one can in principle obtain a radial distribution function of the scattering centers, provided that the scattering phase shifts are known. EXAFS is uniquely suited to analyze the local atomic structure of alloys, since it can focus on the immediate environment around each atomic species, by tuning the photon energy to match a specific elemental core edge, and is sensitive to the local coordination (generally 1-3 coordination shells) rather than to long range order, since EXAFS amplitude decreases with increasing distance as $1/r^2$ [37]-[38].

CdTe and MnTe crystallize, respectively in the zincblende and NiAs (hexagonal) structure, while the ternary alloys exhibit zincblende structure up to x=0.7. Conventional X-ray diffraction shows that the lattice constant of the ternary alloy follows Vegard's law, i.e. it varies linearly with x in the whole single phase region. Conventional diffraction methods, however, provide information only on the average anion-cation distance, and are not suitable to test the microscopic validity of a "virtual crystal" picture of the ternary alloys.

Balzarotti and co-workers [16]-[17] have recently presented a systematic EXAFS analysis of the structural properties of $Cd_{1-x}Mn_xTe$ alloys as a function of concentration. Together with the work of Mikkelsen and Boyce on the *nonmagnetic* $In_{1-x}Ga_xAs$ [30], the investigation of refs. [16]-[17] is a first attempt at characterizing the local microstructure of random ternary semiconductor alloys.

In standard EXAFS experiments single crystals are powdered, supported on thin membranes and exposed to the synchrotron radiation beam. The optical absorption may be measured in transmission or by detecting some by-product of the photoexcitation such as total or partial photoelectron yield, Auger or fluorescence emission [37]-[38]. In the case of $Cd_{1-x}Mn_xTe$, zincblende CdTe and hexagonal MnTe were used as known structural standards to derive and anion-cation (Cd-Te and Mn-Te) scatter functions (amplitude function and phase factor). The radial distribution functions obtained through Fourier analysis of the EXAFS signal above the Te core edges (L_{III} and L_I) show two distinct peaks in the first coordination shell surrounding Te. The x-dependence of the two peaks demonstrates that they correspond to two distinct Cd-Te and Mn-Te bonding distances; a result confirmed by analysis of the EXAFS signal above the Cd (L_{III} and L_I) and Mn (K) edges. The measured Cd-Te and Mn-Te nearest neighbor distances are given in fig. 6 (from ref. [16]-[17]) as a function of concentration x. The solid circles were obtained from a best fit of EXAFS data, open circles are the average values obtained from X-ray diffraction, and open squares are theoretical values calculated from a statistical model proposed in ref. [16] and [17]. The results of fig. 6 indicate that within experimental error, the Cd-Te distance is 2.80 Å at all compositions, and coincides with that observed in CdTe. The Mn-Te bond distance decreases with x from 2.76 to about 2.74 Å, to be compared with the Mn-Te distance in hexagonal MnTe that is 2.92 Å. It is therefore evident that the ternary alloys exhibit substantial local distortion of the zincblende structure.

DISCUSSION

The existence of a dominant Mn 3d feature some 3.4-3.5eV below E_v in the *DOS* of the telluride, selenide and sulphide series is in apparent disagreement with existing theoretical models that forecast a constant 3d ionization energy in the series [10],[32].

Empirical tight binding and LCAO [12] calculations that use the photoemission-determined 3d binding energies as input parameters have been somewhat successful in explaining the properties of these materials, although recent first principle calculations for $Cd_{1-x}Mn_xTe$ [22] show a Mn d band some 2.5eV below the valence band maximum. Also, theoretical results for the diluted impurity limit [24] suggest a 3d impurity binding energy constant relative to the vacuum level rather than to the valence band maximum in the alloy series. This would result from the antibonding nature of the deep gap level with respect to the impurity atom-host orbital combinations [24].

Optical absorption and reflections measurements of $Cd_{1-x}Mn_xTe$ (x>0.4), $Cd_{1-x}Mn_xSe$ (x>0.45), $Zn_{1-}Mn_xTe$ and $Zn_{1x}Mn_xSe$ reveal a transition in the vicinity of 2.2eV. This has been interpreted as an intra-atomic d-d* multiplet transition or as an interband transition from the top of the valence band to the unoccupied lowest energy state of the $3d^6$ multiplet [22],[35]. The lack of observable Zeeman splitting and the absence of associate photoconductivity [35] seems to rule out the second interpretation, and the first one is in disagreement with the photoemission results unless strong final state effects are present. Wei and Zunger [22] have recently proposed, on the basis of first principle calculations for $Cd_{1-x}Mn_xTe$, that the apparent discrepancy is removed if one takes into account

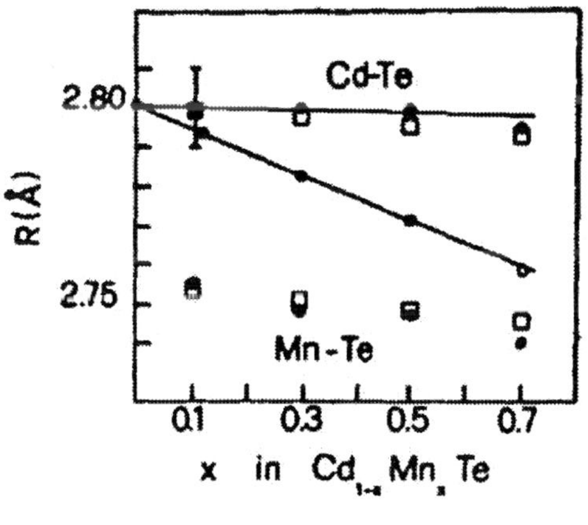

Fig. 6 Average anion-cation nearest-neighbor distances for $Cd_{1-x}Mn_xTe$ vs. concentration. Solid circles: EXAFS data. Open circles: from X-ray diffraction data. Open squares are values calculated from a theoretical model allowing statistical distortion of the anion sublattice. From ref. [15].

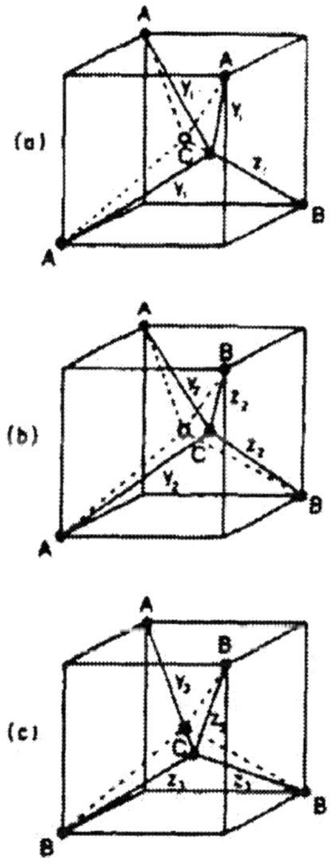

Fig. 7 All possible distortions of the anion sublattice in the tetrahedra with both types of cations. A and B are the cations, C the anion. The open circle and the dashed lines mark the position of the anion and of the bonds when the anion sublattice is undistorted. EXAFS data can be used to derive the average distortion. From ref. [15].

the different electron-electron correlation effects that determine the final state of the photoemission and intrashell optical excitation processes. The location of the final state hole in the localized 3d shell affects the Coulomb repulsion and the exchange splitting and the authors' estimate from a Slater's "transition state" construct that photoemission and intershell optical excitation would be consistent with a 3d ground state multiplet centered some 2.5eV below the top of the valence band (t_+^3 at about 2eV, e_+^2 at about 3.5eV). This suggestion is certainly stimulating, although the well known difficulty encountered by the local density functional formalism in treating highly localized electronic states as well as the poor representation of inter-atomic screening effects given by Slater's transition state construct should inspire some caution.

Whatever the actual magnitude of the electron-electron correlation effects might be, future experimental and theoretical work will have to focus on the apparently constant 3d binding energy in the sulphide, selenide and telluride series, on the observation of a similar binding energy for Fe^{+2} quartet final states [12]-[36] in Fe-based ternary semimagnetic semiconductors, on the compatibility of these findings with the diluted impurity limit [24], and on the composition dependence of the 3d lineshape.

The EXAFS derived experimental input is stimulating a number of theoretical efforts. In particular, Wei and Zunger [22] performed total energy calculations for $Cd_{0.50}Mn_{0.50}Te$ introducing an anion displacement parameter to take into account the local distortion. The equilibrium lattice parameter and the anion displacement calculated from total energy minimization yield values for the Cd-Te and Mn-Te distances in the alloy which are within 1% of the experimental EXAFS values. The total energy calculations point out the existence of an alloy-stabilized [22]-[23] zincblende phase of Mn-Te, with a bond length in agreement with that inferred from a linear extrapolation of the $Cd_{1-x}Mn_xTe$ data. The bond lengths in the ternary alloy remain close to those of the parent binary compounds (zincblende CdTe and the hypothetical zincblende MnTe) since the local distortion of the bond angles allows the system to maintain almost ideal tetrahedral bond lengths [22].

Balzarotti et al. [16]-[17] propose a statistical model to explain how the zincblende structure can accommodate two different cation-anion distances. In analogy with the results of ref. [39] and because of the dominant role of the nearest-neighbor interaction in the alloy they suggest that the cation sublattice remains undistorted, while the anion sublattice is distorted because of the tendency of Te to get closer to the Mn neighbors that to the Cd. The crystal would contain a statistical distribution of tetrahedra where each anion is coordinated to n Mn atoms and 4-n atoms. The three corresponding possible local distortions of the zincblende structure are shown in fig. 7 where A, B and C label the Cd, Mn and Te atoms, respectively. The probability of finding a tetrahedron with Mn atoms is given by a binomial Bernoulli distribution. The Cd-Te and Mn-Te bond distances corresponding to each of the tetrahedra in fig. 7 were derived by minimizing the strain energy needed to change bond lengths through the use of bond-stretching constants for binary compounds taken from the literature [16]-[17]. From the distances in fig. 7 an average over the tetrahedra distribution was performed and the result is shown with open squares in fig. 6. The agreement with EXAFS-derived experimental data is remarkable, and suggests that $Cd_{1-x}Mn_xTe$ indeed exhibits a bimodal distribution of near-neighbor distances with distortion localized mostly within the anion sublattice.

The uncertainty on the general applicability of this simple model has been increased recently by theoretical work by Podgorney et al. [20] who found large distortion of the cation sublattice to be consistent with ternary EXAFS data, at least in the case of InGaAs and CdZnTe. It seems to us of paramount importance to extend the local structural studies to include other semimagnetic alloy systems. EXAFS can clearly provide detailed experimental information of the interatomic distances, while at this stage the existing theoretical models could certainly use a measure of experimental verification.

Synchrotron radiation may also provide information on the density of unoccupied electron states above E_F. These should reflect the superposition of anion p-cation s antibonding states and of the Mn 3d-derived partial density of states. For example, calculations for $Cd_{0.5}Mn_{0.5}Te$ forecast a large 3d empty DOS feature within 2eV from Ev [22]; semi-empirical calculations for $Cd_{1-x}Mn_xTe$ (x=0.3 and 0.7) [19] suggest, instead, a sharp 3d DOS features some 3.5eV above the top of the valence band. In fact, there is little direct experimental information available on this region of the electronic structure, which is also the region where existing theoretical methods become less reliable. To test the theoretical models, and to investigate directly the final states involved in optical excitation processes [22],[35], synchrotron radiation X-ray absorption spectroscopy (XAS) from core levels could be applied in parallel with Bremsstrahlung Isochromat Spectroscopy (BIS) [40]. Both techniques have been already applied to a variety of materials such as metals, intermetallic alloys and silicon-metal compounds [40].

XAS at the Mn $L_{2,3}$ edges, for example, should be dominated by transitions to empty d-states due to dipole selection rules. The BIS spectra give information on the total density of states, with cross sections that are essentially time-reversed photoemission cross sections. Correspondingly, recent calculations suggest that the BIS should offer considerable sensitivity to the d contribution [40]. Comparison of XAS and BIS results as a function of composition x should allow the identification of the different contributions to the DOS above E_F.

We emphasize that XAS should give a local projection of the d-DOS weighted by the matrix element in the presence of a core hole. BIS instead, gives a picture of the empty states in the absence of any core hole. In several d-sp alloys such as $PdAl_x$ the similarity of BIS and XAS results [40] indicates that interatomic screening is effective enough to suppress the complication arising from the core hole potential. In the ternary semimagnetic semiconductor series the 3d impurity appears in a varying sp matrix (from the semimetallic HgTe to the wide gap CdS) and at different concentration. This seems an ideal situation to probe interatomic screening and investigate intershell correlation effects [22] while testing the existing band structure calculations.

ACKNOWLEDGEMENTS

The work at the University of Minnesota was performed in collaboration with A. Wall, S. Chang, P. Philip (University of Minnesota), F. Pool, R. Reifenberger and J.K. Furdyna (Purdue University), and supported in part by the Graduate School of the University of Minnesota and by the Microelctronics and Information Science Center of Minnesota. The assistance of the staff of the University of Wisconsin Synchrotron Radiation Center (supported by NSF) is gratefully acknowledged. We are in debt to a number of colleagues who provided us with results of their work prior to publication: C. Balzarotti, M. De Crescenzi, H. Ehrenreich, K. Hass, L. Ley, A. Mycielski, C. Mailhiot, A. Ramdas and A. Zunger.

REFERENCES

1. J. K. Furdyna, J. Appl. Phys. 53, 7637 (1982).
2. N. B. Brandt and V. V. Moshchalkov, Adv. in Phys. 33, 193 (1984).
3. R. R. Galazka and J. Kossut, Narrow Gap Semiconductors: Physics and Applications, in Lecture Notes in Physics Series (Springer, Berlin, 1980), V. 133.

4. L. A. Kolodziejski, R. L. Gunshor, N. Otsuka, X. C. Zhang, S. K. Chang, and A. V. Nurmikko, Appl. Phys. Lett. 47, 882 (1985).
5. R. N. Bicknell, N. C. Giles-Taylor, D. K. Blanks, R. W. Yanka, E. L. Buckland, and J. F. Schetzina, Mat. Res. Soc. Symp. Proc. 37, 35 (1985).
6. G. Y. Wu, D. L. Smith, C. Mailhiot, and T. C. McGill, Appl. Phys. Lett. (in press).
7. C. Mailhiot and D. L. Smith, Phys. Rev. B 33, 8360 (1986).
8. A. Franciosi, C. Caprile, and R. Reifenberger, Phys. Rev. B 31, 8061 (1985); A. Franciosi, R. Reifenberger, and J. K. Furdyna, J. Vac. Sci. Technol. A 3, 124 (1985).
9. A. Franciosi, S. Chang, C. Caprile, R. Reifenberger, and U. Debska, J. Vac. Sci. Technol. A 3, 926 (1985).
10. A. Franciosi, S. Chang, R. Reifenberger, U. Debska, and R. Riedel, Phys. Rev. B 32, 6682 (1985).
11. A. Wall, C. Caprile, A. Franciosi, R. Reifenberger, and U. Debska, J. Vac. Sci. Technol. A 4, 818 (1986).
12. A. Wall, C. Caprile, A. Franciosi, M. Vaziri, R. Reifenberger, and J. K. Furdyna, J. Vac. Sci. Technol. A 4, 2010 (1986).
13. M. Taniguchi, L. Ley, R. L. Johnson, J. Ghijsen, and M. Cardona, Phys. Rev. B 33, 1206 (1986).
14. M. Taniguchi and L. Ley (private communication).
15. A. Balzarotti, A. Kisiel, N. Motta, M. Zimnal-Starnawska, M. T. Czyzyk, and M. Podgorny, Prog. Cryst. Growth and Charact. 10, 55 (1985).
16. A. Balzarotti, M. Czyzyk, A. Kisiel, N. Motta, M. Pdgorny, and M. Zimnal-Starnawska, Phys. Rev. B 30, 2295 (1985).
17. A. Balzarotti, N. Motta, A. Kisiel, M. Zimnal-Starnawska, M. T. Czyzyk, and M. Podgorny, Phys. Rev. B 31, 7526 (1985).
18. B. E. Larson, K. C. Hass, H. Ehrenreich, and A. E. Carlsson, Sol. State Commun. 56, 347 (1985).
19. H. Ehrenreich, K. C. Hass, N. F. Johnson, B. E. Larson, and R. J. Lampert, in Proc. 18th Int. Conf. on the Phys. of Semiconductors, O. Engstrom, ed., World Scientific Publ. Co. (in press).
20. M. Podgorny, M. T. Czyzyk, A. Balzarotti, P. Letardi, N. Motta, A. Kisiel, and M. Zimnal-Starnawaska, Sol. State Commun. 55, 413 (1985).
21. M. T. Czyzyk, M. Podgorny, A. Balzarotti, P. Letardi, N. Motta, A. Kisiel, and M. Zimnan-Starnawaska, Z. Phys. B 62, 153 (1986).
22. S.-H. Wei and A. Zunger, Phys. Rev. B (in press).
23. S.-H. Wei and A. Zunger, Appl. Phys. Lett. (in press).
24. M. J. Caldas, A. Fazzio and A. Zunger, Appl. Phys. Lett. 65, 671 (1984).
25. A. Mycielski, P. Dzwonskowski, B. Kowalski, B. A. Orlowski, M. Dobrowolska, M. Arciszewska, W. Dobrowolski, and J. M. Beranowski, J. Phys. C. 19, 3605 (1986).
26. W. Dobrowolski, K. Dybko, A. Mycilski, J. Mycielski, J. Wrobel, S. Piechota, M. Palczewska, H. Szymczak, and Z. Wilamowski, in Proc. 18th Int. Conf. on the Phys. of Semiconductors, O. Engstrom, ed., World Scientific Publ. Co. (in press).
27. G. Margaritondo and A. Franciosi, Ann. Rev. Mater. Sci. 14, 67 (1984).
28. G. Margaritondo and J. H. Weaver, in Methods of Experimental Physics: Surfaces, M. G. Lagally and R. L. Park, eds., Academic Press, New York, (1983).
29. A. Wall, A. Franciosi, D. W. Niles, C. Quaresima, M. Capozi, P. Perfetti, and R. Reifenberger, Phys. Rev. B (in press).
30. B. Sonntag, R. Haensel, and C. Kunz, Sol. State Commun. 7, 597 (1969).
31. B. A. Orlowski, Phys. Status Solidi B 95, K31 (1979).
32. C. Webb, M. Kaminska, M. Lichtensteiger, and J. Lagowski, Sol. State Commun. 40, 609 (1981).
33. P. Oelhafen, M. P. Vecchi, J. L. Freeouf, and V. L. Moruzzi, Sol. State Commun. 50, 749 (1984).
34. A. Wall and A. Franciosi (unpublished).

35. Y. R. Lee, A. K. Ramdas, and R. L. Aggarwal, Phys. Rev. B 33, 7383 (1986)
36. A. Wall, S. Chang, C. Caprile, A. Franciosi, F. Pool and R. Reifenberger, J. Vac. Sci. Technol. A (in press).
37. B. K. Teo, EXAFS: Basic Principles and Data Analysis, Springer Verlag, New York (1986).
38. P. A. Lee, P. H. Citrin, P. Eisenberger, and B. M. Kinkaid, Rev. Mod. Phys. 53, 769 (1981).
39. J. C. Mikkelsen and J. B. Boyce, Phys. Rev. Lett. 49, 1412 (1982); and Phys. Rev. B 28, 7130 (1983).
40. See for example: D. D. Sarma, F. U. Hillebrecht, M. Campagna, C. Carbone, J. Nogami, I. Lindau, T. W. Barbee, L. Braicovich, I. Abbati, and B. De Michelis, Z. Phys. B 59, 159 (1985), and G. Rossi, P. Roubin, D. Chandesris, and J. Lecante, Surf. Sci. (in press).

ELECTRONIC THEORY OF Mn - ALLOYED DILUTED MAGNETIC SEMICONDUCTORS

H. EHRENREICH, K.C. HASS*, B.E. LARSON AND N.F. JOHNSON
Harvard University, Division of Applied Sciences and Department of Physics, Cambridge, Massachusetts 02138
* Also at MIT, Dept. of Physics, Cambridge, MA 02139

ABSTRACT

Recent calculations of the electronic structure and magnetic interactions in Mn - alloyed II-VI diluted magnetic semiconductors (DMS) are summarized. Detailed band structure results are obtained using an empirical tight-binding, coherent potential approximation approach with input from experiment and local spin density band calculations. The dominant magnetic interactions in these systems result from hybridization between spin-split Mn *d* states and *sp* valence bands. Superexchange between Mn moments is well described by a simple three-level model which yields accurate Mn - Mn exchange constants for a variety of II-VI DMS as well as the rocksalt insulators MnO and α-MnS.

INTRODUCTION

This paper summarizes recent theoretical contributions to the electronic structure and magnetic interactions in Mn - alloyed II-VI diluted magnetic semiconductors (DMS) (e.g. $Cd_{1-x}Mn_xTe$). These materials have generated considerable fundamental and technological interest because of their unusual magnetic properties (e.g. nonmetallic spin glass phase) and because of their novel interplay between semiconductor physics and magnetism (e.g. enhanced band edge Zeeman splittings) [1]. The detailed microscopic picture reviewed here, which is in good agreement with experiment, is based on a variety of theoretical techniques developed previously for isoelectronic semiconducting alloys and transition metal oxides. Important issues resolved by this work include the effects of Mn *d* electrons on the electronic structure and the origin of Mn - *sp* band and Mn - Mn exchange interactions.

A brief overview of the developments discussed here and their relationship to previous theoretical models for DMS is presented in Fig. 1. The upper two-thirds of this figure, which will be described in detail, begins with the construction of a detailed electronic structure model based on a realistic microscopic Hamiltonian. Calculations of magnetic interactions within this framework provide fundamental support for the phenomenological spin Hamiltonians H_H and H_K used extensively in early theoretical studies of DMS [1]. A modified $\boldsymbol{k}\cdot\boldsymbol{p}$ approach, which combines the Kondo-like interaction H_K with a simple $\boldsymbol{k}\cdot\boldsymbol{p}$ virtual crystal approximation (VCA) treatment of *sp* band edge states, accounts well for the unique magneto-optical and magneto-transport properties of these materials [2]. The spin 5/2 Heisenberg Hamiltonian H_H similarly provides an excellent description of the magnetic properties and phase diagrams of DMS [3].

ELECTRONIC STRUCTURE

The description of the electronic structure of $II_{1-x}Mn_xVI$ alloys, in principle, is a complicated many body problem. Local magnetic moments in these systems result

Mat. Res. Soc. Symp. Proc. Vol. 89. ©1987 Materials Research Society

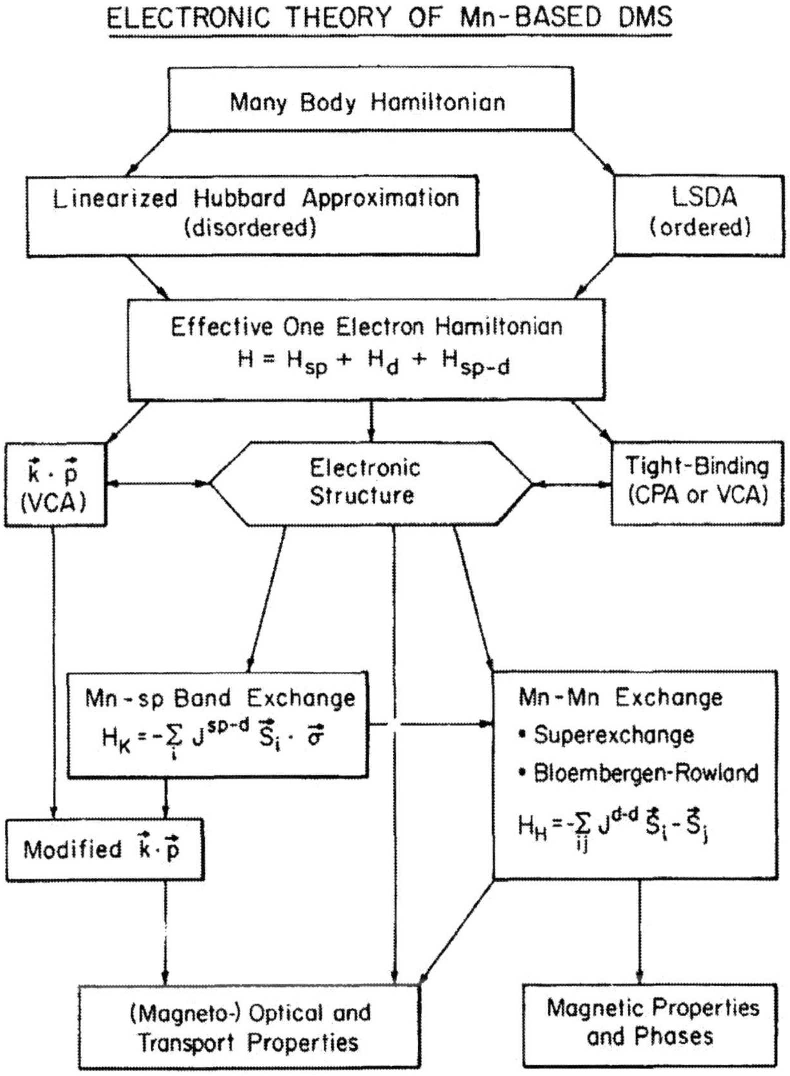

Fig. 1 — Schematic overview of the theoretical developments discussed in the text.

from strong exchange-correlation interactions among the Mn d electrons which strongly favor a spin-polarized d^5 configuration on each Mn site (Hund's rule). We treat these effects by considering the two commonly used approximations shown in the second line of Fig. 1. The linearized Hubbard approximation, on the left, assumes that a Mn d electron of spin σ in orbital α has the energy $\epsilon_d + U_{eff} < n_{\alpha,-\sigma} >$ where $< n_{\alpha,-\sigma} >$ is one if the opposite spin state is occupied and zero if it is unoccupied [4]. This approximation is particularly useful for the alloy tight-binding calculations described below because it is applicable to disordered as well as ordered systems. The second approach, the local spin density approximation (LSDA) [5], involves no adjustable parameters but is easily implemented only in ordered systems. *Ab initio* spin-polarized band calculations based on the LSDA have successfully described the properties of a large number of magnetic compounds [6]. Such calculations are important in the present context for providing band structure information for hypothetical or metastable ordered configurations, particularly tetrahedral MnVI crystals [7-10].

Each of these approximations leads to an effective one electron Hamiltonian which is written schematically in Fig. 1 as the sum of three terms. The first term, H_{sp}, describes itinerant semiconducting *sp* bands. The similar behavior of these bands in $II_{1-x}Mn_xVI$ DMS and in isoelectronic II-VI alloys explains the similar semiconducting properties of these materials in the absence of a magnetic field. The second term, H_d, describes the local aspects of Mn *d* states, including the spin splitting U_{eff} at each site. The final term, H_{sp-d}, describes the hybridization between the two sets of electrons. This effect will be seen to be small compared to U_{eff}, but to represent the dominant source of magnetic interactions in these materials.

We have constructed a realistic electronic structure model based on this Hamiltonian by combining an empirical tight-binding (ETB) [11] description of the various terms with a coherent potential approximation (CPA) [12] treatment of disorder effects (fourth line of Fig. 1). The basis consists of two *s*, six *p* and ten *d* orbitals per cation, and two *s* and six *p* orbitals per anion. We first discuss the unperturbed *sp* band structure which results from H_{sp} acting alone. The effects of Mn *d* electrons will then be examined using information obtained from LSDA calculations and experimental data.

sp Bands

The first detailed calculations of the *sp* band structure of a DMS were performed by Hass and Ehrenreich for $Hg_{1-x}Mn_xTe$ [13]. This work was based on the ETB-CPA formalism developed previously for $Hg_{1-x}Cd_xTe$, which provides a sophisticated means of interpolating between tight-binding parametrizations for the limiting crystals [14]. For HgTe and CdTe in [14], the diagonal tight-binding parameters, which play the most important role, were fit to atomic levels [15], as suggested by Harrison [11]. Off-diagonal parameters (up to second neighbors) and a uniform shift of the diagonal parameters relative to the *sp* bands were chosen to provide the best overall fit to the known HgTe and CdTe band structures. The extension to $Hg_{1-x}Mn_xTe$ was accomplished by considering a hypothetical zincblende (zb) $x=1$ limiting crystal. Diagonal *s* and *p* tight binding parameters for Mn were again chosen from atomic levels on the same absolute energy scale. Off-diagonal parameters for zb-MnTe were assumed to be the same as those of CdTe. This approximation is reasonable in view of the small difference in bond lengths between the two crystals ($\sim$ 2 %) and the fact that off-diagonal parameters in zincblende compounds are relatively insensitive to the chemical nature of the constitutents [11].

Figure 2 shows the resulting placement of the *sp* ETB bands for HgTe, CdTe and zb-MnTe relative to the diagonal *sp* tight-binding parameters assumed in the calculations. (The lower energy Te 5*s* level and associated split-off valence band included in the calculations are not shown here.) The zero of energy corresponds to the valence band maxima which are assumed to be the same in the three crystals because of the large Te 5*p* component of the Γ_8 states. The increase in band gap from HgTe to CdTe results from the much higher cation *s* level in CdTe, which contributes significantly to the conduction band edge Γ_6 state. The still higher *s* level in Mn results in an even larger gap in zb-MnTe. The predicted gap value here of 3.2 eV is in good agreement with the extrapolated value obtained from low temperature measurements on $II_{1-x}Mn_xTe$ alloys [16]. The rise of the cation *s* levels also accounts for the slight decrease in the width of the upper valence region in the series HgTe→CdTe→MnTe.

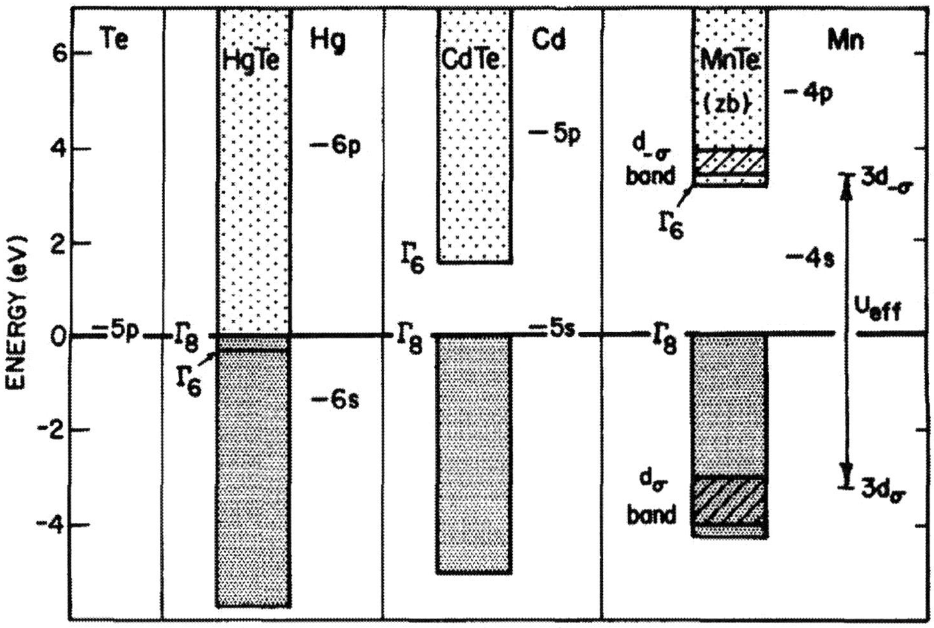

Fig. 2. — Limits of *sp* upper valence and lower conduction bands (darker and lighter boxes, respectively) for HgTe, CdTe and hypothetical zincblende (zb) MnTe relative to atomically-derived *sp* diagonal tight-binding parameters. Details of alignment are discussed in text. The Mn *d* levels are assumed to be spin split by an amount U_{eff} at each site with the occupied states having a spin σ. The resulting regions of large *d* state densities in zb-MnTe are shown as cross-hatched regions labeled *d* "bands."

For many purposes, the behavior of the *sp* bands at intermediate alloy compositions can be reasonably described by a simple VCA averaging scheme. The VCA predicts a nearly linear x dependence of the band features shown in Fig. 2. The additional effects of compositional disorder, included in the CPA, result primarily from the differences between cation s levels in the alloy constitutents. The 1.4 eV difference between Hg and Cd s levels, for example, is known to affect the bowing of band gaps and electron mobilities in $Hg_{1-x}Cd_xTe$, and to give rise to a nearly split valence band density of states near -5 eV [14]. Qualitatively similar behavior is predicted for the *sp* bands of $Hg_{1-x}Mn_xTe$ and $Cd_{1-x}Mn_xTe$ with the most pronounced disorder effects occuring in the former system where the Hg and Mn s levels differ by $\sim$ 3 eV. A consequence of this enhanced disorder is that, for a given band gap, the limiting electron mobility due to alloy scattering in $Hg_{1-x}Mn_xTe$ is estimated to be about a factor of three smaller than in $Hg_{1-x}Cd_xTe$, despite the smaller x value required to produce that gap [13].

Effects of Mn d Levels

The Hubbard-like treatment of Mn d levels described above is illustrated on the right hand side of Fig. 2. A simple perturbative analysis by Hass and Ehrenreich [13] demonstrated that the principal effect of $sp-d$ hybridization is to broaden the spin-split d states into bands, denoted by cross-hatching in Fig. 2. The label σ ($-\sigma$) on the occupied (unoccupied) d band indicates that the associated extended wavefunction for a state of given spin has large amplitude only on those Mn sites whose net spin is in the same (opposite) direction. This basic picture holds regardless of whether the Mn spins are ordered antiferromagnetically, as would be expected for zb-MnTe at low temperatures, or are randomly oriented, as would occur at either higher temperatures or in the spin glass phase of $II_{1-x}Mn_xVI$ alloys.

We obtain ETB parameters describing the location of Mn d levels and the magnitude of $sp-d$ hybridization by fitting to the results of augmented spherical wave (ASW) - LSDA calculations for antiferromagnetic (AF) zb-MnTe [7,9]. The resulting hybridization parameters are in good agreement with previous estimates [13] obtained from Harrison's "universal" tight-binding scheme [11]. The location of Mn d levels relative to the sp bands is less reliable, however, since the LSDA is well known to underestimate quasiparticle excitation energies [17]. Both the ASW results of [7] and [9] and more recent linearized augmented plane wave calculations [10] for AF zb-MnTe place the occupied d states degenerate with the sp valence band near -2 eV and the unoccupied d states in the sp band gap. In our final ETB parametrization shown in Fig. 2 the $3d_\sigma$ level has been shifted to -3.4 eV to agree with the (x independent) location observed in photoemission experiments on $Cd_{1-x}Mn_xTe$ [18]. The $3d_{-\sigma}$ level has been similarly shifted to 3.6 eV to place it about 4.5 eV above the region of the L point of the upper valence band. This placement is consistent with the anomalous structure seen in ellipsometry [19] and reflectivity [20] data for $Cd_{1-x}Mn_xTe$ at 4.5 eV. The magnitudes and signs of the above d level adjustments are consistent with the expected corrections to LSDA eigenvalues [9,17].

The inclusion of d states on the same footing as sp bands in the ETB-CPA requires the consideration of magnetic as well as compositional disorder [21,22]. We model this effect for $II_{1-x}Mn_xVI$ alloys by assuming that the occupied d states have either spin up ($\uparrow$) or spin down ($\downarrow$) at random. The alloy is thus treated as the pseudo-ternary system $II_{1-x}(Mn\uparrow)_{xy}(Mn\downarrow)_{x(1-y)}VI$. The constraint $y=0.5$ is imposed in the ground state calculations described below to ensure zero net magnetization. Variations in the value of y may also be considered to simulate the behavior in an external magnetic field.

Figure 3 shows the calculated CPA total and projected d (shaded) densities of states for $Cd_{1-x}(Mn\uparrow)_{x/2}(Mn\downarrow)_{x/2}Te$ with $x = 0.0$, 0.3 and 0.6. The similar shape of the $x=0.3$ and $x=0.6$ results indicates that the d contribution to the density of states is determined primarily by the local environment of the Mn and is not strongly affected by the presence of magnetic disorder. Compositional disorder resulting from the difference between Cd and Mn s levels, on the other hand, does produce a slight change in the sp contribution to the density of states in the lower valence region shown in the figure. States contributing to the -4.5 eV peak in CdTe are composed primarily of Cd s and Te p orbitals. Upon alloying with Mn, this peak shifts to higher energy and decreases in magnitude while new structure appears in the alloy sp bands near -3.9 eV which is superimposed in Fig. 3 on the much larger Mn d contribution at this energy.

The splitting of the occupied Mn d peak in Fig. 3 is a consequence of $sp-d$ hybridization. The lower energy states, which have t_{2g} symmetry, hybridize more strongly in the tetrahedral environment than those of e_g symmetry and are repelled by the Te p - like upper valence region. The interaction in turn produces an appreciable d admixture throughout the upper valence region, including the valence band maximum. The resulting change in the structure of the density of states near -1 eV has been observed directly in photoemission experiments [18]. The small shift of the valence band maximum is overestimated in the CPA by the neglect of short ranged AF correlations. In contrast to the valence band behavior, the conduction bands in Fig. 3

are relatively unaffected by $sp-d$ hybridization except for a slight broadening of the unoccupied d peak. The d admixture in the conduction band minimum, in fact, is strictly zero by symmetry.

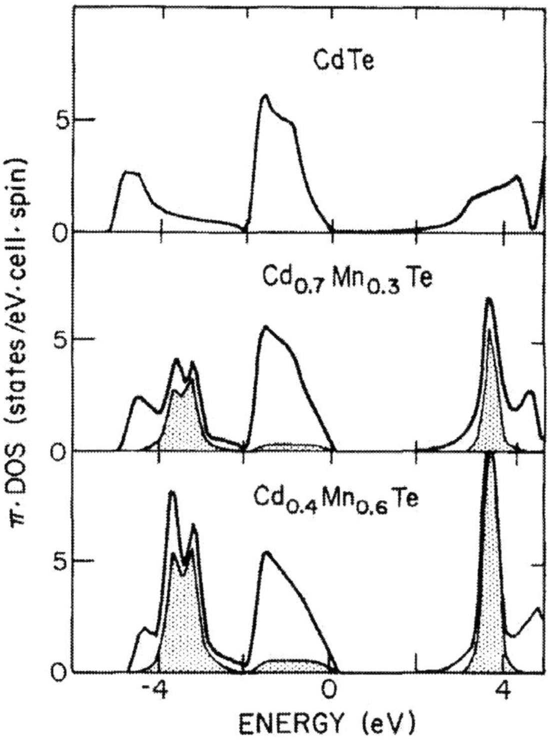

Fig. 3. — Calculated CPA total and projected (shaded) densities of states for $Cd_{1-x}(Mn\uparrow)_{x/2}(Mn\downarrow)_{x/2}Te$ with $x = 0.0$, 0.3 and 0.6.

Much more detailed band structure information can be obtained from an analysis of $\boldsymbol{k}$ dependent properties in the ETB-CPA. Such calculations for $Cd_{1-x}Mn_xTe$ indicate that magnetic disorder causes an appreciable damping of states throughout the region of the upper valence peak in Fig. 3. The damping shows up in the calculations as a broadening of $\boldsymbol{k}$ dependent spectral densities which in the absence of disorder would reduce to series of δ - functions at the band energies. The calculated broadening and peak location of the upper valence band spectral density at L is qualitatively consistent with the anomalous behavior of the E_1 optical transition observed in recent ellipsometry experiments [19]. A more detailed analysis of spectral densities will be particularly important if more extensive angle-resolved photoemission data becomes available.

MAGNETIC INTERACTIONS

The preceding electronic structure model is now used to to examine the principal magnetic interactions in $II_{1-x}Mn_xVI$ DMS along the lines indicated in Fig. 1.

Mn — sp Band Exchange

The Kondo-like interaction H_K between Mn local moments and the spins of band edge electrons and holes is the larger of the two exchange interactions considered here. The relevant exchange constants for most $II_{1-x}Mn_xVI$ DMS have been determined from experiment and found to be $N_0\alpha \equiv J^c_{sp-d} \approx 0.2$ eV for the conduction band edge and $N_0\beta \equiv J^v_{sp-d} \approx -1.0$ eV for the valence band edge [23]. The qualitative difference between these values can be understood in terms of the different effects of $sp-d$

hybridization on the two band edges [8,9,24]. For the conduction band edge, where no d admixture is allowed by symmetry, $N_0\alpha$ is a purely potential exchange interaction; this is a relatively weak effect in these systems, and always ferromagnetic. By contrast, the appreciable d admixture in the valence band edge gives rise, through the Schrieffer-Wolff transformation [25], to the much larger, antiferromagnetic exchange constant $N_0\beta$. Substitution of the above ETB parameters for $Cd_{1-x}Mn_xTe$ in the appropriate Schrieffer-Wolff expression yields an estimate of $N_0\beta$ within 20 % of experiment. Accurate values of both $N_0\alpha$ and $N_0\beta$ in $II_{1-x}Mn_xVI$ DMS have also been extracted from *ab initio* LSDA calculations of band edge spin splittings in hypothetical *ferromagnetic* compounds ($x = 1.0$ and 0.5) [8-10].

Mn — Mn Exchange

The weaker Mn - Mn exchange interaction H_H is known to be short-ranged and antiferromagnetic in $II_{1-x}Mn_xVI$ alloys [3]. Experimentally derived nearest neighbor exchange constants J_{nn}^{dd} are typically of the order of -10 K (-9 x 10^{-4} eV), assuming a total interaction between two spins of $-2J^{dd}(R_{ij})S_i \cdot S_j$ [26]. The hierarchy of Mn - Mn exchange mechanisms has recently been examined quantitatively by Larson, *et. al.* starting from a simplified version (VCA *sp* bands, single $sp-d$ band hopping parameter) of the effective one electron Hamiltonian discussed above [7,9]. The total Mn - Mn interaction in this work is calculated as a fourth order perturbation in H_{sp} with the contributions of various mechanisms classified by intermediate states. Processes which involve the creation of two holes in the *sp* valence band are associated with superexchange [27]. Processes which involve the creation of one electron and one hole are associated with the Bloembergen-Rowland mechanism [28] (RKKY in metals). Two-electron processes are also considered. The approach is superior to that of previous calculations of $J^{dd}(R_{ij})$ in DMS which considered only a second order perturbative treatment of H_K [29]. The retention of only spin degrees of freedom in H_K neglects many of the effects of $sp-d$ hybridization which play an important role in determining $J^{dd}(R_{ij})$.

Numerical calculations based on the above approach indicate that AF superexchange accounts for about 95 % of J_{nn}^{dd} in wide gap Cd and Zn DMS and remains dominant out to about fourth nearest neighbors [7,9]. The Bloembergen-Rowland mechanism makes an increasingly important contribution in narrower gap Hg systems (particularly for more distant neighbors) but is unlikely to ever become dominant. Exchange processes involving holes are more effective than those involving electrons because of the larger density of states and the larger degree of Mn d - sp band hybridization in the *sp* valence band. The net J_{nn}^{dd} values calculated by Larson, *et. al.* for $Cd_{1-x}Mn_xTe$ are -12 K in [7] and -8 K in [9] using a more refined set of parameters. The excellent agreement with experiment (-6.3 K [26]) is a clear indication of the validity of the resulting physical picture and the underlying electronic structure model.

Three—Level Model of Superexchange

We have recently developed a more physically transparent model of superexchange in Mn-based systems which involves only the essential features of the electronic structure [9]. The model contains four parameters: the occupied d level energy ϵ_d, the unoccupied d level energy $\epsilon_d + U_{eff}$, the energy of the valence band maximum E_v, and a Mn d - anion p hybridization parameter V_{pd}. The analytic expression

$$J^{dd}(R) = -2V_{pd}^4 \, [(\epsilon_d+U_{eff}-E_v)^{-2}U_{eff}^{-1} + (\epsilon_d+U_{eff}-E_v)^{-3}] \, f(R) \qquad (1)$$

can be written down immediately in analogy with the full fourth order perturbation theory of [7]. The remarkable feature of this formula is that the function $f(R)$, which describes the dependence on the Mn - Mn separation R, is largely independent of the details of the electronic structure within a closely related class of materials.

For $II_{1-x}Mn_xVI$ DMS, with cubic lattice constant a, $f(R)$ is well approximated out to about fourth nearest neighbors by the Gaussian decay 0.2 exp $(-4.89\ R^2/a^2)$ [7]. Trends in exchange constants upon variations in the group II or group VI element can thus be understood directly in terms of the corresponding trends in electronic structure parameters. By further supplementing this model with the Schrieffer-Wolff expression

$$N_0\beta = -2V_{pd}^2 \, [(\epsilon_d+U_{eff}-E_v)^{-1} + (E_v-\epsilon_d)^{-1}] \qquad (2)$$

one can obtain a consistent description of *both* Mn - *sp* band and Mn -Mn exchange in these materials.

The utility of these expressions is first illustrated by considering those wide gap $II_{1-x}Mn_xVI$ DMS for which reliable experimental values of J_{nn}^{dd} have been reported [26]. The composition x=0.1 is chosen for clarity although none of the parameters or results depend strongly on x. The electronic structure parameters used in the model are given in the left-hand columns in Table I. The E_v - ϵ_d values for the two Cd alloys are taken from photoemission experiments [18]. The value U_{eff} = 7.0 eV in $Cd_{1-x}Mn_xTe$ is chosen from Fig. 2 and the increase in U_{eff} in $Cd_{1-x}Mn_xSe$ is estimated from the smaller dielectric constant in this system. Both E_v - ϵ_d and U_{eff} are assumed to be the same in the corresponding Zn alloys since these quantities are relatively insensitive to the cation species. The values of V_{pd} are determined from Eq. (2) and the experimental $N_0\beta$ values in the table [23].

Table I — Electronic structure parameters used as input in three-level model and associated experimental (exp.) and theoretical (th.) exchange constants for a variety of Mn-based DMS and rocksalt insulators.

	$E_v-\epsilon_d$	U_{eff}	V_{pd}	$N_0\beta$ (exp.)	J_{nn}^{dd} (th.)	J_{nn}^{dd} (exp.)
$Cd_{0.9}Mn_{0.1}Te$	3.4 eV	7.0 eV	0.88 eV	−0.88 eV	−8.0 K	−6.3 K
$Cd_{0.9}Mn_{0.1}Se$	3.4	7.6	1.01	−1.11	−9.0	−7.9
$Zn_{0.9}Mn_{0.1}Te$	3.4	7.0	0.99	−1.12	−13.0	−8.8
$Zn_{0.9}Mn_{0.1}Se$	3.4	7.6	1.06	−1.22	−11.0	−13
α-MnS	0.5	8.0	1.08	-	−1.7	−4.4
MnO	−2.5	9.0	1.84	-	−5.0	−7.2

The resulting theoretical values of J_{nn}^{dd} obtained from Eq. (1) are compared to the corresponding experimental values [26] in the right-hand columns of the table. The simple model not only yields absolute values of the exchange constants within 50 % of experiment for each DMS considered, but it also accounts for the observed trends towards larger J_{nn}^{dd} in $Cd_{1-x}Mn_xSe$ compared to $Cd_{1-x}Mn_xTe$ and when Cd is replaced by Zn. Both of these trends result from an increase in V_{pd} which we attribute to a decrease in the Mn - anion bond length. The failure of the theory to account for the experimentally observed increase in J_{nn}^{dd} in $Zn_{1-x}Mn_xSe$ compared to $Zn_{1-x}Mn_xTe$ may be an artifact of our assumed constancy of E_v - ϵ_d and U_{eff} in Cd and Zn alloys.

The three-level model of superexchange outlined here is expected to have more general applicability to a variety of Mn - based non-metals. The function $f(R)$ will differ somewhat depending on the symmetry and *sp* bandwidths in a given class of materials but within each class $f(R)$ is expected to remain largely independent of the actual atomic constituents [9]. We illustrate this point by considering the Mn - based rocksalt insulators MnO and α-MnS. By coincidence, the appropriate $f(R)$ for Mn nearest neighbors in these materials (~ 0.017) turns out to be the same as in $II_{1-x}Mn_xVI$ DMS, despite the difference in symmetry and smaller *sp* bandwidths in the rocksalt systems [9]. The values of $E_v - \epsilon_d$, U_{eff} and V_{pd} for MnO and α-MnS given in Table I were obtained in [9] from a variety of theoretical ingredients including LSDA calculations [30] and tight-binding "scaling laws" [11]. The resulting theoretical values of J_{nn}^{dd} are again in reasonable agreement with experiment [31]. More importantly, the increase in J_{nn}^{dd} from α-MnS to MnO is well accounted for in the model and related directly through Eq. (1) to changes in the corresponding electronic structure parameters.

ACKNOWLEDGEMENTS

We are grateful to A. E. Carlsson and R. J. Lempert for their collaboration on some of the calculations reviewed here [7-9,21]. This work was supported by the Joint Services Electronics Program (N00014-84-K-0465), the Defense Advanced Research Projects Agency (through ONR Contract N00014-86-K-003) and the National Science Foundation (DMR 85-14638).

REFERENCES

1. N.B. Brandt and V.V. Moschalkov, Adv. in Phys. **33**, 193 (1984); J.K. Furdyna, J. Appl. Phys. **53**, 637 (1982) and references therein.
2. J. Kossut, Phys. Stat. Sol. (b) **78**, 537 (1976).
3. S.B. Oseroff and P.H. Keesom, to be published in *Semiconductors and Semimetals*, ed. by R.K. Willardson and A.C. Beer (Academic, New York).
4. J. Hubbard, Proc. Roy. Soc. (London) A **281** 401 (1964).
5. U. Von Barth and L. Hedin, J. Phys. C **5**, 1629 (1972).
6. A.R. Williams and U. Von Barth, in *The Inhomogeneous Electron Gas*, ed. by S. Lundqvist and N.H. March (Plenum, New York, 1983).
7. B.E. Larson, K.C. Hass, H. Ehrenreich and A.E. Carlsson, Solid State Commun. **56**, 347 (1985).
8. K.C. Hass, B.E. Larson, H. Ehrenreich and A.E. Carlsson, J. Mag. Magn. Mat. **54—57**, 1283 (1986).
9. B.E. Larson, K.C. Hass, H. Ehrenreich and A.E. Carlsson (unpublished).
10. S.H. Wei and A. Zunger, Phys. Rev. Lett. **56**, 2391 (1986) and unpublished.
11. W.A. Harrison, *Electronic Structure and the Properties of Solids* (Freeman,

San Francisco, 1980).

12. H. Ehrenreich and L.M. Schwartz, in *Solid State Physics*, ed. by H. Ehrenreich, F. Seitz and D. Turnbull (Academic, New York, 1976) Vol. 31, p. 149.
13. K.C. Hass and H. Ehrenreich, J. Vac. Sci. Technol. A **1**, 1678 (1983).
14. K.C. Hass, H. Ehrenreich and B. Velický, Phys. Rev. B **27**, 1088 (1983).
15. C.E. Moore, *Atomic Energy Levels* (U.S. G.P.O., Washington, D.C., 1949); F. Herman and S. Skillman, *Atomic Structure Calculations* (Prentice-Hall, Englewood Cliffs, 1963).
16. N.T. Khoi and J.A. Gaj, Phys. Stat. Sol. (b) **83**, K133 (1977); R. Brun de Re, T. Donofro, J. Avon, J. Magid and J.C. Wooley, Nuovo Cim. **2*D***, 1911 (1983).
17. A.E. Carlsson, Phys. Rev. B **31**, 5178 (1985).
18. M. Taniguchi, *et. al.*, Phys. Rev. B **33**, 1206 (1986); A. Franciosi, *et. al.*, *Proc. of the* 18th *Int. Conf. on the Phys. of Semicond.*, Stockholm, 1986 and references therein.
19. P. Lautenschlager, S. Logothetidis, L. Viña and M. Cardona, Phys. Rev. B **32**, 3811 (1985).
20. T. Kendelewicz, J. Phys. C **14**, 6407 (1981).
21. H. Ehrenreich, K.C. Hass, N.F. Johnson, B.E. Larson and R.J. Lempert, *Proc. of the* 18th *Int. Conf. on the Phys. of Semicond.*, Stockholm, 1986.
22. B. Velický, J. Mašek, V. Cháb and B.A. Orlowski, Acta Physica Polonica A**69**, 1059 (1986).
23. D. Heiman, Y. Shapira and S. Foner, Solid State Commun. **51**, 603 (1984) and references therein.
24. A.K. Bhattacharjee, G. Fishman and B. Coqblin, Physica **117&118*B***, 449 (1983).
25. J.R. Schrieffer and P.A. Wolff, Phys. Rev. **149**, 491 (1966).
26. B.E. Larson, K.C. Hass and R.L. Aggarwal, Phys. Rev. B **33**, 1789 (1986); Y. Shapira, S. Foner, D.H. Ridgley, K. Dwight and A. Wold, Phys. Rev. B **30**, 4021 (1984); L.M. Corliss, J.M. Hastings, S.M. Shapiro, Y. Shapira and P. Becla, Phys. Rev. B **33**, 608 (1986).
27. P.W. Anderson, in *Solid State Physics*, ed. by F. Seitz and D. Turnbull (Academic, New York, 1963) Vol. 14, p. 79; B. Koiller and L.M. Falicov, J. Phys. C **8**, 695 (1975).
28. N. Bloembergen and T.J. Rowland, Phys. Rev. B **97**, 1697 (1955).
29. G. Bastard and C. Lewiner, Phys. Rev. B **20**, 4265 (1979); V.C. Lee and L. Liu, Phys. Rev. B **29**, 2125 (1984).
30. T. Oguchi, K. Terakura and A.R. Williams, Phys. Rev. B **28**, 6443 (1983).
31. J.S. Smart, in *Magnetism*, ed. by G.T. Rado and H. Suhl (Academic, New York, 1963) Vol. 3, p. 63.

BAND STRUCTURE AND ELECTRONIC EXCITATIONS IN $Cd_{1-x}Mn_xTe$

SU-HUAI WEI AND ALEX ZUNGER
Solar Energy Research Institute, Golden, Colorado 80401

ABSTRACT

We have performed spin-polarized, self-consistent local spin density total energy and band structure calculations for the prototype semimagnetic semiconductor alloy $Cd_{1-x}Mn_xTe$. Based on the calculated band structures and taking into account the many body effects of localized states, we propose a schematic energy level diagram to interpret the d→d*, p→d, and photoemission transitions in $Cd_{1-x}Mn_xTe$.

INTRODUCTION

Manganese doped II-VI semiconductors exhibit the interesting combination of magnetism and semiconductivity [1-4]. These systems are distinct from conventional octet isovalent semiconductor alloys in that they include an open-shell Mn d^5 ion, and differ from dilute d-electron impurity systems in that MnC^{VI} compounds show considerable solid solubility in common-anion $A^{II}C^{VI}$ compounds. The $A^{II}_{1-x}Mn_xC^{VI}$ system hence provides a unique link between impurity and alloy physics in semiconductors systems. In this paper we study $Cd_{1-x}Mn_xTe$ using the self-consistent, first principle, spin-polarized, general potential linearized augmented plane wave (FLAPW) method [5] within the local spin density functional formalism.

BAND STRUCTURE

We model the 50%-50% $Cd_{1-x}Mn_xTe$ alloy by a ferromagnetic (F) ordered structure (see the inset to Fig. 1). The equilibrium structural parameters of this compound and its binary constituents have been obtained by a total energy calculation [6]. We find the bond lengths are essentially conserved in the alloy and to be R(Mn-Te) = 2.73 Å and R(Cd-Te) = 2.80 Å. These values are in good agreement with experimental data.

In Fig. 1a and b we show the spin-polarized band structure of F-$CdMnTe_2$ for spins up and down, respectively. We find that the lowest valence band in this system consists primarily of Te s orbitals, with a minor Cd admixture. The next highest band is the Cd d band which peaks at $E_v^{\uparrow} - 9.2$ eV and $E_v^{\downarrow} - 7.5$ eV [where $E_v^{\uparrow}$ and $E_v^{\downarrow}$ denote the valence band maxima (VBM) for spin up and spin down, respectively]. The upper valence band complex has Mn d and Te p characteristic; these differ widely for spin up and spin down. Regarding its d components, we find that the spin-up Mn d band is occupied and centered at $E_v^{\uparrow} - 3.7$ eV, whereas the spin down d band is empty and centered at $E_v^{\downarrow} + 2.9$ eV. The +4.9 eV separation between them constitutes the effective d band exchange (x) splitting $\Delta_x(d)$. We find, however, that another important exchange splitting exists in the problem—the p-d exchange splitting $\Delta_x(pd) = E_v^{\downarrow} - E_v^{\uparrow}$ of the top of the valence bands for spin up and spin down—and that it is <u>negative</u>: the top of the valence band for spin-up ($E_v^{\uparrow}$) is 1.7 eV <u>above</u> the top of the valence band for spin down ($E_v^{\downarrow}$). The origin of this anomaly has been discussed in terms of a p-d repulsion mechanism [6].

Mat. Res. Soc. Symp. Proc. Vol. 89. © 1987 Materials Research Society

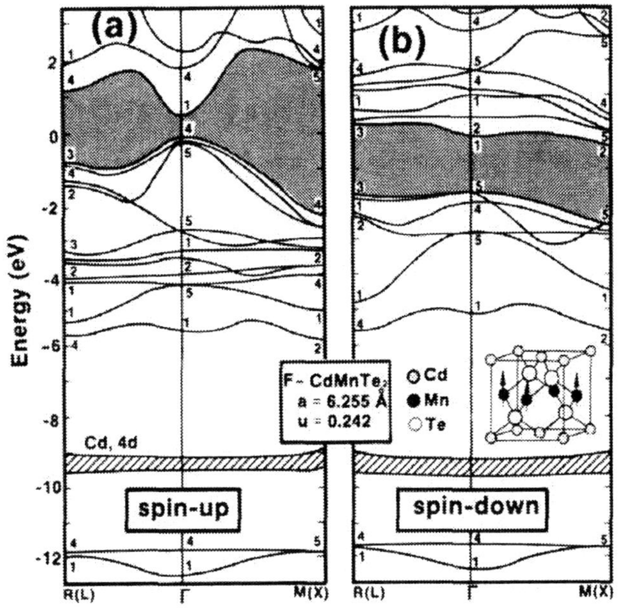

Fig. 1: Electronic band structure of F-CdMnTe$_2$ (a and b). The zero of the energy is at $E_v^{\uparrow}$. Symbols in parentheses are the points of fcc BZ. The band-gap regions are shaded.

We find that the introduction of Mn into CdTe has little effect on states that are localized on the Te and Cd atoms (e.g., Te 5s, Cd 4d states). Furthermore, the energy levels of F-CdMnTe$_2$ are very close to the average of the corresponding energy levels of the end-point compounds CdTe and F-MnTe (see Fig. 2), implying a small optical bowing in CdMnTe$_2$.

To see the effects of the AF ordering of Mn atoms on the electronic band structure, we performed a band structure calculation for the zincblende MnTe in the first kind of antiferromagnetic (AF) order, (inset to Fig. 3).

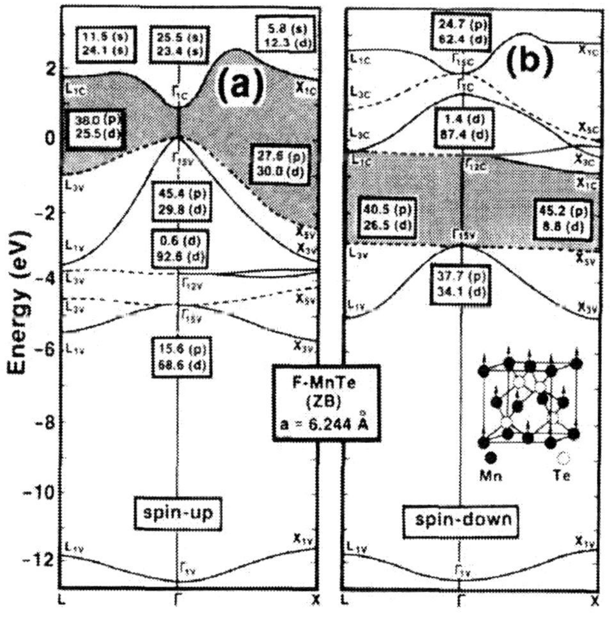

Fig. 2: Electronic band structure of F-ZB-MnTe (a,b). The inset depicts the spin ordering of this phase. Numbers in the boxes are the percentage ℓ character within the muffin-tin spheres [5] for Te and Mn atom.

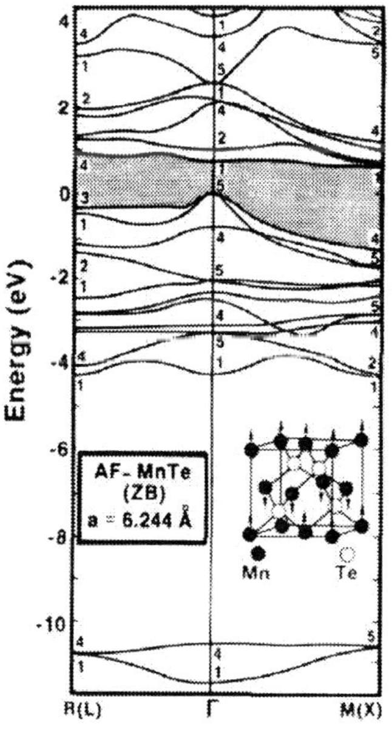

Fig. 3: Electronic band structure of AF-ZB-MnTe. The zero of the energy is at the VBM.

Figure 3 depicts the band structure of the AF-ZB-MnTe at a = 6.244 Å. We find the occupied Mn d band to be centered at $E_v - 2.5$ eV whereas the empty d band is centered at $E_v + 2.2$ eV; hence, the d-d exchange splitting is $\Delta_x(d) \approx +4.7$ eV. The direct band gap at Γ is 1.13 eV. Our calculation indicates that AF-ZB-MnTe is a semiconductor rather than a metal, because the intra-atomic exchange splitting of the Mn ion is large enough to keep the unoccupied spin states above the top of the Te p bands. The AF ordering increases the band gap. In contrast with the ferromagnetic case, there is no negative p-d exchange splitting in the AF spin arrangement. This is so because in the AF phase the Te p levels are repelled equally by d↑ and d↓.

Comparing the band eigenvalues of F-MnTe and AF-MnTe in the ZB structure, we see that: (i) for all the states except the Mn 3d, the average of spin-up and spin-down eigenvalues of F-MnTe is very close to the value of corresponding states in the AF-MnTe; (ii) the exchange splitting of Mn 3d states is smaller in AF phase than in the F phase. This means that both spin-up and spin-down Mn 3d states are closer to the VBM in the AF phase, increasing thereby the p-d hybridization. Our comparison indicates that it is possible to predict the general features of the band structure of a more complicated AF arrangement from that of the simpler F phase by a proper averaging of the corresponding states. This method is adopted in this study to find the location of Mn 3d↑ states of spin disordered $Cd_{1-x}Mn_xTe$ from the calculated F-$CdMnTe_2$ band structure. We find it to be at about $E_v - 2.5$ eV. Our results for F-MnTe and AF-MnTe are generally in good agreement with previous calculation [7], except for an upwards shift of their s-like conduction bands, presumably because of their omission of relativistic effects.

A MODEL FOR THE ELECTRONIC TRANSITIONS

Considerable controversy exists in the literature regarding the nature of the optical transitions in $Cd_{1-x}Mn_xTe$ for $x > 0.4$, at the 2.0-2.3 eV energy range. These sharp transitions have been interpreted as intra-atomic d→d* multiplet transitions of the $^6A_1 \to {}^4T_1, {}^4T_2$ type [3] as well as interband transitions [4]. The similarity of the transition energies in the alloy to those of the impurity systems [8], as well as the near composition-independence of the transition energy has been used to support the interpretation of d→d* excitations. Based on our calculated band structure calculations it appears to us that much of the confusion in this area results from the failure to properly recognize the strong dependence of the system's energy on its electronic configuration and occupations. Such strong dependences (i.e., correlation effects) characterize systems that sustain localized states [8,9,11]; they vanish in the one-electron band structure model for <u>extended</u> states. We illustrate this point in the schematic energy level diagram depicted in Fig. 4.

The ground state of the neutral system involves the Mn^{2+} formal oxidation state in which the e_+ and t_+ states are fully occupied. We consider here only low-energy excitations; hence, we simplify in Fig. 4 the notation to include only the e_+, t_+, and e_- levels. Hence, the ground state of the neutral system (Fig. 4b) is denoted as $[Mn^{2+}, d^5, e_+^2 t_+^3 e_-^o, {}^6A_1]$.

(i) <u>Ionization</u>; When we ionize the Mn atom in a photoemission experiment, the two limiting situations (in crystal field language) involve ionizations from either the t_+ or the e_+ orbitals, i.e., from the top (Γ_{15}) or the bottom (Γ_{12}) of the d↑ band. In the former case, the final state can be denoted formally as $[Mn^{3+}, d^4, e_+^2 t_+^2 e_-^o F^1, {}^5T_2]$ (where the final state orbital F—a conduction or vacuum level is occupied), whereas in the latter case we have $[Mn^{3+}, d^4, e_+^1 t_+^3 e_-^o F^1, {}^5E]$. Our band structure calculation indicate that

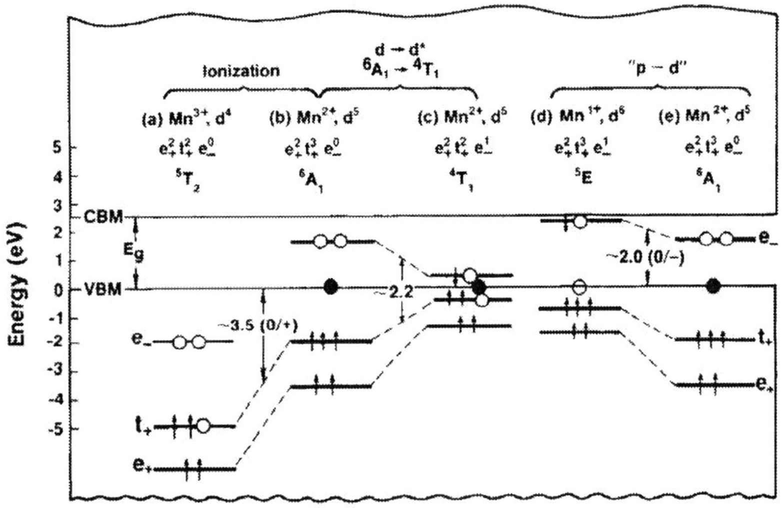

Fig. 4: Schematic energy level diagram depicting various transitions in Mn-rich $Cd_{1-x}Mn_xTe$. Open circles denote holes.

there is a significant difference between the two cases: whereas the t_+ orbital is hybridized with the host crystal, the e_+ orbitals are essentially non-bonding on the symmetry basis. As we ionize t_+^3, (Fig. 4b) to t_+^2 (Fig. 4a), the orbital energy drops, as we have relieved part of the Coulomb repulsion and reduced the exchange splitting. We can model the transition energy using Slater's "transition-state construct" by considering the eigenvalue difference at an intermediate occupation. The ionization energy relative to the valence band maximum E_v in this donor-like transition is

$$\Delta E_t(0/+) = E[Mn^{3+}] - E[Mn^{2+}] \cong \{\varepsilon_v - \varepsilon_{t_+}\}_{e_+^2 t_+^{2.5} F^{0.5}} . \quad (1a)$$

The observed value [10] of Mn d states is at ε_v - 3.5 eV. In our calculation $\varepsilon_{t_+} \cong \varepsilon_v$ - (2.0-2.5)eV in the initial state, hence the relaxation shift (a correction to Koopmans theorem) in this transition is $\Delta_t \cong$ -(1-1.5) eV. The reason that this shift is so much smaller than in the free Mn^{2+} ion [11] is related to the effective screening of the t_+ hole in the solid: similar calculations for 3d impurities in semiconductors [11] have shown that the hybridized host crystal resonances respond to the creation of a hole in a localized t orbital by increasing the amplitude of their wavefunction on the impurity site, thereby returning to it much of its charge lost in the ionization process (the "self-regulating response" [11]). Hence, whereas in occupation number space we have a $d^5 \to d^4$ transition, in coordinate space we have a $L^M d^5 \to L^{M-1} d^5$ transition, where the ligands (L, initially with M electrons) have lost approximately one electron and the impurity site remained essentially neutral ("charge transfer" excitation [11]). This process is not to be confused with ordinary dielectric screening (returning $1-1/\varepsilon$ electrons to the ionized site, where ε is the dielectric constant) which occurs on a far larger distance scale (a few, not one bond length), and is unrelated to the details of the hybridization (i.e., e-type or t-type holes).

At the extreme limit when an e_+ electron, rather than a t_+ electron is ionized, we have instead of Eq. (1a)

$$\Delta E_e(0/+) \cong \{\varepsilon_v - \varepsilon_{e+}\}_{e_+^{1.5} t_+^3 F^{0.5}} . \tag{1b}$$

Since the e_+ orbital is non-bonding, the self-regulating response is not operative (only ordinary dielectric screening is available), hence the relaxation correction Δ_e is expected to be much larger. Rigid band theory is entirely inadequate to explain such transitions. Here, the transition is $L^M d^5 \to L^M d^4$ both in occupation number space and in coordinate space. The recently observed resonant photoemission transition [10] at $\sim\varepsilon_v$-7 eV is hence interpreted as this type of excitation.

We have so far considered the limits of pure t_+ and e_+ excitations. In actuality, the $d\uparrow$ band is dispersed and contains a continuous mixture of orbital representations. In addition, alloy disorder acts to further intermix such states, implying coupling of $^5T_2$ and ^5E-type final states. Further quantitative (supercell) studies are necessary to clarify this point.

(ii) Intra d excitations; When we excite $[Mn^{2+}, d^5, {}^6A_1]$ into its lowest energy many-electron state we remove an electron from t_+ of $[Mn^{2+}, d^5, {}^6A_1]$ and place it in the lowest unoccupied Mn orbital, which is e_-, thereby creating $[Mn^{2+}, d^5, e_+^2 t_+^2 e_-^1, {}^4T_1]$ (higher multiplets, e.g., $^4T_2$, ^4E, $^4A_1$ are possible, too). In this transition (Fig. 4c) the exchange splitting in the maximum spin (S = 5/2) $^6A_1$ configuration is reduced (S = 3/2) in the $^4T_1$ configuration; hence, the formerly unoccupied e_- orbital moves down in energy. The excitation energy is given by the transition-state construct as

$$\Delta E({}^6A_1 \to {}^4T_1) = \{\varepsilon_{t_+} - \varepsilon_{e_-}\}_{e_+^2 t_+^{2.5} e_-^{0.5}} . \tag{2}$$

We interpret this as the 2.2 eV transition observed [12] in $Cd_{1-x}Mn_xTe$, similar to the values observed in the ZnS:Mn, ZnSe:Mn impurity systems [11]. Note that, since the excited electron still resides in a Mn orbital (e_-), it exerts screening on the system; hence, relaxation effects are smaller than in the ionization experiment where the t_+ electron is removed from the site. Because of the p-d hybridization, this transition also has some Te p character.

(iii) p-d excitations; We can also conceive of a transition in which a valence band electron is added to the lowest unoccupied Mn orbital (a "Te p→Mn d↓" transition; Fig. 4d and e). Here, we create the ground state of Mn^{1+}, i.e., $[Mn^{1+}, d^6, e_+^2 t_+^3 e_-^1, {}^5E]$, plus a hole in the valence band. This acceptor-like transition has the energy

$$\Delta E_e(0/-) = \Delta E(Mn^{2+}/Mn^{1+}) = \{\varepsilon_v - \varepsilon_{e-}\}_{e_+^2 t_+^3 e_-^{0.5} V^{M-0.5}} , \tag{3}$$

where the valence band initially had M electrons. In this case, the e_- orbital of Mn^{2+} moves on account of its increased Coulomb repulsion and its lower spin (S = 4/2), and hence smaller exchange splitting, but not as much as in the $^4T_1$ case (Fig. 4c), which has an even smaller spin S = 3/2. We interpret the ~2.0 eV transition observed in luminescence [2] as this transition (not as a stokes shifted d-d* transition). Because of the large mixing of Te p and Mn d states at the VBM, this transition also has some d-d* character.

A few observations are in order. First, since the p→d transition occurs at slightly lower energies (~2 eV) than the d→d* transition (~2.2 eV, the failure to observe Zeeman splitting for the former [12] may simply be due to masking of the p-d transition by the broad [12] d-d* transition, since the expected Zeeman splitting (~0.05 eV at B=15T) cannot be clearly resolved [12]. Second, the assumption of Vecchi et al.[2] that the Mn d↑ bands occur at the same energy in ionization and d-d* excitation is clearly invalid, in view of the large occupation-dependent relaxation effects.

(iv) The Mott-Hubbard Coulomb Energy; The Mott-Hubbard Coulomb energy is defined as the energy required to ionize an atom and place the electron on a distant identical atom. This corresponds to the difference between acceptor and donor energies, both referred to the same band edge. In the present system these are

$$U(e_-t_+) = \Delta E_e(0/-) - \Delta E_t(0/+) = (\varepsilon_v + 2.0) - (\varepsilon_v - 3.5) = 5.5 \text{ eV} . \quad (4a)$$

$$U(e_-e_+) = \Delta E_e(0/-) - \Delta E_e(0/+) = (\varepsilon_v + 2.0) - (\varepsilon_v - 7) = 9.0 \text{ eV} . \quad (4b)$$

Notice that we are transferring from a t_+ (or e_+) orbital to an e_- orbital; hence, $U(e_-t_+)$ is not the ordinary (diagonal) Mott energy since it receives a contribution from the crystal field and exchange splittings (unlike the d^3-d^4-d^5 case). Similarly, $U(e_-e_+)$ has a larger exchange splitting and relaxation contribution. The large Coulomb correlation energies are the reason why simple, one-electron (occupation-independent) considerations are invalid in this system.

ACKNOWLEDGEMENTS

This work was supported by the Office of Energy Research, Materials Science Division, U.S. Department of Energy, Grant No. DE-AC02-77-CH00178.

REFERENCES

1. J. K. Furdyna, J. Appl. Phys. 53, 7637 (1982).
2. M. P. Vecchi, W. Giriat, and L. Videla, Appl. Phys. Lett. 38, 99 (1981).
3. E. Müller, W. Gebhart, and W. Rehwald, J. Phys. C16, L1141 (1983).
4. E. I. Grancharova, J. P. Lascaray, J. Diouri and J. Allegre, Phys. Stat. Solidi 113B, 503 (1982).
5. S. -H. Wei, H. Krakauer, and M. Weinert, Phys. Rev. B 32, 7792 (1985).
6. S.-H. Wei and A. Zunger, Phys. Rev. Lett. 56, 2391 (1986).
7. B. E. Larson, K. C. Hass, H. Ehrenreich, and A. E. Carlsson, Solid State Commun. 56, 347 (1985); K. C. Hass et al., J. Mag. Magn. Mat. 54-57, 1283 (1986).
8. V. A. Singh and A. Zunger, Phys. Rev. B 31, 3729 (1985).
9. J. Zannen, G. A. Sawatzky and J. W. Allen, Phys. Rev. Lett. 55, 418 (1985).
10. M. Taniguchi, L. Ley, R. L. Johnson, J. Ghijsen, and M. Cardona, Phys. Rev. B33, 1206 (1986).
11. A. Zunger in Solid State Physics, Edts. H. Ehrenreich and D. Turnbull, (Academic Press, New York, 1968), 39, 275 (see Sec. IV).
12. Y. R. Lee, A. K. Ramdas and R. L. Aggarwal, Phys. Rev. B 33, 7383 (1986).

GROUND AND EXCITED ELECTRONIC ENERGY SURFACES OF THE MnS_4 CLUSTER IN $ZnS:Mn^{2+}$

J.W. RICHARDSON* AND G.J.M. JANSSEN**
* Purdue University, Dept. of Chemistry, West Lafayette, IN 47907
** University of Groningen, Laboratory of Chemical Physics, Department of Chemistry, Groningen, The Netherlands

ABSTRACT

Gaussian-based Self Consistent Field (SCF) MOs are obtained at various nuclear geometries for the MnS_4 cluster in the external potential of cubic ZnS. Electronic relaxation and d-shell electron correlation effects are then included. For T_d symmetry, quartet d-d excitation energies calculated as functions of R(Mn-S) qualitatively resemble the simple CF diagram. While separations between successive quartet levels agree closely with experiment, the threshold is about 0.5 eV high; much of this discrepancy is removed by including additional correlation effects from charge-transfer states. Large Jahn-Teller (JT) splittings of $^4T_1$ and $^4T_2$ levels are found with D_{2d} distortion. Difficulty in accurately evaluating force constants interferes with predicting the corresponding deformations and stabilization energies. Estimates, encorporating the observed Stokes' shift and JT stabilization energy, are that R(Mn-S) decreases by ~0.1 Å and the S-Mn-S dihedral angle increases by ~8°, in the lowest quartet level.

INTRODUCTION

The composition-independent electronic absorption processes in II-VI:Mn^{2+} diluted magnetic semiconductor systems, which emerge when the band-to-band absorption exceeds ~2.4 eV, are now commonly ascribed to intrinsic sextet-quartet d-d transitions of essentially localized Mn^{2+} ions. Some features in emission have similarly been explained.

Semiempirical fittings of crystal field parameters (Dq, B, and C, among others) to the absorption spectrum have proved difficult [1,2,3]. In the prototypical $ZnS:Mn^{2+}$ using free-ion values for B and C produces the observed spread of quartet levels in the spectrum [4] with Dq ~ 350 cm^{-1}. Fitting the threshold of absorption to the lowest quartet, however, requires a Dq three times larger; but then the overall spread of the quartet levels is too great and their order is altered. Only with severe reductions of B,C can a fit be obtained and then with ambiguity in assigning some of the higher-energy bands. Orthorhombic (e-mode) Jahn-Teller (JT) distortion of the lowest (a^4T_1) excited level has been identified [5]. This level is also the origin of the emission whose maximum lies ~0.22 eV below the lowest maximum in absorption [6,7] and may possibly also account for the emission near 1.9 eV [8]. Unlike other II-VI:Mn^{2+} systems, no near-infrared (~1.2 eV) emission has been reported.

The high density of orbitally degenerate quartet levels and the potentially large number of vibronic coupling constants involved makes difficult a semiempirical characterization of such distortions in the lowest quartet Adiabatic Potential Energy Surface (APES) [9].

As an alternative, <u>ab initio</u> theoretical calculations of the ground and lower excited d-d states of the MnS_4^{6-} cluster have been made for combinations of a_1 (stretching) and e (bending) distortions from the equilibrium geometry. The level of accuracy attained in some other simple transition-metal cluster systems by the methods used herein suggests that physically meaningful conclusions may thus be drawn about the assignments of levels and the distortions occurring. As well, a test is provided for the validity of the cluster model (central transition-metal ion plus

Mat. Res. Soc. Symp. Proc. Vol. 89. ©1987 Materials Research Society

its nearest neighboring anions) in quantitatively characterizing optical processes in these systems.

THEORETICAL COMPUTATIONAL MODEL

The reference system considered here is a MnS_4^{6-} cluster, having T_d symmetry and R_o(Mn-S) = 4.5917 au = 2.43 Å, embedded in a fixed external electrostatic potential field, V_{ext}, generated by point charges of +2e and -2e at the Zn and S sites of a surrounding infinite cubic ZnS structure. Additional calculations in D_{2d} symmetry are reported for combinations of R = (R_o, $R_o \pm 0.3$ au) with dihedral angle α = (α_0, $\alpha_0 \pm 10^o$, $\alpha_0 + 20^o$, $\alpha_0 + 50^o$), where α_0 is 109°28' and α is the S-Mn-S angle bisected by the 4-fold inversion axis of the distorted MnS_4 cluster.

Large gaussian bases for Mn (Wachters'(14,9,5) set [10] plus diffuse d [11] and p functions) and S (Roos-Siegbahn (10,6) set [12]) are used to represent the SCF MOs calculated by the SYMOL program at the University of Groningen. All electrons are explicitly included; orbital wavefunctions and total energies are calculated independently for various states of the open-shell "d" configurations e^xt^y, where x + y = 5 and e and t represent the 2- and 3-fold degenerate MO's whose major constituent is a Mn 3d orbital and whose degeneracies are partially lifted under orthorhombic distortion. Excitation energies reported here were calculated using orbital energies and electronic interaction integrals obtained at each geometry from the SCF solution to a hypothetical state, called H, whose energy is the average of the $^6A_1$ ground state and all 24 excited quartets [$^4T_1$ (3x3) + $^4T_2$ (3x3) + $^4A_1$ + $^4A_2$ + ^4E(2x2)].

Including effects of electron correlation is essential for accuracy in calculating the spectrum. Three types of correlations have been included. (a) Correlations among the 3d electrons of a free Mn^{2+} ion are introduced by the Correlation-Energy-Correction (CEC) method [13]. (b) Additional correlations specific to the reduced symmetry T_d or D_{2d} of the MnS_4 cluster are introduced by Configuration Interaction (CI) among the "Ligand-Field" states (LF-CI) of the e^xt^y configurations. In the context of crystal field theory, these CI terms mix states of the strong-field representation, based upon cluster MOs, and produce the intermediate field regime. (c) Only briefly mentioned below are further correlations specific to the Mn-S bonding-type interactions. They have been treated to a limited extent by allowing a number of S to Mn charge transfer states to be variationally mixed into the LF-CI state functions. Effects of spin-orbit interaction are neglected.

GROUND STATE PROPERTIES

Not unexpectedly, the total energy of the e^2t^3-$^6A_1$ ground state of MnS_4^{6-} shows a theoretical equilibrium value of R(Mn-S) well beyond the presumed true value of 2.43Å. In the absence of V_{ext}, the equilibrium symmetry is found to be T_d; including V_{ext}, however, shows a surprising tendency toward orthorhombic distortion, with a shallow minimum found at $\alpha \sim \alpha_0 + 20^o$. Both effects must result from omitting the non-Coulombic repulsion potentials of next-neighboring ions from V_{ext}.

Thus the total energy of the cluster obtained in the present calculational model does not provide reliable theoretical estimates of a_1 or e mode vibrational potentials for the ground state and, where needed, empirical information must be sought. The calculations do provide some indication, however, at least of a large anharmonicity in the e-mode vibration tending toward a flattened tetrahedral arrangement.

Many other electronic characteristics of the cluster are insensitive to V_{ext}. For example, either with or without V_{ext}, the open-shell SCF e and t MOs obtained show very little back-transfer of electron density from Mn AOs onto the S atoms. One relevant measure of this transfer is the calculated reduction of the two-electron coulombic and exchange interaction integrals from the ion values. Calculations on the hypothetical H state, for example, yield a reduction of 5%, which appears consistent with the highly localized nature of the Mn^{2+} ions in the solid. Thus vertical d-d excitations do not produce significant change in the net electronic charge density within the MnS_4 cluster, lending support to the cluster model for further theoretical analysis of the resulting changes in energy.

TETRAHEDRAL EXCITED STATES

Vertical excitation energies, calculated at R = 4.2917, 4.5917, and 4.8917 au in tetrahedral symmetry are listed in Table I. Allowing the $^6A_1$ state to relax electronically (by independently reoptimizing its SCF total energy) increases the values given by 0.07 eV. Several features of these results are relevant here.

The overall spread of the seven lowest calculated transitions (~0.85 eV) and the intervals between them agree with the qualitative CF diagrams for Dq ~ 350 cm^{-1} and closely with experiment. Energies of the a^4T_1 and a^4T_2 levels, and also $^4A_1$ and a^4E, show distance dependence consistent with shifts in band positions produced by hydrostatic pressure, seen in the similar system $ZnSe:Mn^{2+}$ [14]. At this level of calculation, the $^4A_1$ level remains slightly below (~0.02 eV) a^4E at the three distances shown, in disagreement with several semiempirically-based arguments [2,3]. The fourth band seen at 2.89 eV correlates with this calculated b^4T_1 level and the band(s) at 3.19 eV with b^4T_2 and b^4E; this is the ordering also found in the CF diagram at Dq ~ 350 cm^{-1}.

The major discrepancy between this level of calculation (SCF + LF-CI + CEC + relaxation and experiment is that the calculated quartet levels lie almost uniformly 0.5 eV too high above the $^6A_1$ ground state. Preliminary analysis of results obtained by also including a limited category of charge-transfer states to mix with the LF-CI mainfold [type (c) CT-CI correlations described earlier] indicates that such interactions depress the

Table I. Lower vertical excitation energies (in eV) of tetrahedral MnS_4^{6-} with V_{ext}, LF-CI, and CEC; from SCF on hypothetical H state.

Excitation $^6A_1 \rightarrow$	R(Å) 2.27	2.43	2.59	Expt. (a,b)	Error (c)
a^4T_1	2.66	2.84	2.96	2.34	0.50
a^4T_2	2.88	2.99	3.06	2.49	0.50
$^4A_1$	2.96	3.04	3.08	2.67	0.38
a^4E	2.99	3.06	3.10		
b^4T_1	3.49	3.48	3.47	2.89	0.59
b^4T_2	3.45	3.59	3.67	3.19	0.44
b^4E	3.59	3.67	3.72		

a From data compiled in Ref. [2]. The maximum at 3.19 eV consists of two bands, at 3.15 and 3.19 eV, according to Ref. [16]. The lowest doublet, $^2T_2$, may occur near 2.81 eV according to Ref. [6].
b Assignments of only the lowest three bands necessarily are those indicated in the first column. See text.
c Listed error is for the calculations at R = 2.43 Å.

quartet levels relative to the ground $^6A_1$ by ~0.25 eV, eliminating half the discrepancy [15]. Additionally, these interactions affect the higher quartets more than the lower. In particular, the relative positions of b^4T_1 and b^4T_2 are interchanged (agreeing with the currently more popular assignments) and the $^4A_1/a^4E$ degeneracy is almost restored.

In sum, these additional CT-CI correlations preserve the spread of the spectrum found near Dq ~ 350 cm^{-1} in the CF diagram but tend toward producing the order of levels and threshold of absorption near Dq ~1000 cm^{-1}, all in accord with observation. Thus, the cluster model appears capable of yielding near quantitative assessments of many spectral features associated with the Mn^{2+} center. It is probable that much of the residual discrepancy in the d-d absorption threshold could be eliminated if the remaining LF-CI states were also included here, as it was in some similar cases [15].

ORTHORHOMBIC DISTORTIONS OF LOWER QUARTETS

The extended CT-CI calculations have been done only for the T_d cluster at R = 2.43 Å, thus the effect of allowing for these additional correlations with other geometries is not now known. Assuming for the present that they operate uniformly over modest changes in R and α, the LF-CI + CEC calculations based on SCF MOs calculated for such distortions may be used to extract some useful insights about the lowest four quartet APESs arising from a^4T_1, which gives $^4A_2$ + ^4E, and a^4T_2, which gives $^4B_2$ + ^4E, in D_{2d} symmetry. Calculated energies of these states, relative to the ground state at the same geometry, are given in Table II.

Inspection of these results indicates normal JT splitting of a^4T_1 and a^4T_2 for α close to α_0. The two ^4E levels, however, show very considerable interactions between each other and particularly with higher ^4E levels at larger angles. The lowest ^4E, in fact, turns downward and closely parallels $^4B_2$ for $\alpha > \alpha_0 + 10^o$. Strong non-linearity is also seen in the variation of $^4B_2(a^4T_2)$ with α; beyond (unrealistically) large values of α (~$\alpha_o + 40^o$) it even crosses below $^4A_2(a^4T_1)$.

These results also predict that minima in the APESs of the three lowest quartets lie at shorter values of R(Mn-S) and larger values of dihedral angle α than in the ground state. These minima could be located by assuming that

$$E^\star(R,\alpha) = E^o(R,\alpha) + \Delta E^\star(R,\alpha), \tag{1}$$

where $E^\star$ and E^o are total energies of an excited quartet state and $^6A_1$, respectively, and $\Delta E^\star$ is the calculated vertical excitation energy (Table II), each at R,α. There is not available sufficiently accurate theoretical or experimental information (force constants and anharmonicities) about $E^o(R,\alpha)$ to satisfactorily locate the minimum in Eq. (1). Some significant conclusions can still be obtained, however, making use of other experimental data.

Using the observed Stokes' shift, -0.22 eV, to interpolate among the calculated possibilities indicated in Table II , suggests that any combination of $\delta R(Å) = R - 2.43$ and $\delta\alpha(^o) = \alpha - \alpha_0$ approximately satisfying the relation $\delta\alpha - 107\delta R \sim 20$ locates a minimum in the lowest quartet ($^4A_2$) APES. Since $\delta\alpha > 0$ and $\delta R < 0$, the limiting pairs of values are $(\delta R,\delta\alpha)$ = (-.19 Å,0°) and (0,+20°).

Furthermore, the following biquadratic function

$$\begin{aligned}\Delta E^\star(R,\alpha) = {} & [-0.01921\delta\alpha^2 - 0.01601\delta\alpha + 0.10448] \\ & + [0.02528\delta\alpha^2 + 0.02510\delta\alpha + 0.01813]\delta R \\ & - [0.01674\delta\alpha^2 + 0.01814\delta\alpha + 0.01266]\delta R^2 \end{aligned} \tag{2}$$

fits the $\Delta E^\star(R,\alpha)$ values for the $^4A_2$ APES with $\alpha \leqslant 20^o$ where, here, E

Table II. Vertical excitation energies (in eV) above $^6A_1$ ground state as functions of R and α. Calculated from SCF on hypothetical H state, with V_{ext}, LF-CI, and CEC.[a]

R(Å)	α_0-10°	α_0	α_0+10°	α_0+20°	α_0+50°
2.27	2.760	2.664	2.521	2.336	2.098
	2.911	2.882	2.770	2.528	2.011
	2.621	2.664	2.740	2.594	2.101
	2.891	2.882	2.844	2.849	2.780
		-.18	-.32	-.51	-.83
2.43	2.903	2.843	2.751	2.627	2.488
	3.009	2.992	2.921	2.762	2.407
	2.802	2.843	2.908	2.967	2.898
	2.953	2.992	2.955	2.796	2.438
		0.00	-.09	-.22	-.44
2.59	2.997	2.960	2.901		
	3.069	3.058	3.012		
	2.922	2.960	3.012		
	3.005	3.058	3.028		

a In each group of entries the first four are $\Delta E(R,\alpha)$ for $^4A_2$ (top), $^4B_2$, ^4E and ^4E (bottom). For $\alpha_0 \pm 10°$, the characters of the upper and lower ^4E levels are mainly a^4T_1 and a^4T_2, respectively. At larger values of α, they are strongly mixed with other ^4E levels. Where given, the fifth entry is the Stokes' shift predicted, if the minimum in the lowest APES wave were at that combination of R and α. [Stokes shift = $\Delta E(R,\alpha) - \Delta E^o(2.43,\alpha_0)$, assuming that the calculated relaxation energy of $^6A_1$ is independent of R and α.]

is in hartree, δR in bohr, and $\delta\alpha$ in radians.

From Equations 1 and 2 it is readily deduced that the a_1 stretching and e bending force constants for the $^4A_2(a^4T_1)$ APES are reduced from ground state values, e.g., by $\Delta k_a^* = 0.025$ H bohr^{-2} (= 0.4 x 10^5 dyne cm^{-1}) at $\alpha = \alpha_0$; $\Delta k_e^* = 0.038$ H rad^{-2} (= 0.028 x 10^5 dyne cm^{-1}) at R = 2.43Å; and $\Delta k_e^* = 0.050$ H rad^{-2} if R decreases to 2.27Å. If rough estimates of ground state values in dyne cm^{-1} are $k_a^o \sim 5 \times 10^5$ and $k_e^0 \sim 0.1 \times 10^5$, then the fractional reduction for the bending e mode may be substantial. Similarly, if the true equilibrium ground state geometry is T_d at R = 2.43 Å, the linear JT coupling constants for the a^4T_1 level correspond to $(\partial\Delta E^*/\partial R) = +0.01813$ H a_o^{-1} = 0.93 eV Å^{-1} and $(\partial\Delta E^*/\partial\alpha) = -0.01601$ H rad^{-1} = -0.0076 eV deg^{-1}.

The reduction of the excited state force constant, $\Delta k^*(e)$, is significant in understanding relative values of the Stokes' shift, ΔE_{SS}, and the JT stabilization energy of a^4T_1, ΔE_{JT}, which is here assumed to be totally due to the e mode. Now let $(\delta R,\delta\alpha)$ correspond to a minimum in the resulting $^4A_2$ APES; note that the ground state APES at that point has risen above its minimum at $\delta R = 0$, $\delta\alpha = 0$. Define $\Delta E_{SS} = \Delta E^*(\delta R,\delta\alpha) - \Delta E^*(0,0)$, $-\Delta E_{JT} = E^*(\delta R,\delta\alpha) - E^*(\delta R,0) = (1/2)k_e^*\delta\alpha^2$, assume $E(^6A_1) = (1/2)k_a^o\delta R^2 + (1/2)k_e^0\delta\alpha^2$, and consider some possible deformations. For instance, if δR = 0, then $2\Delta E_{SS} = (k_e^0 + k_e^*)\delta\alpha^2 = (2k_e^0 - \Delta k_e^*)\delta\alpha^2$ but $-2\Delta E_{JT} = k_e^*\delta\alpha^2 = (k_e^0 - \Delta k_e^*)\delta\alpha^2$. This gives

$$-\Delta E_{JT}/\Delta E_{SS} = (k_e^0 - \Delta k_e^*)/(2k_e^0 - \Delta k_e^*) \quad (3)$$

Taking the upper estimate of $\Delta E_{JT} = -330$ cm^{-1} from studies [5] on ZnS:Mn^{2+} and ΔE_{SS} = 0.22 eV gives $\Delta k_e^*/k_e^0 \sim 0.77$; using the calculated value Δk_e^*

$= 0.038$ H rad^{-2} produces an estimated $k_e{}^o = 0.05$ H rad^{-2} $= 0.037\times10^5$ dyne cm^{-1} and $k_e{}^* = 0.008\times10^5$ dyne cm^{-1}.

These values are totally unrealistic since the extremely small value obtained for $k_e{}^*$ produces an impossibly large distortion. This situation is rectified when the change in R(Mn-S) is also included. It also contributes to the Stokes' shift, but not to the ΔE_{JT} cited for the e-mode deformation. Extending the previous argument, ΔE_{ss} now becomes

$$\Delta E_{ss} = \frac{1}{2}(2k_a{}^o - \Delta k_a{}^*)\delta R^2 + \frac{1}{2}(2k_e{}^o - \Delta k_e{}^*)\delta\alpha^2 \qquad (4)$$

Assuming that $\Delta k_a{}^*$ is negligible and that the earlier relation $\delta\alpha$(deg) - 107R(Å)~20 holds, numerical investigation reveals the following approximate location of the minimum in terms of $x = \Delta k_e{}^*/k_e{}^o$

x	δR(Å)	δα(deg)
0.0	-.13	+6.5
0.2	-.11	+8.2
0.4	-.10	+9.0
0.6	-.08	+12.0

Reasonable estimates for the relaxed $^4A_2(a^4T_1)$ state are that R(Mn-S) decreases by ~0.10Å and the S-Mn-S angle α increases by ~8 to 9°.

ACKNOWLEDGMENT

This investigation was supported in part by the Netherlands Foundation for Chemical Research (SDN) with financial aid from the Netherlands Organization of the Advancement of Pure Research (ZWO).

REFERENCES

1. See, for example, L.E. Orgel, J. Chem. Phys. 23, 1004 (1955), for the basic crystal field picture.
2. D. Curie, C. Barthou, and B. Canny, J. Chem. Phys. 61, 3048 (1974).
3. A. Fazzio, M.J. Caldas, and A. Zunger, Phys. Rev. B 30, 3430 (1984).
4. D.S. McClure, J. Chem. Phys. 39, 2850 (1963).
5. R. Parrot, C. Naud, C. Porte, D. Fournier, A.C. Boccara, and J.C. Rivoal, Phys. Rev. B 17, 1057 (1978).
6. D.W. Langer and S. Ibuki, Phys. Rev. 138, A809 (1965).
7. H.E. Gumlich, R.L. Pfrogner, J.C. Shaffer, and F.E. Williams, J. Chem. Phys. 44, 3929 (1966).
8. O. Goede and D.D. Thong, Phys. Stat. Solidi (b) 124, 343 (1984).
9. A.D. Liehr, J. Phys. Chem. 67, 389 (1963) gives all vibronic coupling constants through third order.
10. A.J.H. Wachters, J. Chem. Phys. 52, 1033 (1970).
11. P.J. Hay, J. Chem. Phys. 66, 4377 (1977).
12. B. Roos and P. Siegbahn, Theor. Chim. Acta 17, 209 (1970).
13. L. Pueyo and J.W. Richardson, J. Chem. Phys. 67, 3577 (1977). Note that the CEC correction makes use of the differences between the empirical and theoretical SCF sextet-quartet excitation energies of the free Mn^{2+} ion.
14. S. Ves, K. Strössner, W. Gebhardt, and M. Cardonna, Solid State Comm. 57, 335 (1986).
15. G.J.M. Janssen, Doctoral Thesis, University of Groningen, 1986.
16. A. Mehra, J. Electrochem. Soc. 118, 136 (1971).

MAGNETORESISTANCE AND HALL EFFECT NEAR THE METAL-INSULATOR TRANSITION OF $Cd_{1-x}Mn_xSe$

Y. SHAPIRA*
Francis Bitter National Magnet Laboratory and the Department of Physics
Massachusetts Institute of Technology, Cambridge, MA 02139

ABSTRACT

A large magnetoresistance was observed in n-type $Cd_{1-x}Mn_xSe$ ($x \leq 0.10$) at low temperatures. The concomitant changes in the Hall coefficient as a function of H were also studied. These magnetotransport effects are caused by the s-d interaction. Some specific magnetoresistance mechanisms are discussed. The most likely dominant mechanisms are related to the giant spin splitting of the conduction band.

INTRODUCTION

From previous works on magnetic semiconductors [1] one could have expected that the s-d interaction in dilute magnetic semiconductors (DMS's) would lead to large magnetotransport efffects at low temperatures. However, to obtain a more precise knowledge of the magnetoresistance (MR) and Hall effect in DMS's, and to unravel the mechanisms responsible for them, requires a considerable experimental and theoretical effort. Such efforts have been made only in the last few years. Among the various DMS's which were investigated, the $Cd_{1-x}Mn_xSe$ system has received a great deal of attention. Magnetotransport studies in this system were carried out in Poland [2-4], at IBM [5], and at MIT and Brown University [6-8]. This talk summarizes the results obtained by the MIT-Brown group; it is not intended to be a review of all the magnetotransport works on $Cd_{1-x}Mn_xSe$. However, some comparisons with the results obtained in Poland and at IBM will also be made.

The MIT-Brown measurements were made on three series of Ga-doped single crystals which were grown by a modified Bridgman method [9]. The three series corresponded to Mn concentrations x = 0.01, 0.05, and 0.10. Each series consisted of several n-type samples with room-temperature carrier concentrations n_{RT} in the range 10^{17} to 10^{18} cm^{-3}. This range of n_{RT} includes the characteristic concentration $n_c(x)$ at which the metal-insulator (M-I) transition occurs at zero magnetic field. Resistivity and Hall data were taken from 0.3 to 300 K in magnetic fields up to 80 kOe. The studies which were carried out in Poland were on samples with x = 0.05, and with n_{RT} well below n_c [2], near n_c [3], and somewhat above n_c [4]. The work at IBM was on Ga-doped samples with x = 0.4 and $5\times10^{17} \leq n_{RT} \leq 1.5\times10^{18}$ $cm^{-3} < n_c(x=0.4)$. Thus, the experimental data obtained to date span a wide range of n_{RT}, which covers hopping conduction at low temperatures, metallic conduction, and conduction near the M-I transition. It is remarkable that many of the qualitative features of the MR at low temperatures are similar in all these conduction regimes. This similarity is yet to be explained in a convincing manner; the theoretical approaches to date are piecemeal approaches, each applicable to a specific conduction regime.

THE M-I TRANSITION AT ZERO FIELD

Figure 1 shows the resistivity ρ at zero field for samples with x = 0.10 as a function of 1/T. Based on this figure the value of n_c for x = 0.10 was estimated to be near 6×10^{17} cm^{-3}. Similar plots for x = 0.05 and

Mat. Res. Soc. Symp. Proc. Vol. 89. ©1987 Materials Research Society

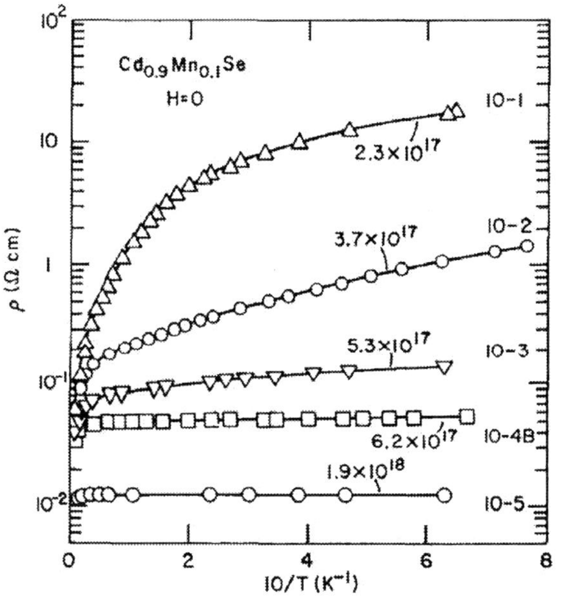

Fig. 1. Temperature dependence of the zero-field resistivities of samples with x = 0.10. The room temperature carrier concentration n_{RT} for each sample (in electrons/cm^3) is indicated.

0.01 show that for samples grown in the same manner, n_c decreases as x decreases. For x = 0.01, n_c was estimated to be near 3×10^{17} cm^{-3}. The latter value is close to that calculated for the parent compound (CdSe) using the Mott relation

$$n_c \cong (0.26/a_H)^3 , \tag{1}$$

where a_H is the effective Bohr radius for the donor. For x = 0.05, we obtained $n_c \cong 5 \times 10^{17}$ cm^{-3}, which is consistent with data obtained in Poland for In-doped samples with the same x. The latter data show that a sample with $n_{RT} = 8.4 \times 10^{17}$ cm^{-3} is on the metallic side of the M-I transition [4], while a sample with $n_{RT} = 4 \times 10^{17}$ cm^{-3} is on the insulating side [3]. It should be mentioned, however, that the value of n_c depends not only on x but also on the sample-preparation technique [8], presumably because n_c also depends on compensation. The increase of n_c with increasing x found for x $\leq$ 0.10 is consistent with the IBM data on Ga-doped samples with x = 0.4. For this much higher Mn concentration, $n_c > 1.5 \times 10^{18}$ cm^{-3}.

The increase of n_c with increasing x is tentatively attributed to two causes. First, the effective mass m* in the conduction band increases slightly with x [10], which should reduce the Bohr radius a_H and, therefore, increase n_c. Second, one expects that as x increases there will be an additional scattering due to positional disorder in the crystal (the data in Ref. 8 show that the room-temperature mobility decreases with increasing x), and there will also be some increase in the scattering at low temperatures due to the s-d interaction with the more numerous Mn spins. One specific s-d scattering mechanism which was mentioned by Sawicki et al. [4] is that due to magnetic polarons. This will be discussed later. Any additional scattering promotes localization, and it therefore increases n_c.

MAGNETORESISTANCE AT LOW TEMPERATURES

A. General Qualitative Features

Figures 2-4 show examples of the transverse MR at low temperatures for samples with different Mn concentrations. These examples are typical in the sense that each figure shows the qualitative features which were observed in

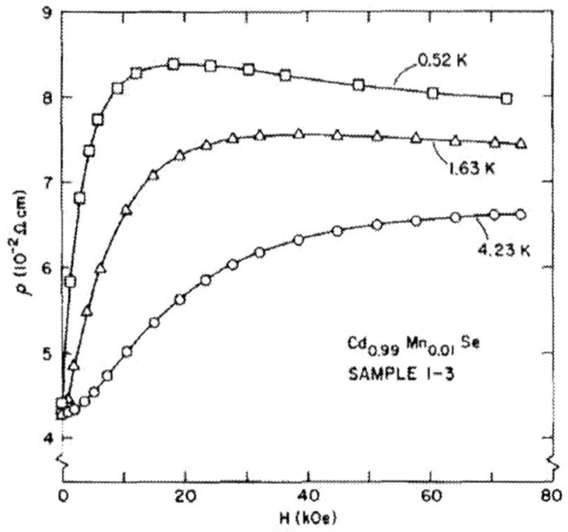

Fig. 2. Transverse magnetoresistance of sample 1-3 (x=0.01, n_{RT}=4.3x10^{17} cm^{-3}) at low temperatures.

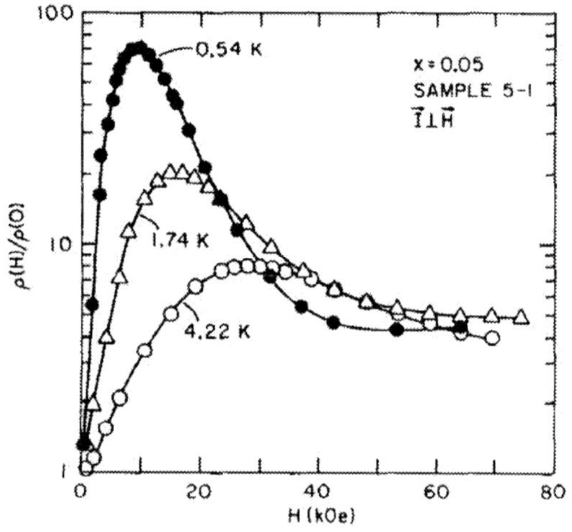

Fig. 3. Transverse magnetoresistance of sample 5-1 (x=0.05, n_{RT}=3.2x10^{17} cm^{-3}) at low temperatures.

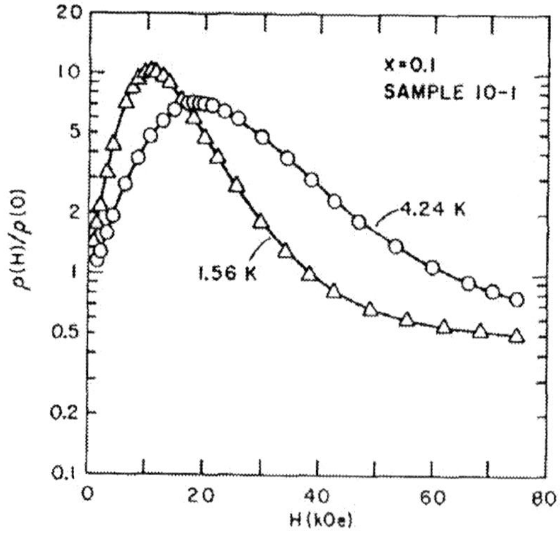

Fig. 4. Transverse magnetoresistance of sample 10-1 (x=0.10, n_{RT}=2.3x10^{17} cm^{-3}) at low temperatures.

other samples with the same x. However, the magnitude of the MR and the detailed shape of the MR curve at a given temperature depend not only on x but also on n_{RT}. The data in Fig. 3 correspond, in fact, to the largest positive MR which we have observed in any sample.

Qualitative features which are observed in all our samples include the following. (i) At low H the MR at T > 0.3 K is always positive. (ii) For samples with x = 0.05 and 0.10 the MR goes through a maximum, and it then decreases at higher fields. (iii) For x = 0.01 the resistivity for $T \gtrsim 2$ K increases monotonically with H, i.e., there is no resistivity maximum below 80 kOe. However, well below 2 K a weak maximum in the MR is also observed for x = 0.01.

Features (i) and (ii) were also observed in Poland in samples with x = 0.05 and $n_{RT} \ll n_c$ [2], and at IBM in samples with x = 0.4 and $n_{RT} < n_c$ [5]. In addition, feature (i) was observed in Poland [4] in a sample with x = 0.05 and n_{RT} slightly above n_c. The data in this case did not extend to sufficiently high fields to observe the resistivity maximum. However, recent data [3] for a sample with n_{RT} slightly below n_c indicate that the resistivity at higher fields decreases with increasing H. These recent data also show that at least in some samples the MR at low H turns from positive to negative when T is lowered to below 0.2 K, i.e., feature (i) disappears at very low temperatures.

B. Dominant Role of the s-d Interaction

The MR effects in Figs. 2-4 are far too large to be ascribed to the classical MR due to the direct action of the Lorentz force. Instead, there is a strong evidence that the MR is primarily caused by the s-d interaction between the spins of the electrons near the conduction band edge and the spins of the Mn ions. As discussed later the MR due to the s-d interaction is expected to be a function of the magnetization M. The evidence which implicates the s-d interaction as the dominant cause of the MR consists of the following correlations between the behaviors of the magnetization and the MR as a function of temperature and magnetic field. (1) A decrease of T leads to a more rapid rise of the MR at low H, as can be seen in Figs. 2-4. This correlates well with the larger low-field differential susceptibility dM/dH at lower T. (2) At the highest fields (near 80 kOe) the MR varies only slowly with H, which correlates well with the slow variation of the M with H at these fields [11]. (3) For x = 0.05 and 0.10 the magnetization M_{max} at the resistivity maximum of a given sample is approximately independent of temperature, whereas the field H_{max} at the maximum decreases appreciably with decreasing T. This is illustrated by the results in Fig. 5. Part (a) of this figure shows that H_{max} is much lower at 1.59 K than at 4.22 K, while part (b) shows that the normalized magnetization M/M_s at the maximum is nearly the same at both temperatures. (The normalization factor M_s, which is called the technical saturation value [12], is constant for a given x. The values for M are based on the results in [11].)

C. Some Qualitative Trends

1. **Temperature dependence.** For x = 0.05 and 0.10 the field H_{max} at the resistivity maximum decreases with decreasing T. In addition, there is a change in the relative magnitudes of the rise and fall of the resistivity ρ in fields below and above H_{max}. As T decreases, the fall of the resistivity above H_{max} (relative to the rise at lower fields) becomes more pronounced. This is illustrated by the data in Fig. 4. The same trend can also be seen in the data of Dietl et al. [2] for x = 0.05 and $n_{RT} \ll n_c$, and in recent data for $n \cong n_c$ [3] which show that at ultra-low temperatures (T < 0.2 K) the MR is negative even at low H.

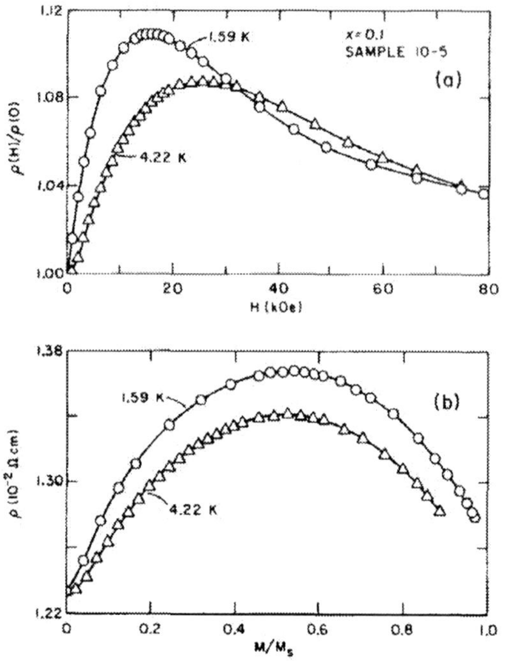

Fig. 5. Transverse MR of sample 10-5 (x=0.10, n_{RT}=1.9 x10^{18} cm^{-3}). (a) MR as a function of H. (b) Resistivity as a function of reduced magnetization M/M_s.

2. Dependence on carrier concentration. For a given x, the magnetization M_{max} at the resistivity maximum increases as n_{RT} increases. This increase, however, is quite slow. For example, for x = 0.10 a change of n_{RT} from 2.3x10^{17} cm^{-3} (sample 10-1) to 1.9x10^{18} cm^{-3} (sample 10-5) increases M_{max} by only 35%. The change of n_{RT} also changes the relative magnitudes of the rise and fall of ρ in fields below and above H_{max}. The fall of ρ above H_{max} is (relatively) more pronounced for lower n_{RT}. For example, the fall of the resistivity at high fields is more pronounced in sample 10-1 (Fig. 4) than in sample 10-5 (Fig. 5). The same trend was also seen by Dietl et al. [2] in samples with $n_{RT} << n_c$. In the latter case, however, the trend stopped at very low values of n_{RT} due to the onset of another magnetoresistance effect which is associated with the shrinking of the donor wavefunction in a magnetic field.

Only limited data are available for samples which are well on the metallic side of the M-I transition. These data suggest that the MR in samples with $n_{RT} >> n_c$ is much smaller than for samples with $n_{RT} \lesssim n_c$.

3. Dependence on Mn concentration. For comparable values of n_{RT}/n_c the fall of ρ above H_{max} (relative to the rise at lower fields) is less pronounced for x = 0.05 than for x = 0.10. This trend persists as x decreases from 0.05 to 0.01; for x = 0.01 a decrease of ρ at high fields occurs only at T $\lesssim$ 2 K, and even at these temperatures this decrease is fairly small (Fig. 2).

The normalized magnetization M_{max}/M_s at the resistivity maximum [12] depends on x. When x $\lesssim$ 0.1, M_{max}/M_s decreases with increasing x, for samples with comparable n_{RT}/n_c. The data which were obtained at IBM [5] suggest, however, that this trend does not persist when x is much higher than 0.1.

HALL EFFECT

Hall data were taken on all samples at 4.2 K. Some data at lower temperatures were also taken for the samples with x = 0.01. The data were analyzed on the assumption (justified in [8]) that the anomalous Hall term was small compared to the ordinary term. The results can be summarized as follows.

1. The positive MR below H_{max} is accompanied by an increase in the magnitude of the Hall coefficient R. At the same time the Hall mobility $\mu = |R/\rho|$ also decreases. An example is shown in Fig. 6.
2. The decrease of ρ in fields above H_{max} is always accompanied by an increase in the Hall mobility. For most values of x and n_{RT} the concomitant percentage change in R is small compared to the percentage change in μ. The exceptions are samples with x = 0.10 and $n_{RT} \lesssim n_c$, for which the percentage decrease in $|R|$ is comparable to the percentage increase of μ.

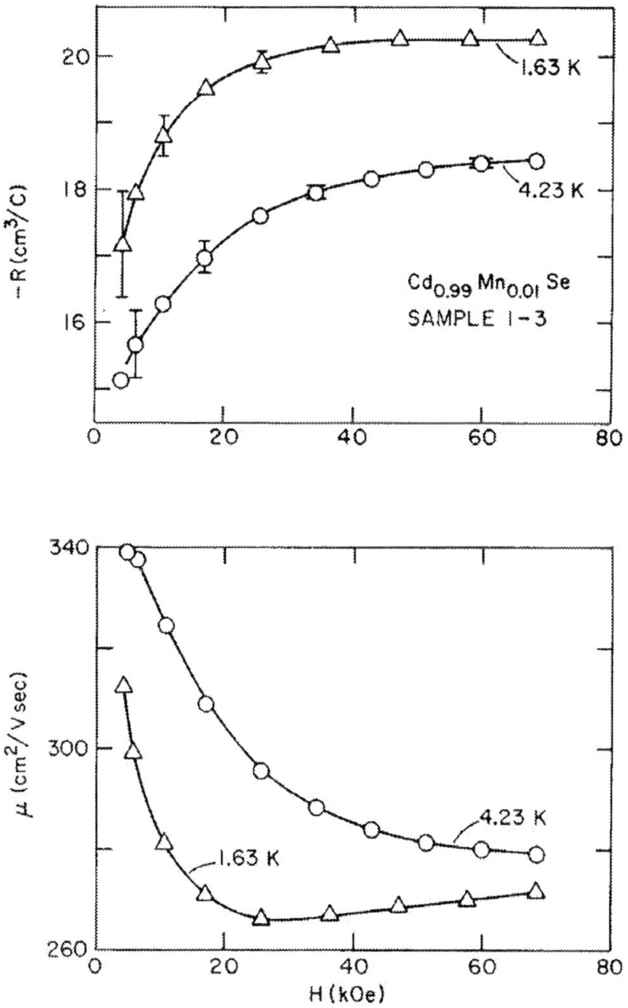

Fig. 6. H-dependence of the Hall coefficient R, and Hall mobility μ, for sample 1-3.

DISCUSSION

The large MR which is observed in n-type $Cd_{1-x}Mn_xSe$ is believed to be caused by the s-d interaction. However, the detailed mechanisms are still a matter of discussion. In an early paper by Dietl et al. [2] an attempt was made to explain the MR in insulating samples on the basis of a hopping model. For more highly conducting samples other explanations have been proposed [3-8]. Here we focus on some possible MR mechanisms which were considered by our group in an attempt to understand the data in our samples. For these samples n_{RT} was comparable to or larger than n_c, and the Hall mobility at 4.2 K was of order 10^2 cm^2/V s. Such a relatively high Hall mobility suggests that the low-temperature conduction was by states which were either above the mobility edge or not too far below it. Accordingly, MR mechanisms which involve hopping conduction between highly localized states will not be considered here.

The following facts suggest that several competing mechanisms contribute to the low-temperature MR. (1) The MR for $x > 0.01$ and $T > 0.3$ K is not a monotonic function of H, i.e., ρ first increases and then decreases with increasing H. (2) In many samples the decrease of ρ above H_{max} is mainly due to an increase of μ and not to a decrease of $|R|$, whereas the positive MR below H_{max} is always accompanied by a large increase of $|R|$. Thus, it is likely that the dominant MR mechanism (or mechanisms) below H_{max} is different from that (those) above H_{max}.

Two classes of MR mechanisms will be considered: (i) Those which are consequences of the giant spin splitting of the conduction band. In principle, the spin splitting can lead to either a positive or a negative MR. (ii) The negative MR due to the H-dependence of the scattering of conduction electrons by Mn spins.

A. Spin Splittings of the Conduction Band

The conduction band of a DMS, and the tail states below the band, undergo a giant spin splitting in the presence of a magnetic field [13]. This is shown schematically in Fig. 7. In $Cd_{1-x}Mn_xSe$ the spin splitting δ can be represented as

$$\delta = \delta_x(M/M_s) , \tag{2}$$

where δ_x is 5.7 meV for x = 0.01, 19 meV for x = 0.05, and 26 meV for x =

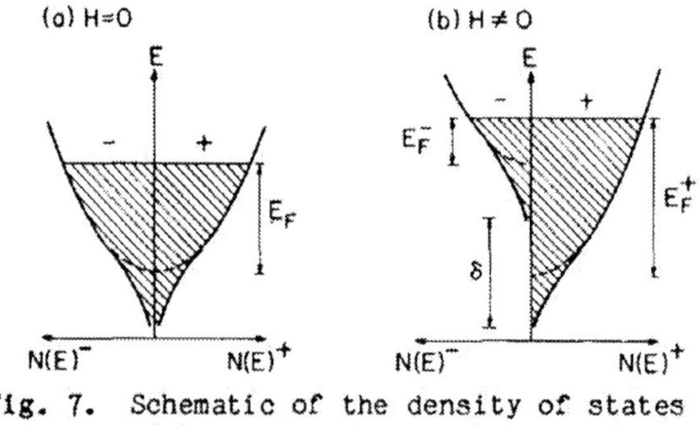

Fig. 7. Schematic of the density of states N(E) as a function of energy E for the conduction band for H = 0 and H ≠ 0. δ is the spin splitting, and E_F is the Fermi energy measured from the bottom of the parabolic portion of the band. Shaded areas represent occupied states at low T.

0.10 [8]. Physically, δ_x is approximately equal to the largest spin splitting (at the highest fields) which was achieved in the low-temperature experiments on $x \leq 0.10$. Estimates show that this largest spin splitting is comparable to the Fermi energy E_F at H = 0, when n_{RT} is in the range 10^{17}-$10^{18}cm^{-3}$. ($E_F \cong 18$ meV when $5x10^{17}$ electrons/cm^3 are in the conduction band.)

An important consequence of the giant spin splitting is the redistribution of electrons between the two spin subbands [14]. Because δ_x is comparable to E_F, this redistribution is substantial, i.e., at the highest fields the number of electrons in the majority-spin (+) subband is substantially higher than that in the minority-spin (-) subband. The MR mechanisms which are considered below are consequences of this electron redistribution.

1. **Screening radius.** The conduction electrons screen the Coulomb potential of an ionized impurity. The scattering from such an impurity depends on the screening radius r_s. In the Thomas-Fermi approximation r_s depends on the density of states at the Fermi level, summed over both spin subbands. It has been shown [14] that for a simple parabolic band this approximation leads to an increase of r_s with increasing band splitting. Ignoring other effects of the electron redistribution, the increase of r_s should lead to a positive MR. However, theoretical estimates by Gan and Lee [15] suggest that this effect is too small to account for the large MR in $Cd_{1-x}Mn_xSe$. A similar conclusion was also reached by the group at IBM [5] from an analysis of the measured magnetocapacitance.

2. **Electron-electron exchange and correlation.** Exchange and correlation effects between electrons lead to an H-dependence of the scattering from ionized impurities. When the conduction band splits, the electron cloud which screens an ionized impurity has a net (+) spin polarization. Exchange and correlation effects then cause the scattering of an incoming (+) electron to be stronger than that of a (-) electron. This was discussed by Kim and Schwartz for the case of a ferromagnetic metal [16], and by Gan and Lee [15] in the context of DMS's. The H-induced changes of the mobilities in the two spin subbands due to this effect can be much larger than those caused by the change in the Thomas-Fermi screening radius, and they can lead to a large MR. The sign of the MR at low fields depends on the choice of parameters.

3. **Changes in the Fermi energies $E_F^{\pm}$.** The electron redistribution which accompanies the giant band splitting raises the Fermi energy E_F^+ in the majority-spin subband and lowers E_F^- in the other spin subband. The conductivity in either subband depends on the difference between the Fermi energy and the mobility edge for the subband, i.e., as this difference increases the conductivity of the subband increases. It was shown by Fukuyama and Yosida [17] that when the contributions of both subbands are added, the changes in $E_F^{\pm}$ (ignoring changes in the mobility edges) lead to a negative MR. This negative MR can be appreciable, particularly on the insulating side of the M-I transition. It is likely that the Fukuyama-Yosida mechanism plays a role in the fall of the resistivity in fields above H_{max}.

B. Scattering by Mn Spins

The s-d interaction leads to a scattering of electrons by Mn spins. Two types of such scattering are: 1) spin-disorder scattering from spins which are not near donors or acceptors, and 2) coherent scattering from ferromagnetic spin clusters associated with bound magnetic polarons (BMP's) or other types of magnetic polarons (MP's).

Spin-disorer scattering was discussed by Haas [18], among others. This scattering leads to a negative MR [14]. However, estimates show that in the samples under consideration, spin-disorder scattering is too weak to be of importance [15,19].

Coherent magnetic scattering from BMP's was discussed by Nagaev and coworkers [1,20], among others. When the ferromagnetic alignment inside the BMP is much larger than outside the BMP, this scattering can be much stronger than spin-disorder scattering. Sawicki et al. [4] have suggested that in $Cd_{1-x}Mn_xSe$ samples which are near the M-I transition, coherent scattering from MP's associated with quasi-localized electrons makes an important contribution to the resistivity below 1 K. While this may be the case, we doubt that coherent scattering from such MP's played an important role in the MR which we have observed, at least not in the experiments above 1 K. The reason is that the expected Mn-spin alignment in the MP's associated with quasi-localized electrons should be very small. The arguments for the last conclusion are: (a) The ferromagnetic alignment of the Mn spins increases with electron localization. Thus, the ferromagnetic alignment in MP's associated with quasi-localized electrons should be smaller than in BMP's associated with localized electrons. (b) Even in the BMP's associated with localized electrons near hydrogenic donors, the ferromagnetic alignment at temperatures above 1 K is very small [21].

The conclusion that BMP's (or MP's) did not make an important contribution to the MR in our samples is also supported by another argument. The MR associated with BMP's is known to be negative. Such a negative MR was reported for many magnetic semiconductors, e.g., [22,23]. The question then is whether the negative MR above H_{max} is due to BMP's. As Figs. 3 and 4 indicate the magnitude of the decrease of ρ above H_{max} is in many cases larger than the zero-field resistivity. Thus, even if the resistivity at $H = 0$ were entirely due to BMP's, the disappearance of this resistivity at high fields would have accounted for only a small fraction of the decrease of ρ above H_{max}.

More complicated scattering mechanisms which involve a combination of Coulomb and s-d scattering were also considered by Nagaev and coworkers [1]. However, as discussed in Ref. 8 we do not believe that they are of major importance in the present context. Thus, our present belief is that the dominant mechanisms responsible for the MR in our experiments are consequences of the giant spin splitting of the conduction band (and of the band tail); the scattering from Mn spins is of lesser importance.

ACKNOWLEDGEMENTS

I am grateful to P.A. Lee, Z. Gan, and P.A. Wolff for many useful discussions. This work was supported by the National Science Foundation Grants Nos. DMR-8504366 and DMR-8511789. The crystal growth work in Professor Wold's laboratory at Brown University was supported by NSF Solid State Chemistry Grant DMR-8601345. Crystal-growth facilities were provided by Brown University's Materials Research Laboratory, which is supported by the National Science Foundation.

REFERENCES

* Collaborators: D.H. Ridgley, R. Kershaw, K. Dwight, and A. Wold (Brown University), and N.F. Oliveira, Jr. (MIT and the University of Sao Paulo).

1. E.L. Nagaev, Physics of Magnetic Semiconductors, (Mir, Moscow, 1983).
2. T. Dietl, J. Antoszewski, and L. Swierkowski, Physica B 117-118, 491 (1983).
3. T. Wojtowicz, T. Dietl, M. Sawicki, W. Plesiewicz, and J. Jaroszynski, Phys. Rev. Lett. 56, 2419 (1986); M. Sawicki, T. Wojtowicz, T. Dietl, J. Jaroszyniski, W. Plesiewicz, and J. Igalson (preprint).

4. M. Sawicki, T. Dietl, J. Kossut, J. Igalson, T. Wojtowicz, and W. Plesiewicz, Phys. Rev. Lett. 56, 508 (1986).
5. J. Stankiewicz, S. von Molnar, and W. Giriat, Phys. Rev. B 33, 3573 (1986).
6. Y. Shapira, D.H. Ridgley, K. Dwight, A. Wold, K.P. Martin, J.S. Brooks, and P.A. Lee, Solid State Commun. 54, 593 (1985).
7. Y. Shapira, D.H. Ridgley, K. Dwight, A. Wold, K.P. Martin, and J.S. Brooks, J. Appl. Phys. 57, 3210 (1985).
8. Y. Shapira, N.F. Oliveira, Jr., D.H. Ridgley, R. Kershaw, K. Dwight, and A. Wold, Phys. Rev. B 34, 4187 (1986).
9. D.H. Ridgley, R. Kershaw, K. Dwight, and A. Wold, Mater. Res. Bull. 20, 815 (1985).
10. J. Stankiewicz and L. David, J. Appl. Phys. 56, 3457 (1984).
11. D. Heiman, Y. Shapira, S. Foner, B. Khazai, R. Kershaw, K. Dwight, and A. Wold, Phys. Rev. B 29, 5634 (1984).
12. For a definition of M_s see Y. Shapira, S. Foner, D.H. Ridgley, K. Dwight, and A. Wold, Phys. Rev. B 30, 4021 (1984).
13. J.K. Furdyna, J. Appl. Phys. 53, 7637 (1982).
14. Y. Shapira and R.L. Kautz, Phys. Rev. B 10, 4781 (1974); R.L. Kautz and Y. Shapira, in Proceedings of the 20th Annual Conference on Magnetism and Magnetic Materials, AIP Conf. Proc. No. 24, edited by C.D. Graham, G.H. Lander, and J.J. Rhyne (AIP, New York, 1975), p. 42.
15. Z. Gan and P.A. Lee (unpublished).
16. D.J. Kim and B.B. Schwartz, Phys. Rev. B 15, 377 (1977).
17. H. Fukuyama and K. Yosida, J. Phys. Soc. Jpn. 46, 102 (1979).
18. C. Haas, Phys. Rev. 168, 531 (1968); IBM J. Res. Dev. 14, 282 (1970); CRC Crit. Rev. Solid State Sci. 1, 47 (1970).
19. Y. Ono and J. Kossut, J. Phys. Soc. Jpn. 53, 1128 (1984).
20. A.P. Grigin and E.L. Nagaev, Pis'ma Zh. Eksp. Teor. Fiz. 16, 438 (1972) [Sov. Phys.-JETP Lett. 16, 312 (1972)].
21. D. Heiman, P.A. Wolff, and J. Warnock, Phys. Rev. B 27, 4848 (1983).
22. Y. Shapira, S. Foner, N.F. Oliveira, Jr., and T.B. Reed, Phys. Rev. B 10, 4765 (1974).
23. S. von Molnar, A. Briggs, J. Flouquet, and G. Remenyi, Phys. Rev. Lett. 51, 706 (1983).

DC- AND FIR-MAGNETO-TRANSPORT IN Zn(1-x)Mn(x)Se

W. Erhardt*, M. von Ortenberg*, A. Twardowski**, and M. Demianiuk***
* Physikalisches Institut der Universität Würzburg,
Röntgenring 8, D-8700 Würzburg, F. R. Germany
** Institute of Experimental Physics, University of Warsaw,
Hoza 69, Pl-00-681 Warsaw, Poland
*** Institute of Technical Physics, Military Technical Academy,
Pl-00-908 Warsaw, Poland

ABSTRACT

DC- and FIR-magneto-transport measurements on Zn(1-x)Mn(x)Se for $x<0.05$ are interpreted in terms of nonresonant spin-flip hopping and resonant spin-flip conductivity. In combination with magnetization data the exchange-matrix element of the conduction band is determined to $\alpha = 0.32$ eV.

INTRODUCTION

Zn(1-x)Mn(x)Se is a semimagnetic wide-gap material of the Zincblende structure. Due to the exchange interaction of the statistically distributed, localized spins of the paramagnetic component Mn++ the spin properties of the quasi-free electrons are essentially modified. Whereas most of the semimagnetic materials investigated so far are belonging to the family of narrow-gap semiconductors, the results on wide-gap materials are not so numerous [1-5]. Therefore the principal objective of the present investigation was the study of semimagnetic effects in a typical wide-gap representative and relate the results to those of the narrow-gap materials.

EXPERIMENTAL

Pure Zn(1-x)Mn(x) is so highly resistive, that an artificial population of the conduction band becomes necessary. Despite to the pronounced exciton effects [6], we did not succeed in populating the condcution band with a sufficient carrier concentration by interband excitation. As alternative method we applied a Zn-diffusion-doping technique and the present results are based on samples treated in this way, which reduced the room-temperature resistivity to the order of k-Ohm*cm. As experimental techniques for the investigation of the electron properties we applied DC-magneto transport and submillimeter-magneto spectroscopy, both as a function of temperature, and magnetization measurements. We like to point out, that very often only the combination of DC- and high-frequency experiments allow a more comprehensive understanding of the microscopic interactions involved.

DC-Transport

The temperature dependence of the resistivity, as plotted in Fig. 1 for a 1%-Mn sample, reveals clearly that the transport mechanism at low temperature is dominated by MOTT's law [8]:

$$R = R_0 \exp((T_0/T)^{**}0.25) \qquad (1)$$

For MOTT's temperature parameter we obtain the very small value of T_0 = 3 K . The unscreened BOHR radius in ZnSe is of the order of a = 33 Å, using $\varepsilon = 9.2*\varepsilon_0$ and $m* = 0.147*m$ [9], so that the density of states at the FERMI

Mat. Res. Soc. Symp. Proc. Vol. 89. © 1987 Materials Research Society

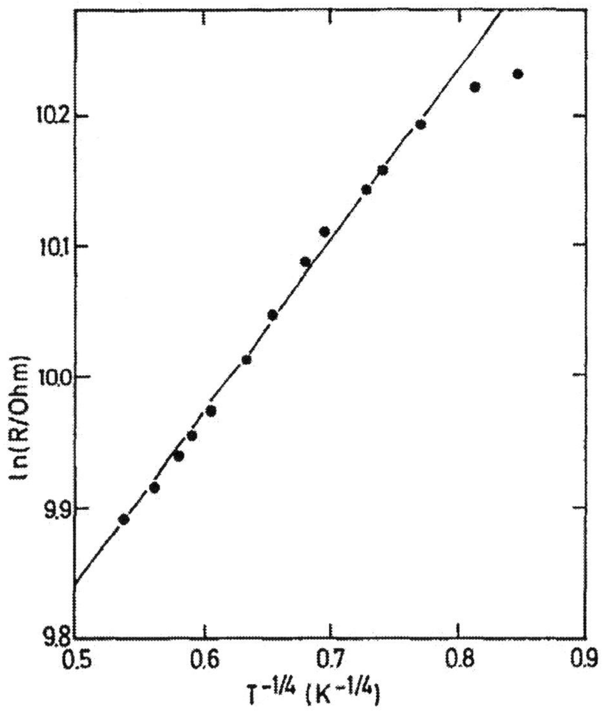

Fig.1 The temperature dependence of the resistivity demonstrates MOTT's variable-range hopping.

level would result in a value:

$$N(E_F) = \frac{18.1}{kT_0*a^3} = 2*10^{21}\ (\text{meV*ccm})^{-1} \tag{2}$$

The transverse magneto resistance of all samples exhibits a pronounced structure as shown representatively in Fig. 2 for the same sample as before. Corresponding data for Cd(1-x)Mn(x)Se were reported by DIETL et al. for T = 1.5 K, showing a similar negative differential magneto resistance, which was qualitatively explained by the bound magnetic polaron [10]. The interesting feature of our results, however, is the pronounced temperature shift of the maximum in the magneto resistance indicating the influence of the s-d-exchange interaction on the hopping process. A similar, however much less pronounced effect on the magneto-resistance has been observed for Ge and Si [11]. For these materials the theoretical understanding was given by MOVAGHAR and SCHWEITZER [12,13] by a theory of spin-flip hopping, which was later modified by OSAKA [14]. According to this theory the hopping conductivity is produced by "normal" and "anomalous" transfers in that sense, that the "anomalous" hop involves a simultaneous spin-flip of the electron, whereas the spin of the "normal" transfer is conserved. Due to occupation an "anomalous" nearest-neighbour hop is more probable than a "normal" second-nearest-neighbour transfer. As result the net conductivity is strongly determined by spin-flip hopping. The spin-flip process itself is produced by the transverse part $\vec{B}_\perp$ of the internal field $\vec{B}_i$:

$$\vec{B}_i = \vec{B}_d + \vec{B}_{hyp} + \vec{B}_{ex} \tag{3}$$

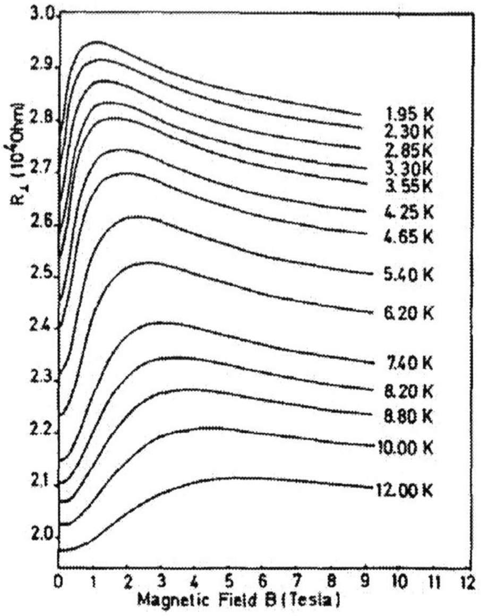

Fig. 2 The magneto-resistance shows a strong temperature dependence

as a superposition of dipolar, hyperfine and exchange fields. Whereas the exchange field is negligible for Si and Ge [12], in our experiment it represents the dominant part of the interaction. The resultant magneto-resistance is given by the following expression [14]:

$$\Delta R/R(B=0) = c(T)*\left\{\frac{B_\perp^2}{B_h^2}\left[1 - \frac{B_h^2}{B^2}\ln(1 + \frac{B^2}{B_h^2})\right] - b^2\ln(1 + \frac{B^2}{B_h^2})\right\} \qquad (4)$$

The temperature parameter T_0 is related to the value used in equ. (1) but not necessarily exactly the same. B is the external magnetic field and b is related to the average g-shift [12]. The "hopping" field B_h is related to the hopping frequency at zero magnetic field in the following way [13]:

$$B_h = B_h^0*\exp(-(T_0/T)**0.25) \qquad (5)$$

and reflects the temperature dependence. It should be noted that in equ. (4) enters only the average of the square of the transverse internal field, which depends, of course, via the exchange interaction also on the external magnetic field. At this point we introduce an essential modification of the MOVAGHAR and SCHWEITZER theory and assume, that the $B_\perp$ is directly related to the exchange coupling matrix element α and has the following form:

$$\overline{B_\perp^2} = B_{\perp 0}^2*(S(S+1) - \overline{<S_z>^2})/S(S+1),\ S = 5/2,\ B_{\perp 0} = x*\alpha/(2*\mu_B) \qquad (6)$$

Here x is the chemical composition parameter. From equ. (6) follows, that the transverse component of the internal field $\vec{B}_\perp$ is essentially determined by the magnetization of the sample showing the usual GAJ-like behaviour [15].

In Fig. 3 we have plotted the simulation using equ. (5) and equ. (6) by solid curves in comparison with the experimental data as represented by dotted curves. The fit values are:

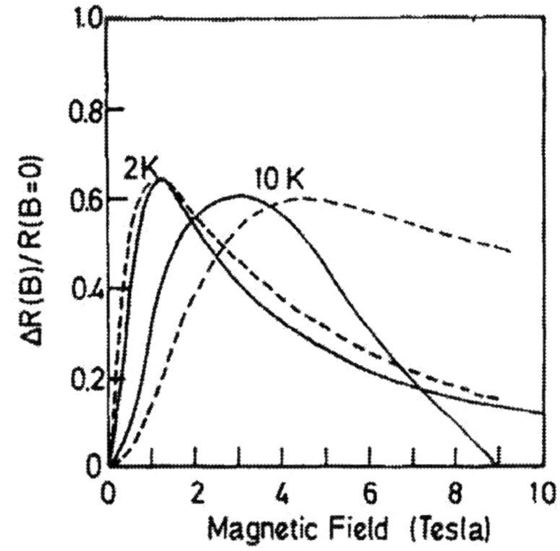

Fig. 3 The simulated magneto-resistance (solid curves) in comparison with the experimental data (broken curves)

$$B_h^0 = 13.2 \text{ Tesla}, \; b = 0.95 \qquad (B_{\perp 0} = 0.01 * \alpha / (2\mu_B) = 27.7 \text{ Tesla}) \quad (7)$$

Where we have used for T_0 = 3K and α = 0.32 eV as derived in the next paragraph.

The temperature shift of the maximumn in the magneto-resistance to higher fields with increasing temperatures is a direct consequence of the growing value of $\vec{B}_\perp$. For a temperature independent $\vec{B}_\perp$ as used by KUIVALAINEN et al. [11] for Si, the maximum is shifted to lower magnetic fields for higher temperatures. Our simulation demonstrates clearly the importance of the spin-flip components of the exchange interaction for the transport process, which are solely responsible for the shift of the maximum to higher magnetic fields with temperature. For higher temperatures the normal hopping transfer without spin-flip becomes nonnegligible, so that the agreement is less satisfactory.

FIR-Magneto Spectroscopy

Whereas DC-transport properties provide essential information on the scattering process involved in the hopping transfer, high-frequency experiments reveal in addition the resonance structure of the dielectric response. Whereas the undoped samples did not show any structure in the FIR-laser-magneto-transmission spectra, after diffusion doping the spectra in Fig. 4 were recorded, here represented for the radiation energy of 5.4 meV. Note that the dominant background modulation corresponds to the structure of the DC-magneto transport and is independent of the radiation frequency. The transmission minimum at zero magnetic field, which has also been observed by DOBROWOLSKA et al. given for Cd(1-x)Mn(x)Se 17 , cannot be explained by impurity transitions, because the photo-conductivity signal for energies up to 14 meV shows no impurity excitation, but can be quantitatively explained by a thermo-modulation of the DC-magneto-resisitance. The background modulation of the transmission spectra can be explained by the dielectric response due to nonresonant, high-frequency hopping-transfer including spin-flip according to MOVAGHAR and SCHWEITZER 12 , assuming that the magnetic field dependence of the conductivity is not essentially changed in the FIR.

Superimposed on the broad "hopping modulation" of the transmission spectra we observe a weak, but strongly temperature dependent resonance. Note

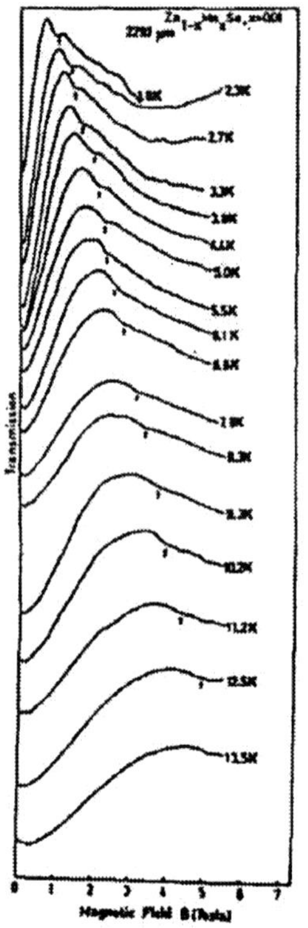

Fig. 4 The FIR magneto-transmission spectra

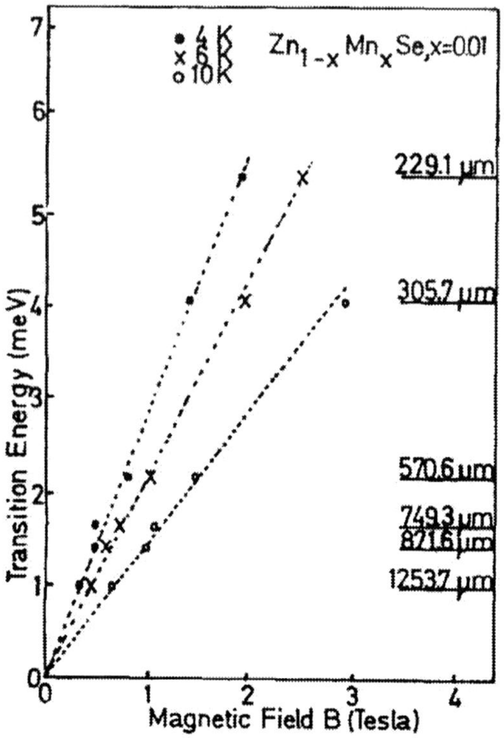

Fig. 5 The resonance position scales with frequency

that the resonance field is changed by 500% by increasing the temperature from 2 K to 13 K. Similar spectra have been observed by DOBROWOLSKA et al. [16, 17] in Cd(1-x)Mn(x)Se and have been explained as electric-dipole induced spin-flip of a donor bound electron. In contrast to the data on Cd(1-x)Mn(x)Se our resonance positions scale and extrapolate to zero energy for zero magnetic field as shown in Fig. 5. This means that no zero-field spin-splitting of a free or bound magnetic polaron is present. This result is quite interesting, because our DC-data have underlined the importance of the spin-flip part. We also like to point out, that no indication of a cyclotron resonance is observed in our spectra. Evidently the mobility of the hopping process in our samples is so low, that the cyclotron motion is nearly completely suppressed in contrast to the spin precession, where no transfer in space is involved.

Applying the KOSSUT-GALAZKA [2] theory of the exchange interaction we are able to derive from both the spin-resonance and magnetization data the exchange coupling matrix element of the conduction band directly without interference of the valence band to $\alpha = 0.32$ eV. This value is in rather good agreement with the value $\alpha = 0.26$ eV as obtained from the giant exciton splitting by TWARDOWSKI et al. [15].

SUMMARY

The DC- and FIR-magneto spectra of Zn-doped Zn(1-x)Mn(x)Se are essentially determined by spin-flip processes. Whereas the transport process is dominated by a non-resonant space transfer accompanied by simultaneous spin-flip, for the electric-dipole excited spin-resonance no hopping process is involved and thus yielding a sharp resonance structure.

REFERENCES

1. see references in: N.B. Brandt and V.V. Moshchalkov: Advances in Physics, 33, 193 (1984)
2. J. Kossut, Solid State Commun. 27, 1237 (1978)
3. H. Pascher, E.J. Fantner, G. Bauer, W. Zawadzki, M. von Ortenberg, Solid State Commun. 48, 461 (1983)
4. G. Karczewski and L. Kowalczyk, Solid State Commun: 48, 461 (1983)
5. G. Karczewski and M. von Ortenberg, Proc of the "17th Intnl. Conf. on the Physics of Semicond.", San Francisco, p. 1435 (1984)
6. A. Twardowski, T. Dietl, M. Demianiuk, Solid State Commun. 48, 845 (1983)
7. J. Gautron, C. Raisin, and P. Lemasson, J. Phys. D: Appl. Phys. 15, 153 (1982)
8. see in: H. Böttger an V.V. Bryksin, Hopping Conduction in Solids, Weinheim; Deerfield Beach, Fl.: VCH (1985)
9. H.W. Hölscher, A. Nöthe and Ch. Uihlein, Physica 117B&118B, 395 (1983)
10. T. Dietl: Springer Series in Solid State Sciences 24 "Physics in High Magnetic Fields", ed. S. Chikazumi and N. Miura, Springer-Verlag Berlin Heidelberg New York, p. 344 (1981)
11. P. Kuivalainen, J. Heleskivi, M. Leppihalme, U. Gyllenberg-Gästrin, and H. Isotalo, Phys. Rev. B26, 2041 (1982)
12. B. Movaghar and L. Schweitzer, phys. stat. sol. b80, 491 (1977)
13. B. Movaghar and L. Schweitzer, J. Phys. C: Solid State Phys. 11, 125 (1978)
14. Y. Osaka, J. Phys. Soc. Japan 47, 729 (1979)
15. A. Twardowski, M. von Ortenberg, M. Demianiuk, and R. Pauthenet, Solid State Commun. 849 (1984)
16. M. Dobrowolska, H.D. Drew, J. K. Furdyna, T. Ichiguchi, A. Witowski, and P.A. Wolff, Phys. Rev. Letters 49, 845 (1982)
17. M. Dobrowolska, A. Witowski, J.K. Furdyna, T. Ichiguchi, H.D.Drew, and P.A. Wolff, Phys. Rev. B29, 6652 (1984)

Strain Modification of Alloy Fluctuations in CdTe/(Cd,MnTe) Superlattice Systems

S. A. Jackson

AT&T Bell Laboratories
Murray Hill, New Jersey 07974

and

C. R. McIntyre

Department of Physics
Massachusetts Institute of Technology
Cambridge, Massachusetts 02139

ABSTRACT

We have calculated the effect of strains on the valence band offset and bandgap in the CdTe/(Cd,Mn)Te superlattice system for various growth directions and alloy compositions. We find that strain can modify the effect of alloy fluctuations on band offset and bandgap. The consequences of this for the formation of interface states will be discussed.

In this paper we consider two aspects of the effect of alloy disorder on optical properties of (111) and (001) oriented CdTe/(Cd,Mn)Te superlattices (SLS). First, we revisit the questions, proposed earlier by Zhang et al.[1] and Chang et al.[2] of whether local fluctuations in interface strain (in a superlattice due to lattice mismatch) which accompany concentration fluctuations, can provide an effective localizing potential which can give different binding energies for holes at the interface for (001) vs (111) SLS. Starting from the Schroedinger equation for a particle in a random potential generated by alloy fluctuations in the magnetic layers in a superlattice (SL) or quantum well (QW), we derive a self-consistent expression for the energy which optimizes the density of states (DOS) of an interface exciton state localized by the random potential. This energy is then minimized w.r.t range parameters for motion both perpendicular and parallel to the layers to find the optimal DOS as well as the broadening of the exciton linewidth due to alloy fluctuations. When this is done, we find the distance within which the exciton is localized near the interface and the energy which characterizes this localization. Secondly, we estimate the effect of the random strain field, again due to fluctuations in the interface strain, on the bandgap of the strained material via an effective Debye-Waller factor.

Chang et al.,[2] by assuming that all the valence band offset was due to fluctuations in the heavy hole - light hole band splittings, estimated the binding

Mat. Res. Soc. Symp. Proc. Vol. 89. © 1987 Materials Research Society

energy of a heavy hole in strained CdTe on $Cd_{.76}Mn_{.24}Te$ (since the heavy hole band provides the smallest bandgap) in a fluctuation induced well at the interface with a width of 2a extending across the interface, where a was a mean perpendicular lattice constant. The fluctuation volume was then $V=2aa_B^2$, where a_B is the exciton Bohr radius of the heavy hole. This means, as discussed by Goede et al.,[3] that the mean concentration fluctuation is

$$\Delta=2[x(1-x)/N]^{1/2} \tag{1}$$

where $N=2V/\Omega$ and Ω is the unit cell volume. The depth of the well is given by the effect of concentration fluctuations on one-half the heavy hole - light hole band splitting, i.e.

$$U_0=\frac{b\partial e_T}{\partial x}\Delta \tag{2}$$

for a (001) SL (where ϵ_T is the tetragonal distortion $= e_{zz}-e_{xx}$), and a similar expression with $b\partial\epsilon_T/\partial x$ replaced by $d/\sqrt{3}\,\frac{\partial e_{xy}}{\partial_x}$ for a (111) SL. Here b (=1.1 eV) and d (=5.45 eV) are the usual deformation potential constants for uniaxial strain along (001) and (111) respectively. They find for the binding energies $E_B = 6$ meV for (001) SLS and $E_B = 10$ meV for (111) SLS for x = .24. Since we expect the interface physics to be the same for both the SLS and a QW, we use a QW to do our analysis since it is mathematically simpler. Previous derivations of exciton line broadening in random alloy semiconductors(5) have used the optimal fluctuation technique of Halperin and Lax,(6) Zittarz and Langer,(7) and Lifshitz.(8) One solves the self-consistent equation for the wavefunction of a particle in a random potential (μ = Lagrange parameter),

$$\left[\frac{-\hbar^2}{2m}\nabla^2 - \mu\psi(\underset{\sim}{r})\int |\psi(\underset{\sim}{r}')|^2\, W(\underset{\sim}{r}-\underset{\sim}{r}')\, d^3\underset{\sim}{r}'\right]\psi = E\psi \tag{3}$$

where the random potential

$$V(\underset{\sim}{r}) = \sum_i v(\underset{\sim}{r}-\underset{\sim}{R}_i) = \eta\sum_i \delta(\underset{\sim}{r}-\underset{\sim}{R}_i) \tag{4}$$

is a sum of contributions from each site and

$$W(\underset{\sim}{r}-\underset{\sim}{r}') = \overline{N}\int v(\underset{\sim}{r}-\underset{\sim}{R})\, v(\underset{\sim}{r}'-\underset{\sim}{R})\, d^3R \tag{5}$$

($\overline{N}$ = mean impurity concentration). In equation (4)

$$\eta = \frac{\alpha}{\overline{N}}\sqrt{2x(1-x)} \tag{6}$$

where $\alpha = dE_c/dx$ or dE_v/dx for the conduction or valence band edges. Equation (3) results from optimizing the DOS $= \rho_o \exp\left\{\frac{-(K-E)^2}{S}\right\}$ with respect the wave function $\psi(\underset{\sim}{r})$, where K = the expectation value of the kinetic energy, E is the energy of the state in question and S is the weighted variance of the random potential, i.e.

$$S = \int d^3\underset{\sim}{r}'\, d^3\underset{\sim}{r}\ |\psi(\underset{\sim}{r})|^2\, W(\underset{\sim}{r} - \underset{\sim}{r}')\ |\psi(\underset{\sim}{r}')|^2 \tag{7}$$

For the combined electron-hole problem, if we assume that the exciton equation, due to the Coulomb interaction, has been solved for the relative wavefunction and the exciton binding energy (~13 meV) is larger than typical energies due to alloy fluctuations, then for the center-of-mass (CM) wavefunction, the self-consistent equation is of the same form as in equation (3), but where in equation (4), the coordinate $\underset{\sim}{r}$ becomes the CM coordinate $\underset{\sim}{R} = \frac{m_e \underset{\sim}{r}_e + m_h \underset{\sim}{r}_h}{m_e + m_n}$ and the CM random potential results from averaging the full potential over the relative wavefunction, e.g.

$$\bar{V}_h(\underset{\sim}{R}) = \int d^3\underset{\sim}{r}\ |\phi(\underset{\sim}{r})|^2\, V_h(\underset{\sim}{R} + \frac{m_n}{M}\underset{\sim}{r}) \tag{8}$$

where $\underset{\sim}{r}$ is the relative coordinate. The assumption that any localization energy due to alloy fluctuations is small compared with the exciton binding energy implies that $(R/r) >> \left(\frac{m_{e,h}}{M}\right)^{1/2}$, in which case $\bar{V}(\underset{\sim}{R}) \approx V(\underset{\sim}{R})$ for both electrons and holes. Then in equation (4), $\alpha = \frac{dEg}{dx} = 1.5$ eV for the CdTe/CdMnTe system.

It is known that, under strain, the valence band degeneracy is lifted and an anisotropic hole mass results, leading to an anisotropic (w.r.t. mass) form of equation (3). If we have a SLS or QW grown on a CdMnTe buffer layer, the CdTe layer is strained such that the heavy hole band rises and for a (111) QW (in units of the bare electron mass), $m_{11} \equiv m_z = 0.77$, $m_\perp = 0.44$, while for an (001) QW, $m_z = 0.53$, $m_\perp = 0.143$; for the CdMnTe layers we use for the heavy hole mass, $m_{nh} = 0.5$. Recognizing that the QW makes the z-direction special, we use a wavefunction similar to that used by Wu et al[9], i.e.

$$\psi(\underset{\sim}{R}) = \frac{\eta}{\sqrt{\pi}} \exp[-\eta^2 R_\perp^2/2]\ \frac{\nu^{1/2}}{\pi^{1/4}} \exp[-\nu^2(Z - Z_o)^2/2] \tag{9}$$

We integrate (average) the random potential over the z-wavefunction, then the random potential has the form

$$V(\underset{\sim}{R}) = \eta \sum_i |\phi(Z_i)|^2 \delta^{(2)}(\underset{\sim\perp}{R} - \underset{\sim i\perp}{R}) \tag{10}$$

Then leads to an effectively 2D random potential problem where it is known that the DOS has the form

$$n(E) \sim |E_{eff}|^{3/2} e^{\lambda E_{eff}/\tilde{Q}} \tag{11}$$

($\lambda = .93\ 4\pi$) resulting from solving the 2D self-consistent equation and minimizing the exponent in (11), where

$$\tilde{Q} = \int dz\, \theta[|z| - L/2]\, |\phi(z)|^4 \int d^2\underset{\sim\perp}{R'}\, d^2\underset{\sim\perp}{R}\, |\psi(\underset{\sim\perp}{R})|^2 W(\underset{\sim\perp}{R} - \underset{\sim\perp}{R'})\, |\psi(\underset{\sim\perp}{R'})|^2 \tag{12}$$

and $E_{eff} = E - K_z$, where

$$K_z = \frac{\hbar^2}{2m_z} \int dz\, |\frac{\partial \phi}{\partial z}|^2 \tag{12}$$

The value of E_{eff} which maximizes the DOS is $E_{eff} = -3/2\ \tilde{Q}/\lambda$, which gives $E = K_z - 3/2\ \tilde{Q}/\lambda$. If now this energy is minimized wrt ν for $Z_0 = L/2$ (exciton localized at the interface), we find for a (001) QW, the z-localization distance is $\nu^{-1} = 35Å$ and $E = 1.3$ meV, while for a (111) QW, we find $\nu^{-1} = 28Å$ and $E = 2$ meV for $x = .25$. It is interesting to compare this result with the result for localization distances and binding energies that would result from assuming a strictly (1D) z-dependent potential, where we find essentially the same results.[10]

This confirms the result of Chang et al.[2] that a disorder induced interface well can lead to binding of a heavy hole at the interface with different binding energies for the different SL orientations although the numerical values are not equal to the observed shifts in spectra for (001) and (111) SLS.

Next we take a different approach. As pointed out by Siggia[4] in his work on k·p theory in semiconductor alloys, the bulk bandgap gets renormalized due to a rescaling of the pseudopotential form factors by a Debye-Waller factor, $\exp(-\frac{1}{6}G^2 \langle u^2 \rangle)$, where G is an appropriate reciprocal lattice vector and $\langle u^2 \rangle$ is the mean square deviation of atoms in the semiconductor from their lattice positions. In (Cd,Mn)Te this would effect the bulk band gap and would be the same for both SL orientations. However, the lattice mismatch strains $e_T(x)$ and $e_{xy}(x)$ give rise to an effective randomization of the atom positions in CdTe due to alloy fluctuations, since the lattice mismatch strain shifts the lattice positions, i.e. $\vec{\ell} \rightarrow \vec{\ell'} = \vec{\ell} + \vec{e} \cdot \vec{\ell}$, where $\vec{e}$ is the strain tensor, where $\vec{\ell}$ is the position of an atom in the unstrained lattice, while $\vec{\ell'}$ is the position in the strained lattice. Alloy fluctuations mean that $e_{ij} = e_{ij}^{(0)} + \frac{\partial e_{ij}}{\partial x} \Delta x$ where $e_{ij}^{(0)}$ is the usual strain tensor. Since the change in e_{ij} is random, this gives rise to fluctuations, $\vec{U}_\ell$, in lattice positions in the strained layers, which leads to an effective Debye-Waller factor

$$e^{-W} = \langle e^{-i\vec{G}\cdot\vec{U}_\ell} \rangle = \int d(\Delta x)P(\Delta x)e^{-i\vec{G}\cdot\vec{U}_\ell} \tag{14}$$

where $\vec{U}_\ell = \Delta x \left(\frac{\partial \vec{e}}{\partial x} \cdot \vec{\ell} \right)$ and where $P(\Delta x)$ is the distribution function for alloy fluctuations. If we take, as suggested by Goede et al.[3], a Gaussian form for $P(\Delta x)$, where the mean fluctuation, Δ, is given by equation (1), we obtain a Debye-Waller factor $e^{-c^2\Delta^2}$ where

$$C = C_0 = 2\pi \sum_{i=1}^{3} \ell_i \frac{\partial e_{ii}}{\partial x} \tag{15}$$

for a (001) SL and

$$C = C_0 + 4\pi(\ell_1 + \ell_2 + \ell_3)\frac{\partial e_{xy}}{\partial x} \tag{16}$$

for a (111) SL, where $\ell_1\ell_2\ell_3$ is the number of unit cells in the volume of interest. Since the Debye-Waller factor effects band edge states which are extended, the volume of interest is $\sim a_B^3$, the exciton volume. Then $\ell_1\ell_2\ell_3 = \frac{V}{\Omega}$, or

$\ell_1 = \left(\frac{4}{\Omega}\right)^{1/3} a_B$, $\ell_2 = \left(\frac{4}{\Omega}\right)^{1/3} a_B$ and $\ell_3 = \left(\frac{4}{\Omega}\right)^{1/3} a_B$. For x=.25 we find for a (001) SL that the bandgap shrinks by a factor .997, and for a (111) SL, it shrinks by .918.

This gives a change in bandgap, due to randomization of lattice positions, of $\Delta E_g = E_g - e^{-W} E_g$. If we assume that this change is equally shared by a change in conduction band offset and valence band offset, i.e. $\Delta E_c = \Delta E_v = \frac{1}{2}\Delta E_g$, we have $\Delta E_v \approx \frac{1}{2} W E_g$. We expect the greatest effect on valence band states. If we assume that this change in valence band offset extends over the same z-range as found in the DOS discussion, we have a potential well, as in equation (3), but where the depth is now $U_0 = \frac{1}{2} W E_g$.

This gives rise to a binding energy for heavy holes of E_B=2.4 meV for and (001) SL and E_B=12.8 meV for a (111) SL, comparing favorably with experiment.

REFERENCES

[1] X.-C. Zhang, S.-K. Chang, A. V. Nurmikko, L. A. Kolodziejski, R. L. Gunshor and S. Datta, *Phys. Rev. B* **31**, 4056 (1985).

[2] S.-K. Chang, A. V. Nurmikko, L. A. Kolodziejski and R. L. Gunshor, *Phys. Rev. B* **33,** 2589 (1986) (and references cited therein).

[3] O. Goede, L. John and D. Henning, *Phys. Stat. Sol. B* **89,** K183 (1978).

[4] Eric D. Siggia, *Phys. Rev. B* **10,** 5147 (1974).

[5] S. D. Baranovskii and A. L. Efros, Sov. Phys. Semicond. **13**, 1328 (1978).

[6] B. I. Halperin and M. Lax, Phys. Rev. **148**, 722 (1966).

[7] J. Z. Harz and J. S. Langer, Phys. Rev. **148**, 741 (1966).

[8] I. M. Lifshitz, Sov. Phys. JETP **17**, 1159 (1963).

[9] J-W. Wu, A. V. Nurmikko and J. J. Quinn, Phys. Rev. B **34**, 1080 (1986).

[10] S. A. Jackson and C. R. McIntyre, unpublished.

EXAFS DETERMINATION OF BOND LENGTHS IN $Zn_{1-x}Mn_xSe$

B.A. BUNKER,* W.-F. PONG,* U. DEBSKA,** D.R. YODER-SHORT,** and J.K. FURDYNA**

*University of Notre Dame, Dept. of Physics, Notre Dame, IN 46530
**Purdue University, Dept. of Physics, West Lafayette, IN 47907

ABSTRACT

EXAFS has been used to determine bond lengths in the diluted magnetic semiconductor $Zn_{1-x}Mn_xSe$. These alloys change from a cubic (zincblende) crystal structure to a hexagonal (wurtzite) structure as a function of composition, with the transition at $x \approx 0.3$. Although x-ray diffraction measurements show the lattice parameter to change by approximately 0.2Å through the series and the crystal structure changes from cubic to hexagonal, the nearest-neighbor bond distances show no change within the experimental uncertainty of 0.01Å.

INTRODUCTION

The ternary alloy $Zn_{1-x}Mn_xSe$ belongs to a class of materials known as diluted magnetic semiconductors. Because of the interaction between the magnetic Mn^{+2} ions and the semiconductor matrix, these alloys exhibit unique electrical, magnetic, and optical properties.[1] It has recently been shown that excellent single-crystal samples of $Zn_{1-x}Mn_xSe$ can be grown for a wide range of compositions, ranging from $x = 0$ to $x \approx 0.57$.[2] For Mn concentrations below about 30%, the room-temperature crystal structure is cubic (zincblende), while for higher concentrations it is hexagonal (wurtzite). Precision lattice-constant measurements have been made for these materials,[3] showing that the mean cation-cation distance increases linearly with Mn concentration with no discontinuity or slope change at the zincblende-to-wurtzite transition concentration.

Mat. Res. Soc. Symp. Proc. Vol. 89. ©1987 Materials Research Society

Although the average cation-cation distance follows Vegard's Law, it is not clear what happens to the nearest-neighbor Mn-Se and Zn-Se bond lengths as the Mn concentration is varied. Extended X-ray Absorption Fine Structure (EXAFS) is a near-ideal probe of the local bond lengths because the environment about each atomic species may be studied separately. Earlier EXAFS measurements of $Ga_{1-x}In_xAs$ [4] and $Cd_{1-x}Mn_xTe$,[5] among others, have shown that bond lengths often change little with alloy composition -- much less than expected from lattice parameter changes as determined by x-ray diffraction. The present work differs from these earlier studies in that it is the first EXAFS investigation of bond lengths in an alloy series undergoing a structural phase transition as a function of composition.

EXPERIMENTAL

The samples used for this work were all grown using the Bridgeman growth technique using an rf-induction-heated graphite crucible.[2] For these measurements, the samples were ground to a fine powder sieved through 400 mesh. The EXAFS spectra were obtained in transmission mode at the National Synchrotron Light Source using the X11A beam line. Both the Zn and Mn edges were measured at room temperature for samples of concentration x = 0.15, 0.35, 0.57.

Data analysis followed standard procedures, including isolation of the EXAFS oscillations, Fourier transforming the data to isolate the first coordination shell, and use of the "ratio method".[6] The magnitude of Fourier-transform of the Zn edge of $Zn_{.85}Mn_{.15}Se$ is shown in Fig. 1. EXAFS measurements always reveal bond lengths relative to some known "standard" sample. In the case of the Zn edges, we compared with pure ZnSe, with the known bond length of 2.455Å. For Mn edges, we compared the x=0.35 and x=0.57 alloys with the x=0.15 alloy to determine the distance differences as a function of composition.

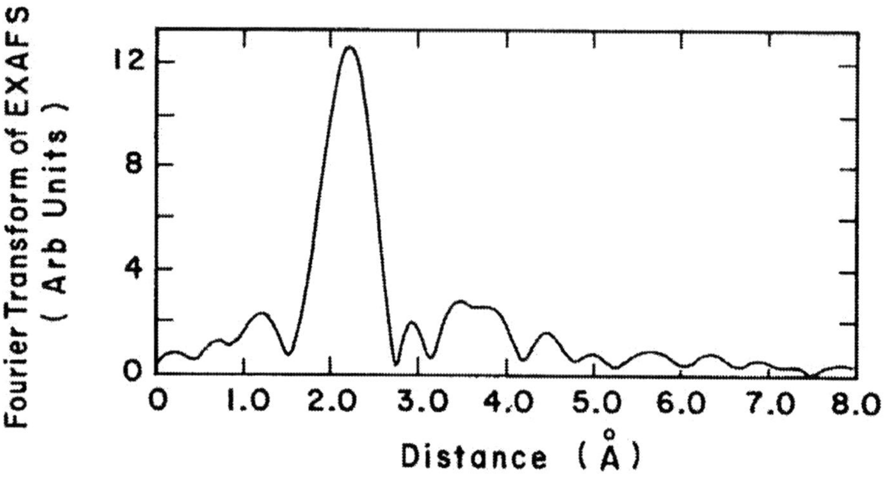

Fig. 1 Magnitude of the Fourier transform of the k-weighted EXAFS oscillations, for the Zn edge of $Zn_{0.85}Mn_{0.15}Se$.

The results of the analysis for bond lengths in $Zn_{1-x}Mn_xSe$ are summarized in Fig 2. As shown, the Mn-Se and Zn-Se bond lengths remain nearly constant (within 0.01Å) independent of the Mn concentration -- even though the structure changes from cubic to hexagonal. As discussed above, the EXAFS results for the Mn-Se bond lengths are relative, showing that there is essentially no change in the bond length as a function of composition. To present the Mn-Se bond length results on this graph, we have set the $Zn_{.15}Mn_{.85}Se$ bond length to the Mn-Se tetrahedral length as determined by extrapolation of the diffraction results. Also shown in Fig. 2 are the results of x-ray diffraction measurements[3] of the mean cation-cation distance. In order to directly compare diffraction results with our bond-length measurements, we have multiplied their results by $(3/8)^{½}$, the ratio of the cation-anion to cation-cation distance in the undistorted tetrahedral crystal.

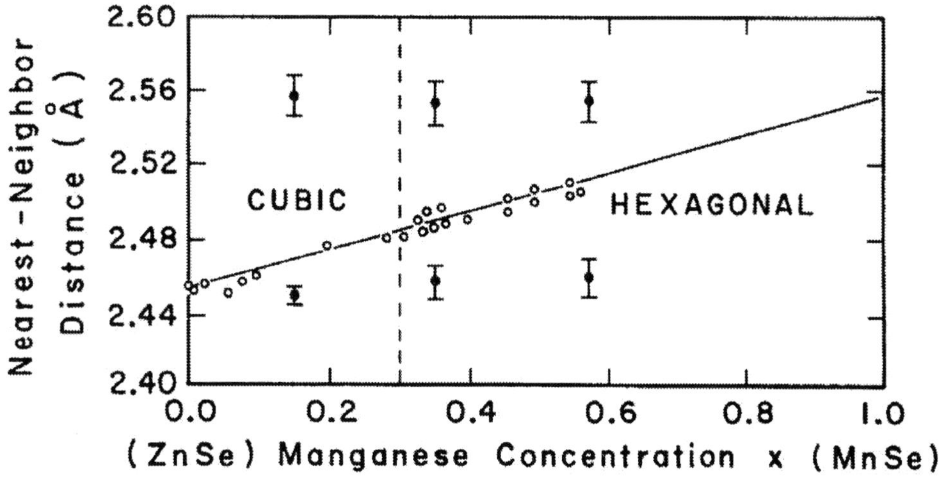

Fig. 2. EXAFS results for Zn-Se bond distance (lower filled circles); EXAFS results for Mn-Se bond distance (upper filled circles); x-ray diffraction determination of cation-cation distance, multiplied by $(3/8)^{½}$ (open circles); and linear fit to diffraction results (dashed line). See text for discussion of absolute bond lengths.

SUMMARY

In conclusion, we have shown that nearest-neighbor bond lengths in the diluted magnetic semiconductor $Zn_{1-x}Mn_xSe$ are essentially constant as a function of Mn composition, even though the crystal structure is cubic for Mn compositions below x = 0.3 and hexagonal for higher Mn compositions. These results show that the local crystal structure is highly distorted from the average "virtual crystal", and this distortion is insensitive to the differences between the cubic and hexagonal phases.

This work was supported in part by ONR #N00014-85-K-0614 and NSF #DMR 83-16988. The X11 beamline at Brookhaven National Laboratory is supported by DOE DMR-DE-AS05-80ER10742.

References

1. J. K. Furdyna, J. Appl. Phys. 53, 7637 (1982).

2. U. Debska, W. Giriat, H.R. Harrison, and D.R. Yoder-Short, J. Cryst. Growth 70,399 (1984).

3. D.R. Yoder-Short,U. Debska and J.K. Furdyna, J. Appl. Phy. 58, 4056(1985).

4. J.C.Mikkelsen, Jr., and J.B.Boyce, Phys.Rev.Lett. 49, 1412 (1982).

5. A. Balzarotti, M. Czyzykand, A. Kisiel, N. Motta, M. Podgorny, and M. Zimnal-Starnawska Phys. Rev. B30, 2295 (1984).

6. E. A. Stern, Contemp. Phys. 19, 289 (1978); P. A. Lee, P. H. Citrin, P. Eisenberger, and B. M. Kincaid, Rev. Mod. Phys. 53, 769 (1981).

Synthesis of the Dilute Magnetic Semiconductor CdMnTe by Ion Implantation of Mn into CdTe

G.H. Braunstein,*[1] D. Heiman,* S.P. Withrow,† and G. Dresselhaus*
*Massachusetts Institute of Technology, Cambridge, MA 02139
†Oak Ridge National Laboratory, Oak Ridge, TN 37831

Abstract

The dilute magnetic semiconductor CdMnTe has been synthesized by ion implantation of Mn into CdTe. Samples of CdTe have been implanted with Mn ions of 60 keV energy to fluences in the range 1×10^{13} cm^{-2} to 2×10^{16} cm^{-2} and subsequently annealed, using rapid thermal annealing, for 10–15 sec at temperatures $300 \leq T_A \leq 730$°C. The successful formation of a near surface layer of CdMnTe is demonstrated by studies of the structural, electronic and magnetic properties of the ion implanted and annealed samples; Rutherford backscattering–channeling analysis of the radiation-induced damage indicates complete recovery of lattice order after annealing at 700°C. Photoluminescence measurements, performed at 2K, reveal an increase in the energy band gap of the ion implanted alloy with respect to CdTe. Application of magnetic fields, up to 8T, produce both the characteristic energy shift of the excitonic recombination peak and polarization of the emitted radiation (in the Faraday configuration) previously observed in bulk-grown CdMnTe material.

Introduction

Ion implantation has evolved as a prominent technique in semiconductor technology. Besides its widespread application for precisely controlled *n*– or *p*–type doping, ion implantation is used to synthesize buried layers of pure species or alloys as well as to create new metastable phases and solid solutions. All these capabilities could be exploited readily within the field of dilute magnetic semiconductors.[1,2]

Only a limited variety of host materials and magnetic species can be prepared as DMS by standard chemical means. Ion implantation allows synthesis of these materials for a much wider variety of magnetic ions and for different physical configurations. Accurate control over the carrier concentration of DMS could be achieved by ion implantation of proper dopants. Thus ion implantation allows for the exploration of new physics and provides promise for novel device applications.

We have recently presented a preliminary study demonstrating the synthesis of DMS using ion implantation, by implanting Mn into CdTe to form CdMnTe.[3] Ion implantation of 1×10^{16} Mn ions/cm^2 at 60 keV into CdTe and subsequent rapid thermal annealing (RTA) at 730°C for 15 sec resulted in the formation of a near surface layer of CdMnTe, as evidenced by:

(a) Rutherford backscattering (RBS)–channeling analysis of the radiation damage which reveals complete recovery of lattice order and

[1] Present Address: Research Laboratories, Eastman Kodak Company, Rochester, NY 14650

Mat. Res. Soc. Symp. Proc. Vol. 89. © 1987 Materials Research Society

(b) Photoluminescence (PL) measurements at 0 and 10 KG applied magnetic fields which show the characteristic magnetic field tuning of the excitonic PL peak as previously observed in bulk grown CdMnTe.[4]

In this report we present more detailed studies of the magnetic field dependence of the photoluminescence, applying magnetic fields up to 8 Tesla. Indeed we observe the characteristic shift of the excitonic recombination and polarization of the emitted radiation with increasing magnetic field in full agreement with DMS theory, providing conclusive support for the synthesis of a DMS by ion implantation. The connection between the structural disorder and the semimagnetic properties of ion implanted CdTe is discussed in more detail elsewhere.[5]

Results and Discussion

Single crystal, (100) oriented CdTe substrates[6] mechanically polished and chemically etched ($\simeq 0.1\%$ bromine–methanol) were used in the present study. The CdTe samples were implanted with Mn ions having an energy of 60 keV to fluences in the range 1×10^{13} cm^{-2} to 2×10^{16} cm^{-2}. The samples were subsequently annealed in an inert gas atmosphere using rapid thermal annealing (RTA) for 10–15 sec at temperatures 300°C $\leq$ T_A $\leq$ 730°C. A flash lamp type annealer was used for $T_A \leq 500$°C. In both cases, the implanted side of the sample was placed on a larger silicon wafer which served both as a cap to avoid contaminants and for equilibrating the sample temperature.

RBS–channeling analysis has been used to study the radiation damage after implantation and after subsequent rapid thermal annealing. These experiments have been reported in detail elsewhere.[3] We focus attention here only on a particular fluence: 1×10^{16} Mn/cm^2 which should result in a concentration of Mn ions high enough to produce observable magnetic effects. Several samples were implanted with 60 keV Mn ions to a fluence of 1×10^{16} cm^{-2} and subsequently annealed, each sample at a different temperature from 300°C to 730°C. The results of the RBS–channeling analysis, partially shown by the spectra in Fig. 1 indicate complete recovery of the crystalline structure after annealing at 730°C for 15 sec. The RBS–channeling analysis also revealed that after annealing, the sample retained at least half of the implanted Mn, which was concentrated in the near–surface region of the sample. This conclusion follows from the measurements of the area of the small but resolvable peak indicated by the arrow in Fig. 1 which corresponds to Mn on the surface of the sample.

Photoluminescence (PL) constitutes a very sensitive technique to monitor the electronic and magnetic properties of DMS.[7] The energy band gap of the alloy increases with increasing Mn concentration resulting in a shift of the excitonic PL peak to higher energies. The magnetic field tuning of the band gap and the magnetic field induced polarization of the recombination radiation are also probed by PL experiments by monitoring the magnetic field induced shift and polarization, respectively, of the excitonic PL peak. In addition, this technique becomes particularly convenient in ion implantation studies where the range over which the implanted ions are distributed is comparable to the skin depth of the probing light. Figure 2 shows the photoluminescence spectra, taken at 2K, for various magnetic fields, for a sample of CdTe implanted with 1×10^{16} Mn/cm^2 at 60 keV and annealed using RTA for 15 sec at 730°C. The narrow peak at about 1.59 eV corresponds at an acceptor bound exciton transition (A°,X) in pristine CdTe.[8] The appearance of this peak in the spectra could be due to inhomogeneities in the ion implanted sample or due to the probing of the region beyond the implant

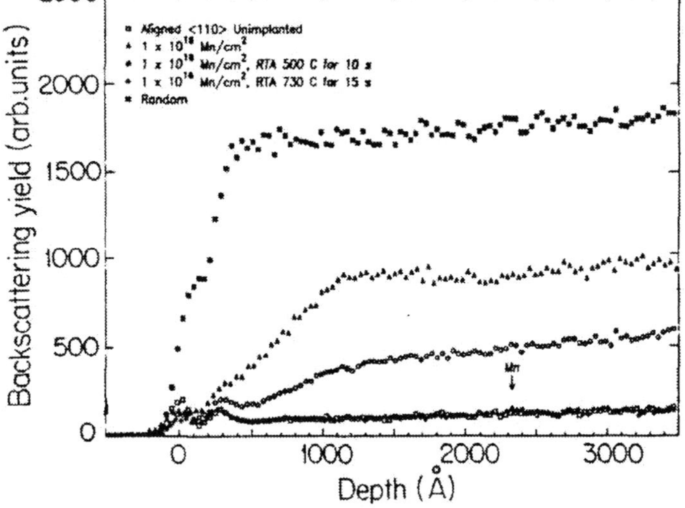

Figure 1: RBS–channeling spectra of CdTe implanted with 1×10^{16} Mn/cm^2 at 60 keV a) pristine, b) as implanted, c) after RTA at 500°C for 15 sec, d) after RTA at 730°C for 15 sec. The arrow indicates a small, barely resolvable peak corresponding to Mn on the surface of the sample annealed at 730°C.

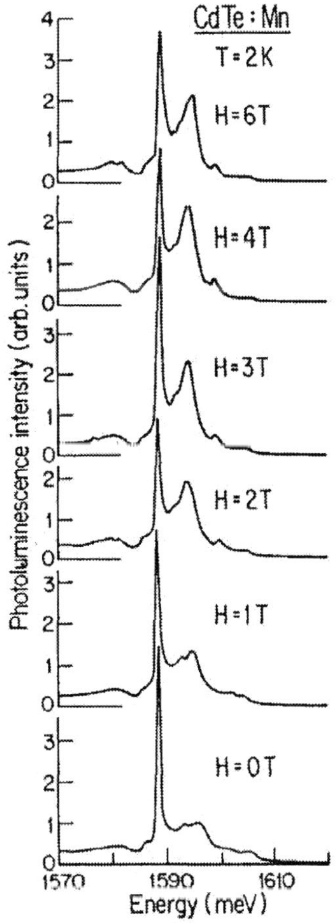

Figure 2: Photoluminescence spectra as a function of photon energy, taken at 2K for various magnetic fields, on a sample of Mn implanted CdTe (1×10^{16} cm^{-2}, 60 keV) annealed using RTA at 730°C for 15 sec.

distribution by the analyzing beam. Using the position of the "CdTe peak" as a reference, the enhancement of the broad and asymmetric peak observed in the spectra of Fig. 2 indicates the formation of a near surface layer of CdMnTe; we denote the broad asymmetric structure of Fig. 2 as the "CdMnTe peak". This "CdMnTe peak" appears at a higher energy than the "CdTe peak" indicating the formation of an alloy with an energy band gap larger than that of pristine CdTe. Also the "CdMnTe peak" shifts to lower energies with increasing magnetic field (for $H \leq 2.5$T). This shift is the result of the large internal exchange field which produces large splittings of the bands and is a clear signature of DMSs.

Upon application of an external magnetic field, the energy band gap of a DMS and consequently also the energy of the PL recombination radiation changes according to the expressions

$$\Delta E_g(x,T,H) = E_g(x,T,H) - E_g(x,T,0) \tag{1}$$

$$\Delta E_g(x,T,H) = AH + BH^2 + \frac{1}{2}xN_0(\alpha - \beta)\langle S_z\rangle \tag{2}$$

The first term has two components: $A = -(g+2K)\mu_B + \frac{1}{2}\hbar\omega_c/H$ due to Zeeman and Landau level splitting, respectively, of the energy bands at the Γ point. Here $g \simeq 1.59$ is the g–factor for conduction electrons in CdTe, K is related to the Luttinger parameter[9,10] K and ω_c is the cyclotron frequency for band electrons, while μ_B is the Bohr magneton. The second term BH^2 in Eq. (2) arises from the diamagnetic contribution to the exciton binding energy. It has been analyzed in detail for CdMnTe by Aggarwal *et al.*[10] The Zeeman splitting, Landau level separation and diamagnetic energy shift are common to any semiconductor, independent of the presence or absence of magnetic ions in the lattice. In the case of DMSs there is an additional contribution to ΔE_g, due to the exchange interaction between localized and itinerant spins, given by the third term in Eq. 2. Here x is the Mn molar fraction, α and β are exchange integrals for conduction and valence bands respectively, N_0 is the number of unit cells per unit volume and $N_0(\alpha - \beta) \cong 1100$ meV.[11] The average spin $\langle S_z\rangle$ follows a Brillouin–like phenomenological function B_S[11]

$$\langle S_z\rangle = S_0 B_S\left[\frac{S g_{\mathrm{Mn}}\mu_B H}{k_B(T+T_0)}\right] \tag{3}$$

where $S = \frac{5}{2}$, and the argument of the Brillouin function is given in square brackets with $g_{\mathrm{Mn}} = 2$ being the Mn^{++} g–factor. Because of the low Mn concentration we take $T_0 \approx 0$ and $S_0 \approx S$ in Eq. (

Figure 3 shows a fitting to the experimental data using Eq. (2). The curve labeled CdTe is a fitting to the energy shift of the "CdTe like" peak and uses only the two first terms of Eq. (2) with A and B as fitting parameters. The curve labeled CdMnTe is a fitting to the energy shift of the "CdMnTe" PL recombination peak. The downshift observed for fields $H \leq 2.5T$, is due to the third term in Eq. (2) and represents a clear indication of the formation of a semimagnetic alloy. The energy upshift, observed in both curves for $H \geq 2.5T$, is due mainly to the Landau level splitting of the bands and the diamagnetic contribution to the exciton energy which, because of the low Mn concentrations involved, become comparable to the exchange splitting in the present study. From comparison with results on chemically synthesized CdMnTe, we obtain an average Mn concentration of $x = 0.002$. This value is at least one order of magnitude smaller than the concentration of Mn initially implanted. Although the RTA should

Figure 3: Photon energy as a function of applied magnetic field in CdMnTe synthesized by implantation of Mn into CdTe (full curve). The PL recombination radiation characteristic of CdTe was also observed in the same sample and is plotted for comparison (dotted line).

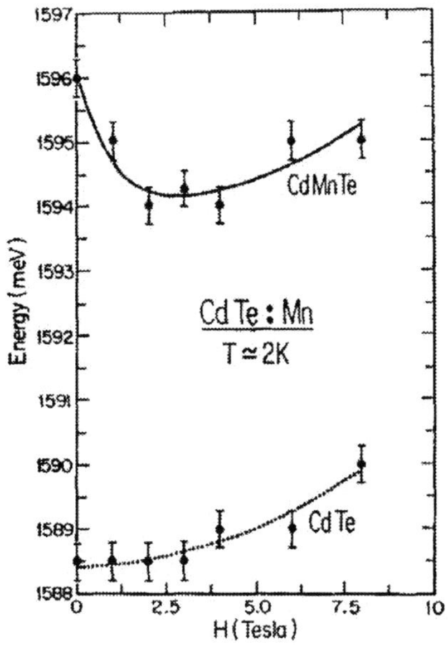

Figure 4: Polarization of the luminescence ρ in the Faraday configuration as a function of applied magnetic field in CdMnTe synthesized by implantation of Mn into CdTe followed by rapid thermal annealing (see text).

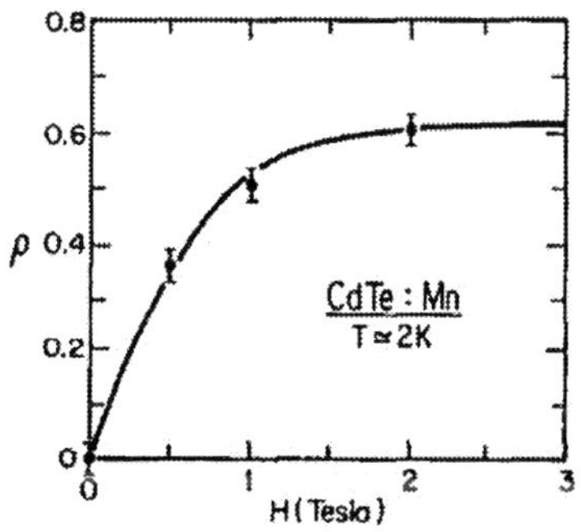

not give rise to a pronounced out-diffusion of Mn under these annealing conditions, some accumulation of Mn in the near-surface region of the sample annealed at 730°C is nevertheless observed in the aligned spectrum and shown by the arrow in Fig. 1. A segregation of Mn to the near-surface region could perhaps provide an explanation for the low efficiency of the ion implantation synthesis. However more work will be required to characterize this behavior more quantitatively.

Another characteristic behavior exhibited by DMS is the polarization of the PL recombination radiation upon application of an external magnetic field. In this connection, we have plotted in Fig. 4 the polarization $\rho = (I_{RCP} - I_{LCP})/(I_{RCP} + I_{LCP})$, of the CdMnTe peak, as a function of magnetic field, where RCP and LCP refer to right and left handed circular polarization, respectively. The experimental data of ρ *vs.* H can be fitted (full line in Fig. 4) with the functional form[12]

$$\rho = \rho_s \tanh\left[\frac{p\mu_B H}{k_B T}\right] \tag{4}$$

where ρ_s is a constant and p is a function of temperature. The curve fitting in Fig. 4 yields $\rho_s = 0.615$ and $p = 3.0$. The value of p is somewhat smaller than in previous studies on higher x samples of bulk grown CdMnTe materials.[12]

In summary we demonstrate that ion implantation can be used to synthesize thin layers of DMS material, thus opening a new approach to the preparation of DMS compounds, including possibly metastable DMS compounds.

Acknowledgments

We would like to thank Prof. M.S. Dresselhaus for helpful discussions. The MIT authors acknowledge financial support from NSF Grant # DMR 83–10482. The work at Oak Ridge National Laboratory was sponsored by the US Department of Energy under contract #DOE–AC05–84OR21400 with Martin Marietta Energy Systems, Inc. The high field experiments were performed at the Francis Bitter National Magnet Laboratory which is supported by the Division of Materials Research of the National Science Foundation.

References

[1] J.K. Furdyna, *J. Appl. Phys.* **53**, 7637 (1982).

[2] N.B. Brandt and V.V. Moshchalkov, *Adv. in Phys.* **33**, 194 (1984).

[3] G.H. Braunstein, G. Dresselhaus, and S.P. Withrow, *Nucl. Instr. and Meth.* B (in press, 1987).

[4] R. Planel, J.A. Gaj, and C. Benoit a la Guillaume, *J. de Physique* **C5**, 39 (1980).

[5] H.J. Jiménez–González, A. Lusnikov, G. Dresselhaus, G.H. Braunstein, and S.P. Withrow, Proceedings of this symposium.

[6] Purchased from II–VI, Inc., Saxonburg, PA 16056, USA.

[7] A. Golnick, J. Ginter, and J.A. Gaj, *J. Phys. C: Solid State Phys.* **16**, 6073 (1983).

[8] K. Zanio, "Cadmium Telluride", in *Semiconductors and Semimetals*, edited by R.K. Willardson and A.C. Beer (Academic Press, New York, 1978) Vol. **13**.

[9] C. Kittel, *Quantum Theory of Solids*, (John Wiley and Sons, Inc., New York, 1963), p. 284.

[10] R.L. Aggarwal, S.N. Jasperson, P. Becla, and R.R. Galazka, *Phys. Rev.* **B32**, 5132 (1985).

[11] J.A. Gaj, R. Planel, and G. Fishman, *Solid State Commun.* **29**, 435 (1979).

[12] D. Heiman, J. Warnock, P.A. Wolff, R. Kershaw, D. Ridgley, K. Dwight, and A. Wold, *Solid State Commun.* **52**, 909 (1984).

THE CONNECTION BETWEEN STRUCTURAL DISORDER AND SEMIMAGNETIC PROPERTIES IN IMPLANTED CdTe

H.J. Jiménez–González, A. Lusnikov, G. Dresselhaus, G.H. Braunstein[1]
Massachusetts Institute of Technology, Cambridge, MA
and S.P. Withrow
Oak Ridge National Laboratory, Oak Ridge, TN

Abstract

The structural disorder and annealing characteristics of dilute magnetic semiconductors synthesized by implantation of Mn ions into CdTe are studied. After implantation but prior to annealing, the implanted Mn ions reside at random sites in a highly disordered CdTe lattice. To separate magnetic and nonmagnetic effects we have implanted CdTe with nonmagnetic Zn ions under the same conditions as the Mn ions (energy 60 keV, fluence $10^{16}cm^{-2}$). Various annealing conditions are applied to monitor the decrease in intensity of the disorder–induced photoluminescence peaks and the increase of the peak intensity of the CdTe–like exciton line is reduced.

Introduction

It was shown previously[1] that dilute magnetic semiconductors (DMSs) can be synthesized by implantation of Mn ions into CdTe (60 keV and fluences up to 1×10^{16} cm^{-2}), followed by rapid thermal annealing (RTA at 740°C for 15 s) to remove the radiation–induced lattice disorder. Evidence for the successful synthesis of a very low concentration DMS $Cd_{1-x}Mn_xTe$ was provided by photoluminescence (PL) studies which indicate an increased band gap and by magnetoluminescence studies which showed the characteristic shift of the excitonic PL peak to lower energies with increasing magnetic field.[1,2] After implantation but prior to annealing, the implanted Mn ions reside at random sites in a highly disordered CdTe lattice. In addition to restoring lattice order, the annealing process allows Mn ions to occupy substitutional Cd sites. The effective Mn concentration in the optical skin depth was estimated by the magnitude of the blue shift in the excitonic PL peak in zero magnetic field. In addition, the implantation of Mn and Te into PbTe at a similar dose to form an n–type semiconductor $Pb_{1-x}Mn_xTe$ has been studied previously.[3]

In the present work, Mn and Zn ions have been implanted into CdTe at an energy of 60 keV and at fluences of $\sim 10^{16}cm^{-2}$. By studying the effect of Zn in CdTe, we can separate the magnetic and non–magnetic effects of implantation. By use of rapid thermal annealing (RTA), the lattice disorder introduced by implantation can be reduced almost completely. The distribution of the implanted species prior to RTA is monitored by SIMS and Auger electron spectroscopy. The effect of RTA on the lattice disorder is studied by PL measurements.

[1]Present Address: Research Laboratories, Eastman Kodak Company, Rochester, NY 14650

Mat. Res. Soc. Symp. Proc. Vol. 89. © 1987 Materials Research Society

Experimental Procedure

Single crystal samples of CdTe grown using the Bridgman technique oriented along the (100) direction were used as the host material for implantation. The samples were $8 \times 4 \times 1$ mm^3 in size, and were polished on one face with alumina powder of 0.05μm grit size. Each piece was chemically etched in a 1% solution of bromine in methanol for 5 minutes, and immediately rinsed in neat methanol.

Three of the samples were implanted with Zn ions at an incident energy of 60 keV and a total fluence of 10^{16}cm^{-2}. The remaining pieces were implanted with Mn at the same energy with fluences from 1×10^{16}cm^{-2} to 2×10^{16}cm^{-2}. Variation of the incident ion energy in principle permits control of the distribution of the implanted ions. However corrections for surface sputtering are important in determining the final depth and distribution of the implanted ions. To study the effect of annealing, the PL and RTA measurements on a test sample were combined in a cycle as follows starting from the as-implanted (unannealed) sample: (a) the PL spectrum was taken in zero field, (b) the PL spectrum was taken in the presence of a magnetic field H, (c) the RTA was performed at the next (higher) temperature T_n for a time Δt, where n identifies the cycle number. With the newly annealed sample, steps (a), (b) and (c) were repeated until some final temperature T_f where an ordered sample was achieved.

All the PL measurements were done at ~10 K. The same annealing time was used ($\Delta t = 10$ s) throughout the experiments to reduce ambiguities in the data interpretation. The annealing temperatures ranged between 400°C and 820°C. The initial and final temperatures of annealing, T_i and T_f respectively, were the same for all Zn implanted samples, though each sample followed a different sequence of annealing steps between T_i and T_f. Relatively large temperature steps were chosen in order to obtain resolvable differences in the PL spectrum at each step.

Prior to implantation, a detailed Rutherford backscattering analysis was performed on one sample (see Fig. 1 in Ref. 2). Along the (110) channeling direction a minimum backscattering yield of ~6.3% was observed relative to the signal obtained from a random direction, in agreement with previous results.[1] The implantation was done at an angle of 6° with respect to the channeling direction, to avoid channeling effects during implantation. To determine the parameters of the implanted ion distribution, a Monte Carlo type calculation was done using the TRIM[4] computer program. After implantation, the channels were found to be effectively destroyed. In order to measure the implantation profile more directly SIMS (secondary ion mass spectroscopy) and Auger electron spectroscopy analysis were employed. The results of the Mn ion distribution determined by SIMS are shown in Fig. 1. To account for the discrepancy between the SIMS data and the TRIM calculation, the effect of sputtering of surface atoms by the implanted ions was modeled.[5] This calculation (to be described elsewhere[5]), provides a natural mechanism for moving the peak of the Mn distribution closer to the surface. Because of this sputtering effect, an upper bound is imposed on the concentration of dopant that can be implanted at a given energy. Though the calculation gives a moderately good overall fit to the experimental SIMS points, it assumes sputtering yields for pure Cd and Te[6] which may over estimate the sputtering of the CdTe surface. Refinements of this model calculation are in progress.

Figure 1: A SIMS profile for CdTe implanted with Mn and having a Mn surface concentration of nearly 3% as determined by Auger. The star points refer to the experimental SIMS measurements and the solid line to the profile calculated for an incident energy of 60 keV by the TRIM program, while the open squares represent a model calculation where surface sputtering was included.[5] All curves are normalized to the same Mn peak concentration.

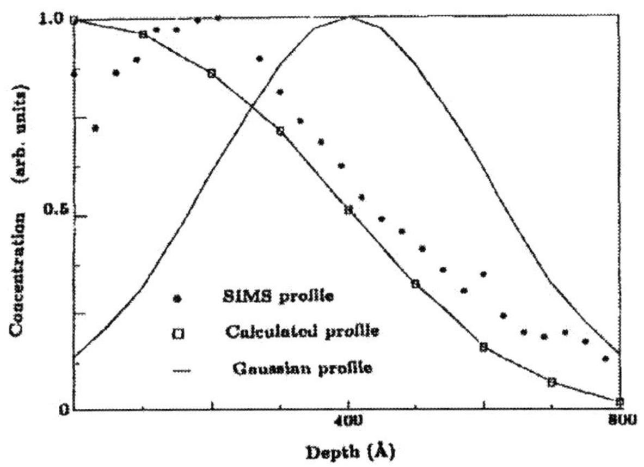

Experimental Results and Discussion

To study the effect of annealing, samples of CdTe were implanted with magnetic Mn ions and nonmagnetic Zn ions at the same energy of 60 keV and to the similar fluence of $\sim 10^{16}$ ions/cm^2. The samples were then characterized by their PL spectra. In Fig. 2, we show the PL spectra of the unimplanted CdTe, and the CdTe samples implanted with Mn and Zn ions but prior to annealing. For illustrative purposes the PL spectra in Fig. 2 are normalized to the same intensities as the main PL for the unimplanted CdTe, though the actual intensities of the PL peaks in the as-implanted and pristine samples are approximately in the ratio of 1:10. The unimplanted sample in Fig. 2 shows the characteristic exciton spectrum for CdTe. Implantation with either Mn or Zn ions at the same energy (60 keV) and fluence (1×10^{16}/cm^2) results in almost the same changes in the PL spectra, including a decrease in the peak of the PL intensity by an order of magnitude, an increase in the PL linewidth by a factor of ~ 6, and a blue shift of the peak PL intensity by ≈ 1.2 meV. Also of interest is the small feature appearing in the PL spectrum of the as-implanted Mn sample at 7870 Å. The SIMS profile in Fig. 1 corresponds to the as-implanted Mn sample shown in Fig. 2, with a total near-surface Mn concentration of nearly 3% according to the indicated distribution as measured by Auger electron spectroscopy. With the preliminary data available thus far, it is not possible to identify the PL lines of the implanted samples as excitonic or band to band transitions or transitions associated with some other lattice defect.

Having established that the non-magnetic Zn and the magnetic Mn give rise to essentially the same shift in the main PL peak(at 7790 Å) for similar implantation conditions, we then examined the effect of rapid thermal annealing (RTA) on the Zn-implanted samples. In planning the experiments, the temperature range $400 < T_a < 820$°C available from the RTA apparatus was expected to satisfy the following limits. The lower value $T_i = 400$°C was expected to yield a small but measurable effect on the PL spectrum and the upper value $T_f = 820$°C was expected to yield a well-ordered sample, yet retaining a significant concentration of the implanted species. Figure 3

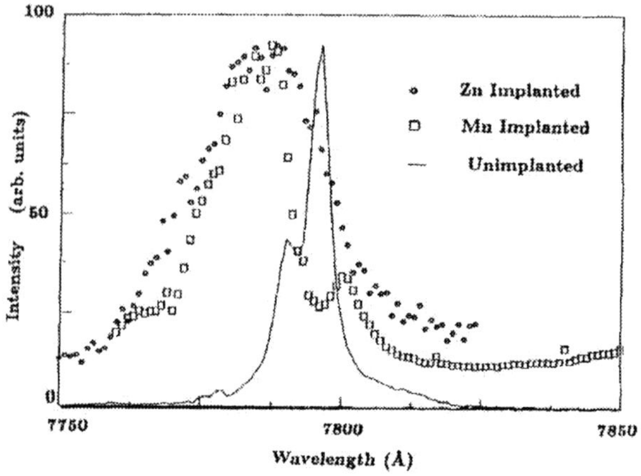

Figure 2: Photoluminescence spectra of unimplanted CdTe,[7] and of CdTe implanted with Mn and Zn (60 keV, $1 \times 10^{16}/cm^2$) but prior to annealing. The spectra are normalized to have the same peak intensity as the main exciton peak in the unimplanted CdTe.

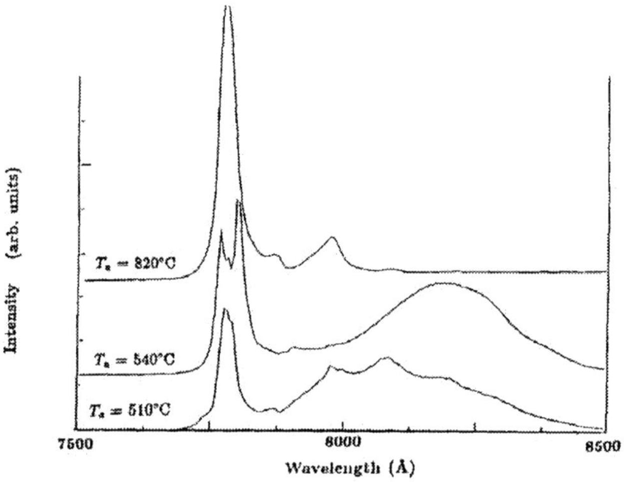

Figure 3: Photoluminescence (PL) spectra on Zn–implanted CdTe samples corresponding to a sequence of annealing steps between T_i and T_f. As the disorder is reduced (T_a is increased), the broad features at long wavelengths decrease in intensity relative to the peaks in the vicinity of 7790Å (see text).

shows the results of the PL spectrum for various annealing temperatures T_a on the same Zn–implanted sample shown in Fig. 2 prior to annealing. It was found that for $T_a = T_i = 400°C$, the broad (FWHM $\sim$ 75 Å) PL peak centered at ~7790 Å before annealing (see Fig. 2) was no longer present, but instead a broad PL peak centered at ~ 7840 Å was observed. The evolution of the broad PL peak for the Zn–implanted sample from Fig. 2 to that in Fig. 3 is presently under investigation. The photoluminescence (PL) spectrum of the disordered material exhibits additional peaks at long wavelengths, most probably related to vacancy–trapped excitons.

In Fig. 3, we focus on the observation of a sharp excitonic peak at ~7790 Å in the PL spectrum of the Zn–implanted sample as the annealing temperature increases. This sharp peak has a threshold value between $400°C \leq T_a \leq 510°C$ (a more precise threshold value is presently under investigation), and the exciton peak grows in intensity as T_a increases and the disorder is reduced. As T_a increases, the features at long wavelengths decrease in intensity relative to the peaks in the vicinity of 7790Å. At the highest annealing temperature T_f (820°C), the peaks for all three Zn–implanted samples were remarkably similar in both position and linewidth. No magnetic field induced shifts were found in any of the PL peaks of the Zn–implanted samples, in contrast to the results for the Mn–implanted samples.

For the present, the origins of the broad PL peaks in the as–implanted Zn–implanted

Figure 4: Shift of the excitonic PL peak versus annealing temperature of Zn (Δ), Mn at $H = 0$ ($\square$), and Mn at $H = 1$ T ($\bullet$) implanted CdTe. The error bars represent a standard deviation calculated from several spectra of the same sample. The dashed line is a visual guide to the figure.

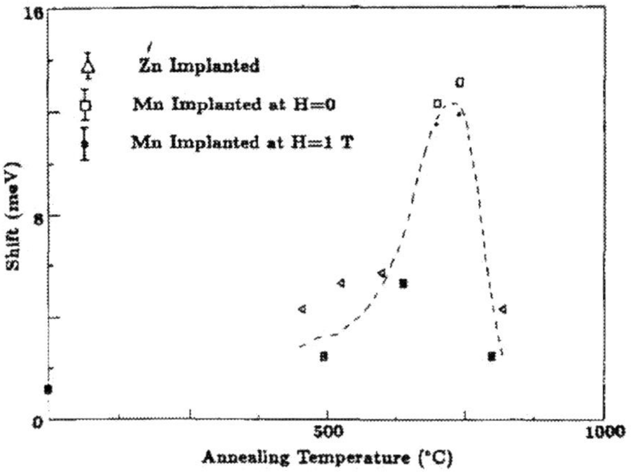

samples and in the annealed samples are not understood, nor is it clear that they have a common origin. However the sharp PL features in the pristine CdTe sample and in the Zn–implanted RTA samples do appear to have a common origin, and are identified with an excitonic PL process. In Fig. 4, a plot of the energy shift of the sharp PL peak ΔE_p (relative to pristine CdTe) versus annealing temperature T_a is shown for Mn and Zn–implanted samples. The T_a dependence of the shift ΔE_p may be indicative of the variation of the band gap with T_a. The decrease in ΔE_p with T_a above $\sim$ 750°C may be due to the loss of Zn and Mn from the optical skin depth of the sample.

The results for the Mn–implanted CdTe at zero field are similar to that reported for Zn. After annealing, the Mn–implanted CdTe shows an increasing shift in the PL peak as the annealing temperature increases as shown in Fig. 4. However, for the two anneals at 740°C and 750°C, the exciton peak showed a much larger shift ΔE_p than for the other annealing temperatures. Unfortunately, no data are yet available for the Zn–implanted samples at $T_a = 740$°C and 750°C, so that a direct comparison of the Zn and Mn–implanted CdTe cannot be made over the entire range of T_a in Fig. 4. Also at these T_a values the exciton peak was found to shift measurably in a 1 tesla magnetic field. Both the PL shift ΔE_p and the magnetic field shift of the PL peak are consistent with a 1% atomic concentration of Mn at the annealing temperature of 740°C. Both of these shifts are also consistent with the results of Braunstein, et al.[2]. Finally at the highest annealing temperature of 820°C for both the Zn and Mn implants the shift decreases indicating an almost complete loss of the implanted species.

Though the energy shifts ΔE_P of the PL lines in the Zn and Mn–implanted samples are similar, the linewidths for the Mn–implanted samples are almost an order of magnitude larger than in the case on Zn implants. Nevertheless in both the Zn and Mn, cases the linewidth decreases with increasing annealing temperature.

The band gap of $Cd_{1-x}Mn_xTe$ synthesized by chemical methods, increases by approximately 14meV for each 1% increase in concentration x.[8] In the case of Mn–implanted CdTe with $T_a = 715$°C,[1] the shift in the main PL peak relative to that for pure CdTe was found to be considerably smaller than expected from the maximum Mn concentration calculated by the TRIM program.[4] Several possible factors may contribute to this discrepancy: out–diffusion of Mn under annealing, limited contribution of interstitial Mn to the change in the energy gap, and sputtering of the sample surface during implantation which reduced the Mn concentration. Model calculations (see Fig. 1) show that sputtering plays a dominant role for the conditions of the present work.[5]

The somewhat different behavior of the PL peak between Zn and Mn is not clearly resolved in the present set of experiments. The annealing range which gave DMS behavior for the Mn–implanted samples was not covered in the Zn–implanted samples. The behavior of the PL peak for the different annealing temperatures suggests that Mn assumes substitutional sites as the annealing temperature increases. Though the number of occupied substitutional sites may be increasing over the entire annealing range, the interstitials are expelled at about 740°C and the substitutional Mn is expelled at about 820°C. Approaches to increase the Mn concentration are under investigation.

In the case of Mn–implanted CdTe we obtain a small blue shift in the PL excitonic line before any annealing is done, a much larger blue shift after annealing, and finally a red shift upon application of a magnetic field. This behavior is clear evidence for the creation of $Cd_{1-x}Mn_xTe$.

These preliminary studies with non–magnetic Zn and magnetic Mn ions show that PL is a powerful method for studying RTA phenomena. The change in the PL spectra as a function of T_a is dramatic. Further work is needed to identify the threshold T_a for the onset of the excitonic peak and to follow the evolution of the broad PL peak in the highly disordered regime. Further studies of the retention of the implanted species as a function of T_a are needed.

Acknowledgments

We would like to thank Prof. M.S. Dresselhaus for helpful discussions. The MIT authors acknowledge financial support from NSF Grant # DMR 83–10482. The work at Oak Ridge National Laboratory was sponsored by the US Department of Energy under contract #DOE–AC05–84OR21400 with Martin Marietta Energy Systems, Inc.

References

[1] G.H. Braunstein, G. Dresselhaus, and S.P. Withrow, *Nucl. Instr. and Meth.* B (in press, 1987).

[2] G.H. Braunstein, D. Heiman, S.P. Withrow, and G. Dresselhaus, Proceedings of this Symposium.

[3] L. Palmetshofer, B. Wakolbinger and E. Zinner, *Nucl. Instr. and Meth.* **209/210**, 725 (1983).

[4] J.P. Biersack and L.G. Haggmark, *Nucl. Instr. and Meth.* **174**, 257 (1980).

[5] A. Lusnikov, (to be published).

[6] H.H. Anderson and H.L. Bay, *Sputtering by Ion Bombardment*, ed. R. Behrish, University of Aarhus–Denmark, p. 75 (1980).

[7] K. Zanio, "Cadmium Telluride", in *Semiconductors and Semimetals*, edited by R.K. Willardson and A.C. Beer (Academic Press, New York, 1978) Vol. **13**.

[8] D. Heiman, P. Becla, R. Kershaw, D. Ridgley, K. Dwight, A. Wold and R.R. Galazka, *Phys. Rev.* **B34**, (1986) (to be published).

SUBMICRON HETEROSTRUCTURES OF DILUTED MAGNETIC SEMICONDUCTORS

R. L. GUNSHOR[*], *L. A. KOLODZIEJSKI*[*], *N. OTSUKA*[**], *S. DATTA*[*], *AND A. V. NURMIKKO*[***]
[*]School of Electrical Engineering, Purdue University, West Lafayette, IN 47907
[**]Materials Engineering, Purdue University, West Lafayette, IN 47907
[***]Division of Engineering, Brown University, Providence, R.I. 02912

ABSTRACT

The successful thin film growth of diluted magnetic semiconductors (DMS) by molecular beam epitaxy has "nucleated" a new field of research in which the DMS material is incorporated in a variety of novel superlattice and quantum well structures. The observation of reflection high energy electron diffraction intensity oscillations in ZnSe and MnSe has enabled the fabrication of ultrathin layered structures involving the "hypothetical" zincblende magnetic semiconductor MnSe. The expected antiferromagnetic ordering of MnSe is increasingly inhibited as the MnSe layer thickness is reduced from ten monolayers to the quasi-2D limit of one monolayer. Further developments include new observations of the epitaxial growth and nucleation of ZnSe, utilizing GaAs epilayers as the substrate material.

INTRODUCTION

ZnSe is one of the more technologically important II-VI compounds because its bandgap corresponds to a wavelength in the blue portion of the visible spectrum. ZnSe has a bandgap approximately twice that of GaAs while possessing a nearly identical lattice constant (0.25% lattice mismatch) and highly compatible thermal expansion coefficients. The prospect of realizing a "perfect" interface allows one to envision many device applications for the ZnSe/GaAs heterojunction. Using GaAs as a substrate material for ZnSe provides for the future integration of ZnSe optical devices on GaAs for optoelectronic applications. In order to investigate the physics and possible applications of multiple quantum well structures and superlattices based on ZnSe, we have focused attention on the wider bandgap diluted magnetic semiconductor material $Zn_{1-x}Mn_xSe$. Not only does incorporation of Mn result in an increase of the ZnSe bandgap, but Mn introduces potentially useful magneto-optical properties such as greatly enhanced Faraday rotation as well as a degree of magnetic field tunability of the direct bandgap. Of immediate importance is the use of Mn as an efficient light emitter for electroluminescent display devices. In this review we address new observations into the nucleation of ZnSe on GaAs where both bulk substrates and molecular beam epitaxial (MBE) epilayers are employed. We will briefly mention the growth and properties of multiple quantum well and superlattice structures employing ZnSe as the well material and $Zn_{1-x}Mn_xSe$ as the barrier material. As an extension of this work we will discuss the implications of incorporating barrier layers of the "hypothetical" zincblende magnetic semiconductor MnSe. The use of reflection high energy electron diffraction (RHEED) intensity oscillations observed in both ZnSe and MnSe, has enabled the growth of superlattice structures with MnSe layer thicknesses down to the quasi-2D limit.

HETEROEPITAXY OF ZnSe ON GaAs EPILAYERS

The various (Zn,Mn)Se epilayers and superlattice structures were grown in a

Mat. Res. Soc. Symp. Proc. Vol. 89. © 1987 Materials Research Society

Perkin-Elmer model 400 MBE using elemental sources. For the ZnSe epilayers, a Zn to Se flux ratio of approximately unity was used, with substrate temperatures ranging from 300 to 400 °C. The layered structures involving the ternary DMS material and the magnetic semiconductor MnSe were all grown at 400 °C with growth parameters previously reported [1]. Conventional chemical etching techniques were used for preparation of the GaAs (100) substrates.

The GaAs epilayers destined for subsequent ZnSe epitaxy were grown in a separate Perkin-Elmer model 400 MBE system. To maintain a contamination-free surface, the GaAs epilayers were passivated with amorphous arsenic before transfer to the II-VI MBE system. The GaAs epilayers with thicknesses ranging from 1.5 to 2.0 μm were grown at a substrate temperature of 600 °C using Ga and As_2 sources. The Ga source shutter was closed to end the growth, at which time the substrate heater was turned off in the presence of a reduced As flux. Following this procedure, Auger electron spectroscopy (AES) revealed only an arsenic peak. The Mo sample holder, to which was mounted the As-passivated GaAs sample, was transferred in air to the II-VI system located in the same room. The amorphous As layer was thermally desorbed in the analytical chamber of the II-VI MBE while being monitored by AES. The desorption of arsenic occurred at approximately 290 °C and ceased shortly after the appearance of a Ga peak. Finally the GaAs sample was transferred to the growth chamber for ZnSe epitaxy.

Reflection high energy electron diffraction was employed to monitor the early stages of growth. The RHEED electron gun operated at 10 kv with a beam current of 1 mA. Intensity measurements were made using a photomultiplier coupled by an optical fiber to an x-y stage mounted at the focal plane of a camera normally used to record RHEED patterns. The RHEED intensity oscillations were observed at the specular spot in the [110] azimuth, with an incident beam angle less than 1 ° (off-Bragg condition).

A number of groups have reported the growth by MBE of ZnSe on GaAs substrates. In all cases the growth occurred by the formation of islands as suggested by the observation of spotty RHEED patterns during the early stages of nucleation [2-5]. Although it has been suggested that the chemical mismatch [5] between this II-VI/III-V heterointerface is responsible for the observed three-dimensional growth, recent observations of the nucleation of ZnSe onto GaAs epilayers seem to contradict this conclusion. A series of ZnSe films were grown on MBE-grown GaAs epilayers, all of which exhibited two-dimensional nucleation. In Fig. 1 is seen the RHEED intensity oscillations recorded during the nucleation of ZnSe on a GaAs epilayer.

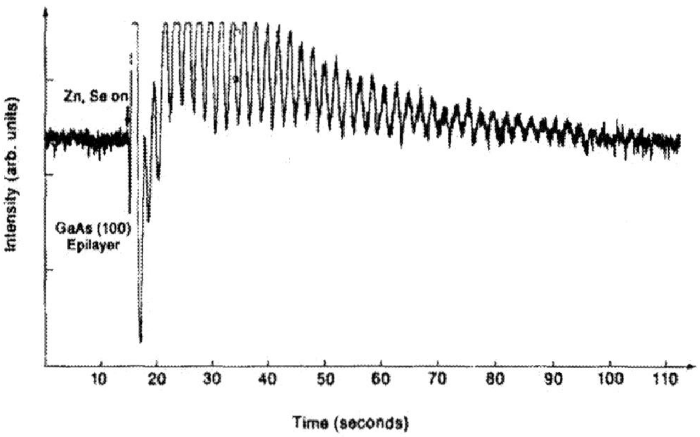

Fig. 1. RHEED intensity oscillations during nucleation of ZnSe on a GaAs epilayer.

Although the visual observation of the time evolution of the RHEED pattern suggested that the early stages of nucleation occurred via a two-dimensional growth mechanism, additional confirmation was provided by the clear presence of the intensity oscillations, which are characteristic of layer-by-layer growth. In contrast the variation of the intensity of the specular reflection during nucleation on a GaAs bulk substrate showed no evidence of oscillatory behavior. As the Zn and Se source shutters were opened to begin growth on the substrate, the intensity rapidly decreased followed by a gradual recovery of the initial intensity (~15 sec.), and a slow decrease to a steady state intensity level slightly lower than the initial GaAs static intensity (~50 sec.). A similar intensity variation has been reported for three-dimensional nucleation of InGaAs on GaAs epilayers [7]. The occurrence of two-dimensional nucleation improves the chances of obtaining a useable heterojunction interface.

The interfacial microstructure of ZnSe grown on GaAs epilayers and substrates were investigated by cross-sectional transmission electron microscopy (TEM). Fig. 2 shows a high resolution electron microscope (HREM) image of the interface between a 1000 Å ZnSe film grown on a 1.5 μm GaAs epilayer.

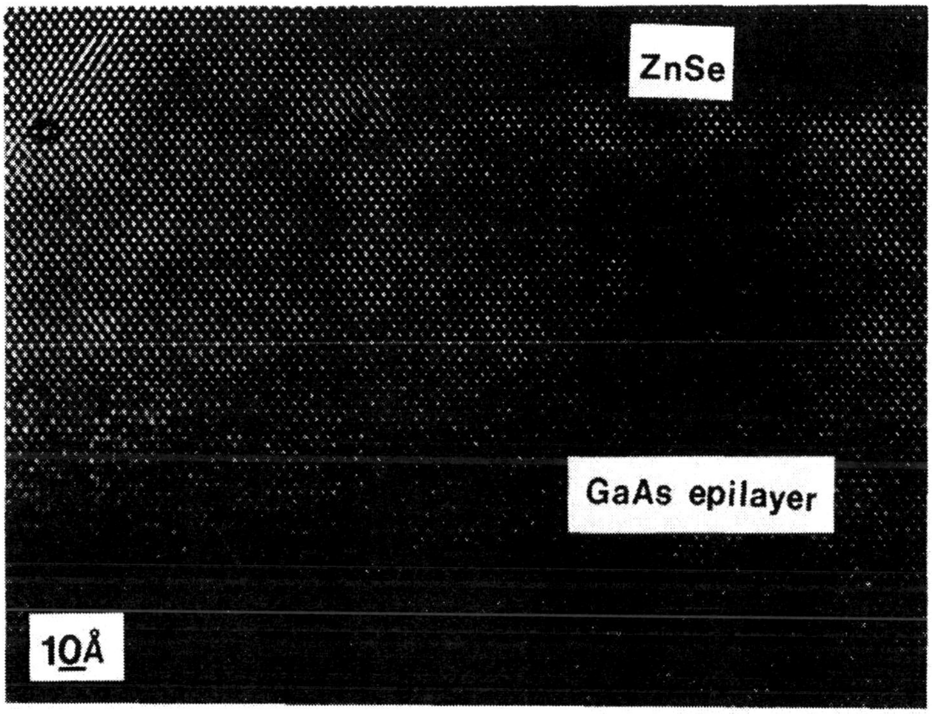

Fig. 2. HREM image of the interfacial region between a 1000 Å ZnSe epilayer and a 1.5 μm GaAs epilayer.

The film exhibits an essentially featureless image of the interface which appears as an atomistically flat boundary over wide areas with a perfectly coherent contact between the two crystals. In contrast, HREM images taken of the interface between ZnSe films grown on GaAs substrates reveal wavy, step-like boundary images indicating the presence of small pits and steps on the surface of the GaAs substrates. Dark field images of the 1000 Å ZnSe film show neither misfit or threading dislocations or stacking faults in observed areas, thus confirming the pseudomorphic character of the layer. TEM images suggest that a major factor influencing the occurrence of two- or three-dimensional nucleation is the relative density of nucleation sites such as surface dislocations and step edges. A GaAs epilayer is expected to have a substantially lower density of surface defects than a conventionally prepared substrate.

The evaluation of strain-shifted photoluminescence and reflection spectroscopy features provide a means for determining the deformation potentials for ZnSe. Fig. 3 shows the photoluminescense measurements of a 1000 Å ZnSe pseudomorphic layer grown on a GaAs epilayer was performed using the UV lines of an argon laser. At liquid helium temperatures the dominant feature is a free exciton at 2.8064 eV, a value 4.4 meV higher than the transition energy in bulk crystals.

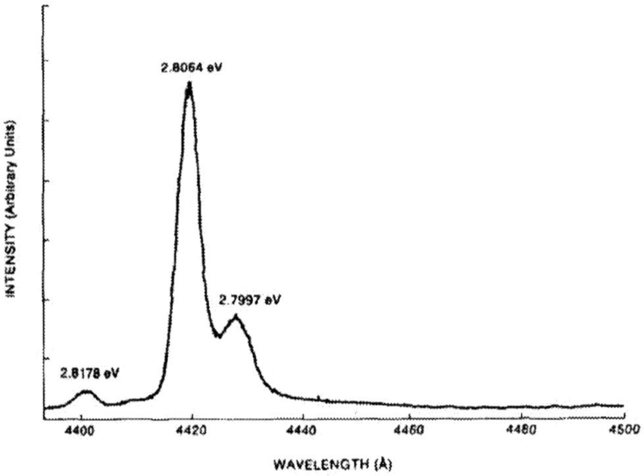

Fig. 3. 8k Photoluminescence spectrum of a 1000 Å ZnSe pseudomorphic epilayer grown on a 1.5 μm GaAs epilayer taken with an excitation density of 3.5 W/cm^2.

The 0.25% lattice mismatch between the ZnSe and GaAs layers determines the degree of strain in the pseudomorphic layer. The strain results in the aforementioned energy shift, as well as removing the light-heavy hole valence band degeneracy. The valence band splitting is revealed by the presence of a second free exciton peak located 11.4 meV higher in energy, and representing a light hole transition. (The feature at 2.7997 eV is usually identified as a neutral donor bound exciton.) Using a Poisson ratio of 1.7 to relate the strain in the plane parallel to the interface with the normal component of strain, one obtains deformation potentials of b = 1.05 eV and a = -4.87 eV. Despite the wide range of deformation potentials reported in the literature, these deformation potentials agree quite well with the most recently reported vallues (a = 1.2 eV and b = -5.4 eV [8]).

MAGNETIC SEMICONDUCTOR SUPERLATTICES OF MnSe/ZnSe

The successful growth of the zincblende (Zn,Mn)Se ternary material in thin film form allowed the fabrication of a variety of multiple quantum well and superlattice structures. In these structures, the ZnSe forms the well while the wider bandgap (Zn,Mn)Se material provides the barrier. The optical and microstructural properties of these multilayer structures have been summarized in Ref. 1. Experiments involving the device potential of these structures have included the demonstration of optically pumped laser oscillations [9], as well as nonlinear excitonic absorption [10].

Attempts at expanding the bandgap energy in zincblende (Zn,Mn)Se barrier layers by increasing the Mn concentration succeeded to the point of actual realization of the hypothetical zincblende magnetic semiconductor MnSe. It is noteworthy that for equilibrium growth of bulk crystals, a pure zincblende crystal structure is obtained only for Mn mole fractions of up to 0.09, with a transition to purely wurtzite crystal structure occurring at x = .30. Extrapolation of the lattice parameter and bandgap data as a function of the Mn content provides a prediction of these material properties for the hypothetical zincblende MnSe; the bandgap is predicted to be 3.4 eV with a lattice parameter of 5.93 Å (1).

Unexpected magneto-optic effects have been observed in superlattice structures consisting of ultrathin (on a monolayer scale) layers of MnSe separated by approximately 45 Å of ZnSe (11). These structures are unique in that the MnSe exists in the metastable zincblende crystal structure whereas bulk crystals of MnSe have the NaCl crystal structure. Reflection high energy electron diffraction intensity oscillations (not previously reported for II-VI semiconductors), were used to control the MnSe layer thickness with a one monolayer resolution. The expected strong antiferromagnetic ordering evidenced in thick MnSe layers ($\sim$ 10 monolayers) is found to be progressively weakened as the layer thickness is reduced to the quasi-2D limit (one monolayer). Although bulk rock-salt MnSe is antiferromagnetic and epilayers of zincblende $Zn_{1-x}Mn_xSe$ exhibit increasing antiferromagnetic ordering for mole fractions of $x \geq 0.20$, the ultrathin layers of zincblende MnSe possess strong paramagnetism. The paramagnetic behavior appears as a red-shift of the excitonic emission energy. The origin of the red-shift is the Zeeman splitting of energy levels in the conduction and valence band(s) arising from an exchange interaction between the magnetic moments of Mn ions and the extended states. For the superlattice structure containing ten monolayers of MnSe (3D) no spectral-red shift was observed, confirming the expected antiferromagnetic ordering. In striking contrast a superlattice consisting of one monolayer of MnSe (quasi-2D) exhibited a large spectral red-shift indicating a strong paramagnetic effect. Figure 4 summarizes the magnetic behavior as deduced from the Zeeman shift of the ground state excitonic transitions observed in photoluminescence with the application of an external magnetic field.

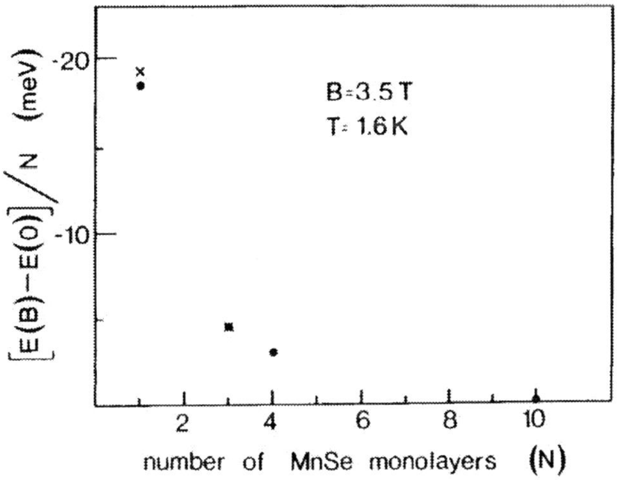

Fig. 4. Red shift of ground state excitonic transition due to an externally applied magnetic field for MnSe/ZnSe superlattices. The energy shift is normalized by the number of MnSe monolayers in a superlattice barrier. The x denotes a structure with interrupted growth at each interface.

Clear evidence of the frustration of magnetic ordering as the 2D limit is approached can be seen. The energy shift is normalized to the number of monolayers to provide a comparison on the basis of equivalent total number of magnetic ions. Further investigation is neded to clarify the precise mechanism responsible for the existence of the paramagnetic behavior exhibited by the ultrathin layers of MnSe.

The authors gratefully acknowledge the following for their contribution to this work: M. R. Melloch, Y. Hefetz, D. Lee, M. Vaziri, S. Bandyopadhyay, M. Yamanishi, G. Studtmann, J. Qiu, C. Choi, Y. Lee, and A. Ramdas. The research at Purdue is supported by Office of Naval Research Contract N00014-82-K0563, and Air Force Office of Scientific Research Grant 85-0185. The work at Brown University was supported by an Office of Naval Research Contract N00014-83-K0638.

REFERENCES

1. L. A. Kolodziejski,, R. L. Gunshor, N. Otsuka, S. Datta, W. M. Becker, and A. V. Nurmikko, IEEE J. Quantum Electronics, QE-22, 1660 (1986).
2. T. Yao, S. Amano, Y. Makita, and S. Naekawa, Jpn. J. Appl. Phys. 16, 1001 (1976).
3. R. M. Park, N. M. Salansky, Appl. Phys. Lett. 44, 249 (1984).
4. L. A. Kolodziejski, R. L. Gunshor, T. C. Bonsett, R. Venkatasubramanian, S. Datta, R. B. Bylsma, W. M. Becker, and N. Otsuka, Appl. Phys. Lett. 47, 169 (1985).
5. T. Yao and T. Takeda, Appl. Phys. Lett. 48, 160 (1986).
6. L. A. Kolodziejski, R. L. Gunshor, M. R. Melloch, N. Otsuka, C. Choi, Y. Hefetz, and A. V. Nurmikko, paper presented at the IV Int'l. Conf. on Molecular Beam Epitaxy, York, England, Sept. 1986.
7. B. F. Lewis, T. C. Lee, F. J. Grunthaner, A. Madhukar, R. Fernandez, and J. Maserjian, J. Vac. Soc. Technol. B2, 419 (1984).
8. A. Blacha, H. Presting, and M. Cardona, Phys. Stat. Sol. (b) 126, 11 (1984).
9. R. B. Bylsma, W. M. Becker, T. C. Bonsett, L. A. Kolodziejski, R L. Gunshor, M. Yamanishi, and S. Datta, Appl. Phys. Lett. 47, 1039 (1985).
10. D. R. Andersen, L. A. Kolodziejski, R. L. Gunshor, S. Datta, A. E. Kaplan, and A. V. Nurmikko, Appl. Phys. Lett. 48, 1559 (1986).
11. L. A. Kolodziejski, R. L. Gunshor, N. Otsuka, B. P. Gu, Y. Hefetz, and A. V. Nurmikko, Appl. Phys. Lett. 48, 1482 (1986).

DILUTE MAGNETIC SEMICONDUCTOR SUPERLATTICES CONTAINING $Hg_{1-x}Mn_xTe$

K.A. HARRIS, S. HWANG, R.P. BURNS, J.W. COOK, JR., AND J.F. SCHETZINA
Department of Physics, North Carolina State University, Raleigh, North Carolina 27695-8202

ABSTRACT

Superlattices containing alternating layers of $Hg_{1-x}Mn_xTe$ and HgTe have been successfully grown by molecular beam epitaxy. These structures are the first superlattices containing layers of a mercury-based dilute magnetic semiconductor to be grown by any thin film technique. The optical and electrical properties of these novel magnetic multilayers are presented and discussed.

I. INTRODUCTION

The pseudobinary II-VI semiconductors in which a fraction x of the group II elemental sites is occupied by a transition metal magnetic ion such as Mn^{++} are referred to as semimagnetic or dilute magnetic semiconductors (DMS). Examples include $Cd_{1-x}Mn_xTe$, $Zn_{1-x}Mn_xTe$, $Cd_{1-x}Mn_xSe$, $Zn_{1-x}Mn_xSe$, $Hg_{1-x}Mn_xTe$, $Hg_{1-x}Mn_xSe$, and $Hg_{1-x}Mn_xS$.

These ternary alloys have attracted much attention because of their novel magnetic, magneto-optic, and magneto-transport properties [1,2] which result from the spin-spin exchange interaction that occurs between the localized magnetic moments associated with the Mn^{++} ions and the conduction band electrons and valence band holes. In $Hg_{1-x}Mn_xTe$, this leads to large positive electronic g factors [3,4], magnetic-field-induced overlap of valence and conduction bands [5,6], and giant negative magnetoresistance at low temperatures [7].

Interest in these materials has been further stimulated by the successful growth of CdMnTe-CdTe and ZnMnSe-ZnSe quantum well structures and superlattices by molecular beam epitaxy (MBE) [8,9]. These new structures exhibit many interesting properties including stimulated emission [10-12] which can be magnetically tuned [13], interfacial localization of excitons [14-15], and quantum size effects similar to those exhibited by non-magnetic multilayer semiconductors [16,17].

Superlattices containing alternating layers of $Hg_{1-x}Mn_xTe$ and HgTe have been successfully grown by MBE. These structures are the first superlattices containing layers of a Hg-based narrow-gap DMS to be grown by any thin film technique. The first type of DMS structure consists of 240 layer pairs of $Hg_{0.97}Mn_{0.03}Te$-HgTe. The band structures of the constituents of the $Hg_{0.97}Mn_{0.03}Te$-HgTe superlattice are shown in Fig. 1(a). As shown in the figure, HgTe has an inverted band structure in which the Γ_6 energy band functions as the light hole valence band and lies below the Γ_8 conduction

Mat. Res. Soc. Symp. Proc. Vol. 89. © 1987 Materials Research Society

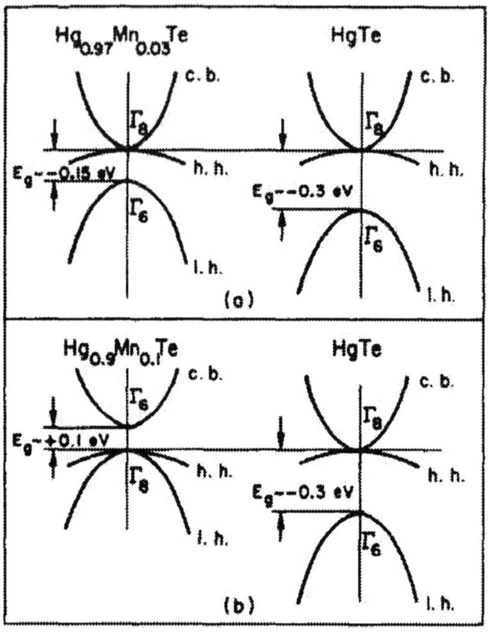

Figure 1. Electronic band structures of constituent layers of (a) $Hg_{0.97}Mn_{0.03}Te$-HgTe and (b) $Hg_{0.9}Mn_{0.1}Te$-HgTe superlattices. The heavy hole bands at the Γ_8 points are shown to be at the same energy, although there may exist a valence band offset which is presently unknown for this structure.

band. This results in a negative energy gap, defined as $E_g = E(\Gamma_6) - E(\Gamma_8)$, with $E_g = -0.3$ eV at 4.2 K. $Hg_{1-x}Mn_xTe$ also exhibits this band structure inversion for $x \leq 0.06$ and, as a result, the $Hg_{0.97}Mn_{0.03}Te$-HgTe superlattice consists of two negative band gap semiconductors.

The second type of DMS structure grown by MBE consists of 200 layer pairs of $Hg_{0.9}Mn_{0.1}Te$-HgTe. Fig. 1(b) shows the band structures of the constituents of the $Hg_{0.9}Mn_{0.1}Te$-HgTe superlattice. For $Hg_{1-x}Mn_xTe$ with $x > 0.06$, the Γ_6 band becomes the conduction band, E_g becomes positive, and the light hole and heavy hole bands are degenerate at the Γ point. In this case, the $Hg_{0.9}Mn_{0.1}Te$ layers, which have a band gap of ~0.1 eV at 4.2 K, serve as barrier layers between HgTe quantum wells, provided the valence offset is < 0.1 eV.

II. EXPERIMENTAL DETAILS

The DMS superlattices were prepared in a Hg-compatible MBE system that was designed and constructed at North Carolina State University [18,19]. They were deposited onto (100) $Cd_{0.98}Zn_{0.02}Te$ substrates which were chemimechanically polished and pre-screened by means of x-ray rocking curve measurements. The substrates were degreased, etched for 30 seconds in a 0.1%

bromine-in-methanol solution, and dipped in concentrated hydrochloric acid to remove the native oxide. The substrates were then quickly inserted into the MBE system and preheated in ultra high vacuum to 400 ^{o}C immediately prior to growth. Two standard effusion cells containing Mn and Te, respectively, were used as vapor sources, while a specially-designed Hg source was employed to provide the high mercury flux required for film growth [18,19]. The source temperatures were adjusted to produce beam equivalent pressures at the substrate of 2 x 10^{-4} Torr for Hg, 2 x 10^{-6} Torr for Te, and either 4 x 10^{-8} or 1.2 x 10^{-7} Torr for Mn during growth of $Hg_{0.97}Mn_{0.03}Te$ or $Hg_{0.9}Mn_{0.1}Te$, respectively. The substrate temperature was maintained at 175 ^{o}C during all depositions. Film thickness measurements and x-ray diffraction techniques were employed to obtain superlattice layer dimensions.

Van der Pauw Hall effect measurements were completed on the superlattices over the temperature range 20-300 K using an applied magnetic field of 3kG. In these experiments, 5mm x 6mm rectangular samples with In-solder corner contacts were employed.

The infrared (IR) transmittance of each superlattice was obtained using a Perkin-Elmer 983-G double beam ratio-recording spectrophotometer which is interfaced to a computer. The transmittance of each substrate (polished on both sides) was recorded prior to film growth and stored in the computer as a base-line correction file. The ultraviolet-visible (UV-VIS) reflectance of each superlattice was measured using a computer-controlled Perkin-Elmer Lambda-9 spectrophotometer equipped with a specular reflectance accessory.

III. RESULTS AND DISCUSSION

Hall effect data for a $Hg_{0.97}Mn_{0.03}Te$-HgTe superlattice consisting of 240 layer-pairs of thickness L_b = 48 A (HgMnTe) and L_z = 48 A (HgTe) is shown in Fig. 2. The superlattice is n-type with a carrier concentration ranging from 4 x 10^{17} cm^{-3} at 300 K to 1.3 x 10^{16} cm^{-3} at 12 K. The corresponding electron mobility ranges from 3 x 10^4 cm^2/V s at 300 K to greater than 1.4 x 10^5 cm^2/V s at 12 K. To our knowledge, this is the highest electron mobility ever obtained for any syuperlattice containing Hg, including those composed of either HgTe-CdTe or HgCdTe-HgTe double layers.

Fig. 3 shows Hall effect data for a $Hg_{0.9}Mn_{0.1}Te$-HgTe superlattice consisting of 200 layer-pairs of thickness L_b = 54 A (HgMnTe) and L_z = 54 A (HgTe). This superlattice is also n-type vioth a somewhat reduced carrier concentration possibly due to quantum confinement effects. The carrier concentration at 300 K is 2.8 x 10^{17} cm^{-3} and falls to below 1 x 10^{16} cm^{-3} at low temperatures. The corresponding electron mobility is also very large and

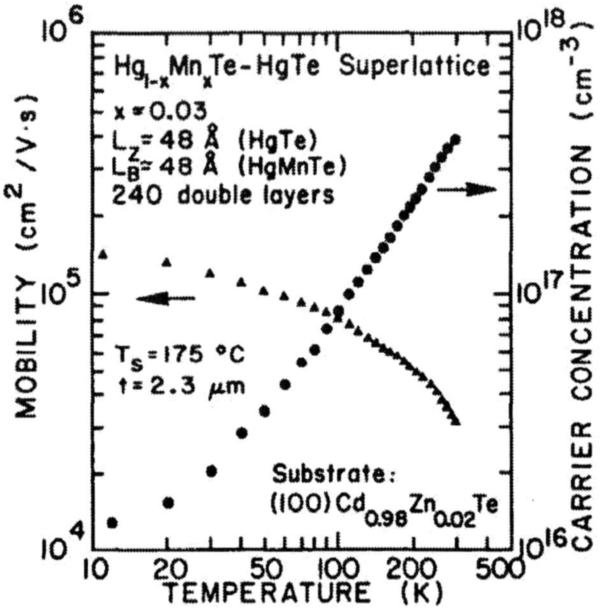

Figure 2. Hall mobility and carrier concentration versus temperature for n-type $Hg_{0.97}Mn_{0.03}Te$-HgTe superlattice.

reaches values greater than 1.1 x 10^5 cm^2/V s at temperatures below 130 K.

Both superlattices exhibit outstanding optical properties. The IR transmittance and UV-VIS reflectance of the $Hg_{0.9}Mn_{0.1}Te$-HgTe superlattice is shown in Fig. 4. The transmittance of the multilayer shows IR absorption out to and beyond 16 μm with soft absorption between 4 and 16 μm, similar to that observed in pure HgTe, which occurs as a result of the small density of states associated with a zero or negative band gap material. The reflectance shows well-resolved peaks at 2.117 eV and 2.773 eV, which are, respectively, the E_1 and $E_1 + \Delta_1$ zincblende transition peaks. The peaks are shifted to higher energies in comparison with pure HgTe, consistent with an increase in Mn content. The reflectance peaks are well formed with no additional structure, indicating high quality layers with little or no stacking faults. The magnitude of the reflectance of both superlattices approaches 40% due to the highly specular surfaces which they possess.

IV. CONCLUSIONS

In summary, novel superlattice structures containing alternating layers

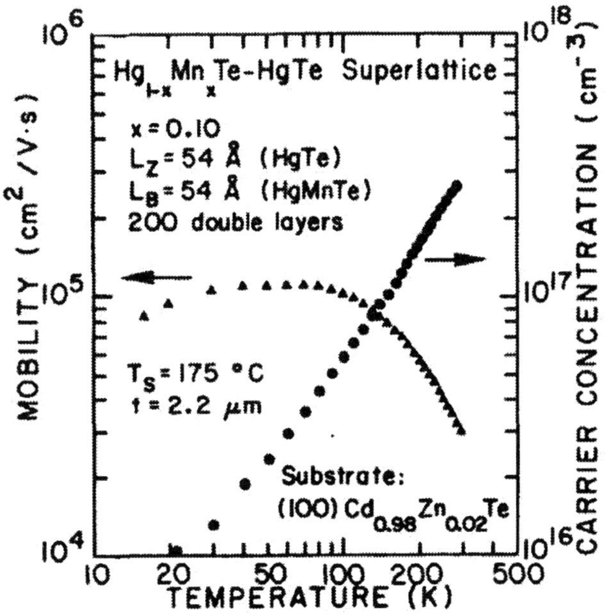

Figure 3. Temperature dependence of the Hall mobility and carrier concentration for n-type $Hg_{0.9}Mn_{0.1}Te$-HgTe superlattice.

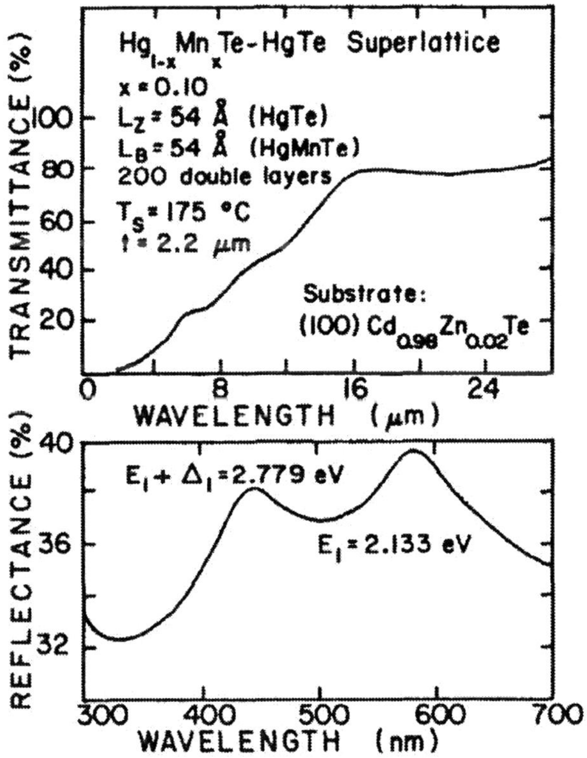

Figure 4. Infrared transmittance and UV-VIS reflectance of $Hg_{0.9}Mn_{0.1}Te$-HgTe superlattice.

of $Hg_{1-x}Mn_xTe$ and HgTe have been successfully grown by MBE. These structures consist of alternating layer pairs of a Hg-based narrow-gap DMS and HgTe. These new multilayers are n-type and exhibit very large electron mobilities. Superlattice transmittances are HgTe-like and UV-VIS reflectances confirm their high quality.

V. ACKNOWLEDGEMENTS

The authors wish to thank M. Chu of Fermionics for providing the CdZnTe substrates. This work was supported by DARPA/ARO contract DAAG29-83-K-0102 and by NSF grant DMR83-13036.

REFERENCES

1. J.K. Furdyna, J. Vac. Sci. Technol. **21**, 220 (1982).
2. J.K. Furdyna, J. Appl. Phys. **53**, 7637 (1982).
3. G. Bastard, C. Rigaux, Y. Guldner, A. Mycielski, J.K. Furdyna, and D.P. Mullin, Phys. Rev. B **24**, 1961 (1981).
4. M. Dobrowolska and W. Dobrowolska, J. Phys. C **14**, 5689 (1981).
5. G. Bastard, C. Rigaux, Y. Guldner, J. Mycielski, and A. Mycielski, J. Phys. (Paris) **39**, 87 (1978).
6. G. Bastard, C. Rigaux, and A. Mycielski, Phys. Stat. Solidi B: **79**, 585 (1977).
7. A. Mycielski and J. Mycielski, J. Phys. Soc. Jpn., **49**, 807 (1980).
8. R.N. Bicknell, R.W. Yanka, N.C. Giles-Taylor, E.L. Buckland, and J.F. Schetzina, Appl. Phys. Lett. **45**, 92 (1984).
9. L.A. Kolodziejski, R.L. Gunshor, T.C. Bonsett, R. Venkatasubramanian, S. Datta, R.B. Bylsma, W.M. Becker, and N. Otsuka, Appl. Phys. Lett. **47**, 169 (1985).
10 R.N. Bicknell, N.C. Giles-Taylor, N.G. Anderson, W.D. Laidig, and J.F. Schetzina, Appl. Phys. Lett. **46**, 238 (1985).
11 R.N. Bicknell, N.C. Giles-Taylor, D.K. Blanks, N.G. Anderson, W.D. Laidig, and J.F. Schetzina, Appl. Phys. Lett. **46**, 112 (1985).
12 R.B. Bylsma, W.M. Becker, T.C. Bonsett, L.A. Kolodziejski, R.L. Gunshor, M. Yamanishi, and S. Datta, Appl. Phys. Lett. **47**, 1039 (1985).
13 E.D. Isaacs, D. Heiman, J.J. Zayhowski, R.N. Bicknell, and J.F. Schetzina, Appl. Phys. Lett. **48**, 275 (1986).
14 X.C. Zhang, S.K. Chang, A.V. Nurmikko, L.A. Kolodziejski, R.L. Gunshor, and S. Datta, Phys. Rev. B **31**, 4056 (1985).
15 J. Warnock, A. Petrou, R.N. Bicknell, N.C. Giles-Taylor, D.K. Blanks, and J.F. Schetzina, Phys. Rev. B **32**, 8116 (1986).
16 L.A. Kolodziejski, T.C. Bonsett, R.L. Gunshor, S. Datta, R.B. Bylsma, W.M. Becker, and N. Otsuka, Appl. Phys. Lett. **45**, 440 (1984).
17 R.N. Bicknell, N.C. Giles-Taylor, D.K. Blanks, R.W. Yanka, E.L. Buckland, and J.F. Schetzina, J. Vac. Sci. Technol. B **3**, 709 (1985).
18 K.A. Harris, S. Hwang, D.K. Blanks, J.W. Cook, Jr., and J.F. Schetzina, J. Vac. Sci. Technol. A **4**, 2061 (1986).
19 K.A. Harris, S. Hwang, D.K. Blanks, J.W. Cook, Jr., and J.F. Schetzina, J. Vac. Sci. Technol. B **4**, 581 (1986).

FAR INFRARED MAGNETOABSORPTION IN $Hg_{1-x}Mn_xTe/HgTe$ SUPERLATTICE

Z. Yang, M. Dobrowolska, H. Luo and J. K. Furdyna
Department of Physics, Purdue University
West Lafayette, IN. 47907

and

K. A. Harris, J. W. Cook, Jr. and J. F. Schetzina
Department of Physics, North Carolina State University
Raleigh, NC. 27695

INTRODUCTION

Hg-based superlattices (SL)--e.g., the HgTe/CdTe SL--exhibit interesting and important physical properties[1-3]. Among these are the tunability of the band gap, achieved by varying the thicknesses of the two constituent materials; and the existence of the valence band offset, on which the details of the valence bands of the SL largely depend[2]. Recently a new type of Hg-based SL, the $Hg_{1-x}Mn_xTe/HgTe$ SL, has been successfully prepared. In addition to the ability of varying its band gap by controlling the Mn content, this SL has the advantage that some of its important parameters, including the band offset, can be tuned by an external magnetic field because of the spin-spin exchange interaction in the $Hg_{1-x}Mn_xTe$ layers. So far, the electronic and optical properties of this type of SL have not been explored.

Far infrared (FIR) magnetoabsorption spectroscopy is ideally suited for probing the details of the energy levels in narrow gap semiconductors and their superlattices, and we have initiated such studies on the $Hg_{1-x}Mn_xTe/HgTe$ SL system. At this early stage we will, for the most part, present the results phenomenologically, with only preliminary attempts at interpretation. Detailed quantitative analysis will follow in future papers, as our understanding of the complex FIR spectra increases with further experimental and especially theoretical work.

SAMPLE AND EXPERIMENT

The $Hg_{1-x}Mn_xTe/HgTe$ SL was grown in the [100] direction on a $Cd_{.96}Zn_{.04}Te$ substrate at North Carolina State University. The sample was n-type in the temperature range from 20K to 300K. The nominal Mn content in the $Hg_{1-x}Mn_xTe$ layer was x=0.1. Recent transmission electron microscope measurements carried out on the sample show a well defined SL structure, with layer thickness of 65Å and 55Å for HgTe and $Hg_{1-x}Mn_xTe$, respectively, and the value of x between 0.07 and 0.10, close to the nominal value[4].

The FIR measurements were performed in Faraday and Voigt geometries, at fixed photon energies between 2.5 and 12.8 meV, in temperatures ranging from 1.6K to 60K, and in magnetic fields up to 5.5 tesla. In the Faraday geometry (**B** normal to the layer plane), cyclotron-resonance-active (CRA) and cyclotron-resonance-inactive (CRI) circular polarizations were used. In the Voigt geometry (**B** in the layer plane), linearly polarized FIR waves were used, with $\mathbf{E} // \mathbf{B}$ (parallel Voigt) or $\mathbf{E} \perp \mathbf{B}$ (perpendicular Voigt geometry). In addition to these standard configurations, experiments were also performed in geometries where the sample was rotated up to 55 degrees from its starting orientation relative to the **B** field.

Mat. Res. Soc. Symp. Proc. Vol. 89. ©1987 Materials Research Society

RESULTS

In this section we will present the results of our magnetotransmission measurements, leaving their interpretation for the next section. The data we have obtained are very complicated and rich in detail. For the sake of clarity, we will present Faraday and Voigt geometry spectra separately since, in addition to differences in selection rules, in a superlattice the magnetic-field-dependent energy level schemes themselves should be different in these two cases.

A. *Faraday geometry*

In the Faraday geometry, magnetotransmission spectra for the CRA polarization show well-defined strong absorption lines, while in the CRI polarization only a few weak lines occur at low temperatures. We shall therefore restrict our attention to the CRA case.

The left hand side of Fig.1 shows typical CRA magnetotransmission spectra obtained for $163\mu m$ at several temperatures. The general feature to be noticed is that the line positions are largely unaffected by the temperature, while the intensities of the lines show a strong temperature dependence. In particular, the line marked A gets weaker when temperature increases, and disappears at T≃30 K; the lines marked B and D are observed between T≃20K and T ≃50K; and the line marked C gains in intensity as the temperature increases, becoming a giant resonance at T=50K and above. Finally at T=1.6K sharp new lines appear (marked E and A' in Fig.1). Figure 2 shows the behavior of these low-temperature lines for several wavelengths. Note the strong dependence on wavelength of the line intensity, different for both lines, suggesting that the two transmissions are of different origin (e.g., inter- and intra-band transition). Figure 3 is a plot of magnetic field positions of lines E and A' vs. photon energy (the so called fan-chart), which extrapolate to the same energy (within the experimental error) at zero magnetic field.

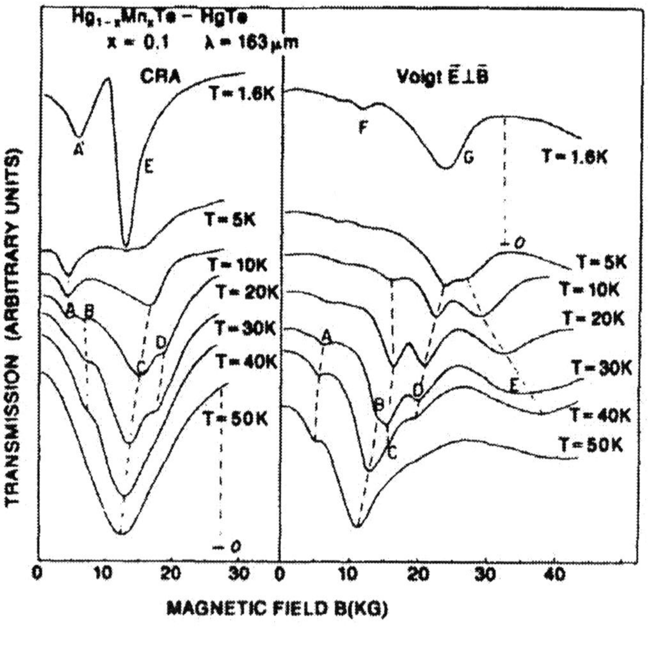

Figure 1

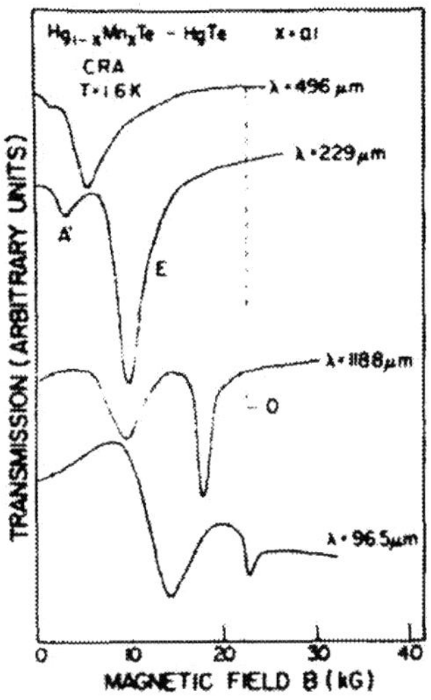

Figure 2

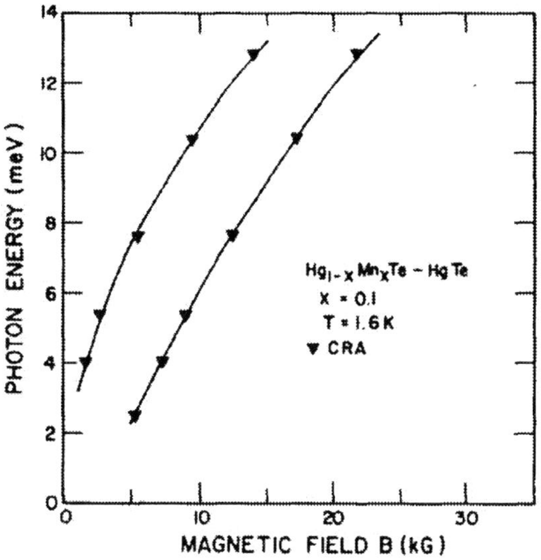

Figure 3

We also performed experiments for **B** tilted relative to the sample layers. This was accomplished by rotating the SL sample, keeping the directions of incident beam and **B** fixed. Figure 4 shows the spectra obtained at several rotation angles at T=1.6K and λ=96.5μm, revealing again a different behavior for the two low-temperature lines. Line A' shifted strongly with the angle, the resonance field of this line following the relation $B(\theta) = B(0)/\cos(\theta)$, where θ is the angle rotated away from the Faraday geometry. Line E, on the other hand, remained at the same position and only decreaesd in intensity.

B. *Voigt geometry*

In the Voigt geometry, the spectra are totally different from those observed in the CRA configuration. The right side of Fig.1 shows typical spectra at λ= 163μm at several temperatures in the perpendicular Voigt (**E**⊥**B**) polarization. Along with the intensity, the positions of the lines now also depend on the temperature. Lines marked A and B in the figure first appear at T≃30K, line B becoming the dominant resonance of the entire spectrum above T=50K. Line C is observed between T=10K and T=40K. Line D and E have the most dramatic dependence of positions on temperature. In the parallel Voigt (**E**//**B**) polarization, the absorption lines (not shown) are shallower than the ones in either CRA or perpendicular Voigt polarizations. As for the Voigt **E**⊥**B** polarization, the positions of lines depend strongly on the temperature. The highest field resonance observed in **E** //**B** is very similar in lineshape and field position to line E in the perpendicular Voigt geometry. Figure 5 shows the fan-charts of line E and its **E**//**B** counterpart at several temperatures. Note that these lines very nearly coincide, except at T=1.6K. An interesting feature of Fig.5 is that the data extrapolate to a finite negative energy at **B**=0, the intercept apparently depending on the temperature.

We also performed measurements when the sample was rotated away from the Voigt geometry, as a function of the rotation angle. As the angle between the **B** field and the layer plane increases, the line marked F quickly disappears and the line marked G shifts its position, following a more complicated pattern than in CRA.

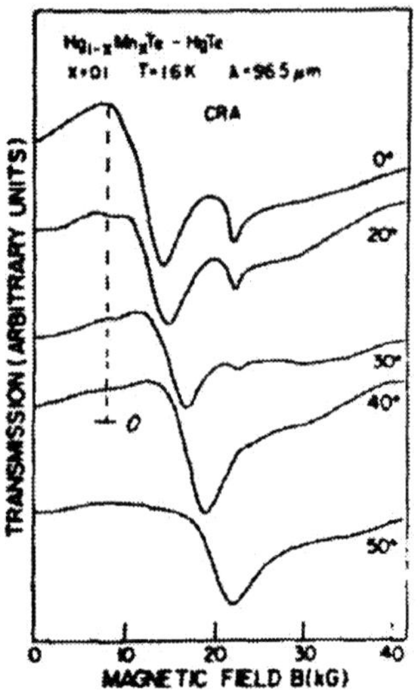

Figure 4

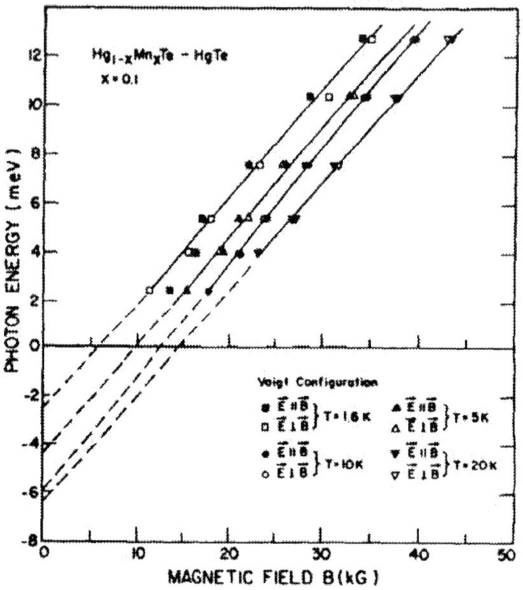

Figure 5

Figure 6 shows the fan-charts for the CRA line A' before and after a 20 degree rotation, and the Voigt **E**⊥**B** line G before and after a 30 degree rotation. The two curves in the case of CRA tend to converge for lower photon energies while, in contrast, the two Voigt curves are nearly parallel to each other. This difference in behavior of CRA and Voigt data is typical of other lines observed in these two geometries.

DISCUSSION

The very fact that in the Voigt geometry magnetoabsorption lines have been observed indicates that the electrons in this SL are not confined to two-dimensional motion in the layers, but sustain a component of motion along the SL growth axis. In other words, the effective mass describing the motion of electrons in the layer plane is of the same order of magnitude as the mass describing the motion along the direction normal to the layers. This is in contrast to the HgTe/CdTe SL of comparable well thickness[5]. The probable reason for this difference is that the magnitude of the barriers which the electrons "see" in our sample is only about a fraction of that in the HgTe/CdTe SL, since $Hg_{.90}Mn_{.10}Te$ has a positive band gap of the order of 100 meV. The effect of this small barrier, however, is still pronounced, resulting in totally different spectra in the CRA and in the Voigt **E**⊥**B** geometries.

Inspection of the data reveals that there are three temperature regions in which the spectra are qualitatively different from one other. At low temperatures (below 2K) the absorption lines differ from higher temperatures, since in that range they most probably involve the transitions from the ground energy level to some excited energy levels. At intermediate temperature (10K $<$ T $<$ 30K) new lines rapidly begin to appear, which may be interpreted as originating from excited energy levels which are gradually becoming populated. The observed strong dependence of line intensities on temperature is suggestive of such thermal redistribution of electrons. Finally, as seen in Fig.1, at temperatures above 40K both CRA and Voigt **E**⊥**B** spectra are dominated by a single strong line.

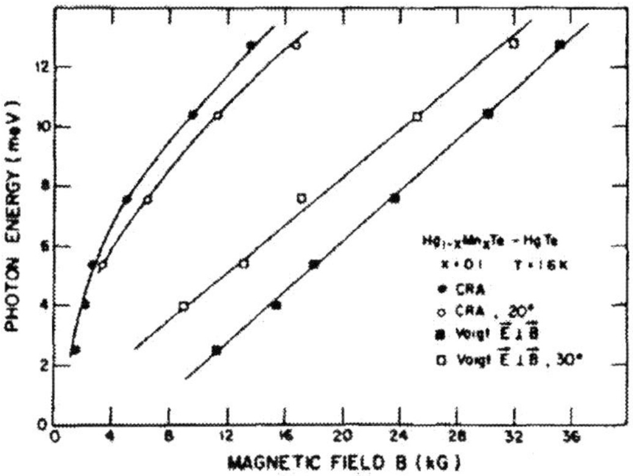

Figure 6

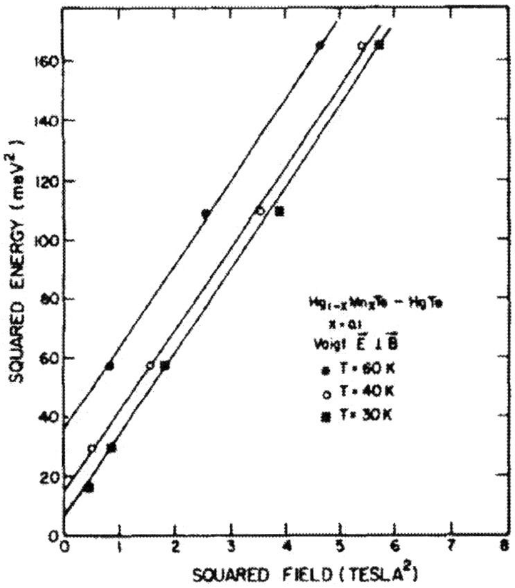

Figure 7

While at this early stage it is difficult to embark on a general interpretation of the observed spectra, we can identify certain characteristic features of the low and high temperature data. As is evident from Fig.4, line E in the CRA data at T=1.6K, λ=96.5μm does not shift its position when the sample is rotated. It is possible that this line is due to the spin-flip transition, since in dilute magnetic semiconductors the spin splitting energy is mainly determined by the exchange interaction between electrons and Mn ions, and is therefore not expected to be affected by the anisotropy of band structure.

In the high temperature data, the identification of the strong line as cyclotron resonance of conduction electrons immediately suggests itself. This interpretation of the CRA spectra gives an effective mass of 0.018m_0. Ths value lies between what it should be for conduction electrons in HgTe and for $Hg_{.90}Mn_{.10}Te$, suggesting (at least qualitatively) that at high temperatures the electrons are no longer confined to the shallow wells, but can sample the properties of the constituent materials during their trajectories. At our shortest wavelengths (96.5μm and 118.8μm) the perpendicular Voigt resonance occurs at fields which yields m^*=0.02m_0, close to that given by the CRA spectra. For longer wavelengths, however, the absorption line shifts to lower magnetic fields, consistent with plasma-shifted cyclotron resonance expected in this geometry for bulk samples. Following standard practice, in Fig.7 we plot the photon energy squared vs. magnetic field squared for the Voigt **E**$\perp$**B** line for two temperatures. The intercept of the straight lines on the energy-squared axis then gives the plasma frequency of the electron gas for the two temperatures.

We feel that it would be premature to attempt a more detailed interpretation at this early stage. Progress in this area will require experiments on superlattices involving additional values of x and additional well and barrier dimensions. Such experiments are then likely to shed considerable light on the physics of this interesting system, including valuable information on the question of band offset in type-III superlattices.

This work was supported by NSF Grants DMR-8313036 and DMR-8600014 andby DARPA/ARO Contract DAAG 29-83-k0102.

REFERNENCES

1) J. N. Schulman and T. C. McGill, Appl. Phys. Lett. **34**, 663 (1978)

2) J. M. Berroir, Y. Guldner, J. P. Vieren, M. Voos, J. P. Faurie Phys. Rev. **B34**, 891 (1986)

3) S. P. Kowalczyk, J. T. Cheung, E. A. Kraut and R. W. Grant Phys. Rev. Lett. **56**, 1605 (1986)

4) N. Otsuka, private communication

5) Our own FIR experiments on HgTe/CdTe superlattices have shown no absorption lines in the Voigt geometry, and so far we have not seen any reports claimimg that such absorption lines have been observed.

$Hg_{1-x-y}Mn_xCd_yTe$ ALLOYS FOR 1.3-1.8 μm PHOTODIODE APPLICATIONS

S.H. SHIN,* J.G. PASKO,* D.S. LO* W.E. TENNANT,* J.R. ANDERSON,**
M. GORSKA,** M. FOTOUHI** and C.R. LU**
* Rockwell International Science Center, 1049 Camino Dos Rios, Thousand Oaks, CA 91360
** Department of Physics and Astronomy, University of Maryland, College Park, MD 20742

ABSTRACT

HgMnCdTe/CdTe photodiodes with responsivity cutoffs of up to 1.54 μm have been fabricated by liquid phase epitaxy (LPE). The mesa device structure consists of a boron-implanted mosaic fabricated on a p-type $Hg_{1-x-y}Mn_xCd_yTe$ layer grown on a CdTe substrate. A reverse breakdown voltage (V_B) of 50 V and a leakage current density of 1.5 x 10^{-4} A/cm² at V = -10 V was measured at room temperature (295K). A 0.75 pF capacitance was also measured under a 5 V reverse bias at room temperature. This device performance based on the quaternary HgMnCdTe shows both theoretical and practical promise of superior performance for wavelengths in the range 1.3 to 1.8 μm for fiber optic applications.

INTRODUCTION

The recent development of optical fibers with low attenuation and minimum dispersion in the 1.3 to 1.6 μm region has motivated research on optical sources and photodetectors for these longer wavelengths. The problems encountered in fabricating long-wavelength avalanche photodiodes (APDs) have been primarily related to materials. Initial devices fabricated from Ge or III-V alloys have exhibited a combination of high leakage currents, soft breakdowns and relatively low gains [1]. These high leakage currents degrade the signal-to-noise performance of a receiver and restrict the usable gain to low values.

A fundamental limitation of long-wavelength materials for fiber optic communications is associated with the ionization rates for electrons (α) and holes (β). According to McIntyre's rule [2], APD noise performance improves by over a factor of 10 with a β/α ratio of only 5. The excess noise in APDs is greatly reduced when the carrier with the largest ionization coefficient is injected into the high field region. Consequently, it is essential to identify materials or device structures which have very different electron and hole ionization coefficients. This condition is met in Si APDs where $\alpha/\beta \geq 20$; in most long-wavelength materials, however, α and β are comparable ($\beta/\alpha \leq 2$), e.g., 1.4 for Ge and only 2.5 for InGaAs.

In this paper, we propose high-performance photodiodes from HgMnCdTe alloys as a possible alternative to group IV or III-V devices for long-wavelength APD applications. Recent work by Becla and Wong [3] has shown good photodiode performance in $Hg_{1-x}Mn_xTe$ and $Hg_{1-x-y}Mn_xCd_yTe$ with energy gaps in the range 100-770 meV at 77K. Moreover, HgTe-based alloy systems offer the same potential advantages for low-noise APDs as does the GaAlSb system, a resonantly enhanced ionization ratio β/α [4]. The β/α ratio of $Hg_{0.3}Cd_{0.7}Te$ is about 10 at a wavelength of 1.3 μm [5]. Since the band structure and electrical properties of $Hg_{1-x}Mn_xTe$ are quite similar to the HgCdTe system, a similar resonance effect is expected at a longer wavelength of λ = 1.8 μm. Furthermore, recent theoretical calculations [6] showed that the addition of Mn in the HgTe alloy might strengthen the Hg-Te

Mat. Res. Soc. Symp. Proc. Vol. 89. ©1987 Materials Research Society

bond, resulting in improved material quality. These results suggest that the resonance condition can be obtained at 1.55 μm with the quaternary alloy $Hg_{1-x-y}Mn_xCd_yTe$ by an appropriate choice of x and y.

Figure 1 shows the variation of the bandgap (E_0) and spin-orbit splitting (Δ_0) as a function of Cd and Mn composition for the ternary $Hg_{1-x}Cd_xTe$ and $Hg_{1-x}Mn_xTe$ systems. The $Hg_{1-x}Cd_xTe$ and $Hg_{1-x}Mn_xTe$ systems provide an $E_0 = \Delta_0$ resonance at 1.3 μm and 1.8 μm, respectively, at room temperature.

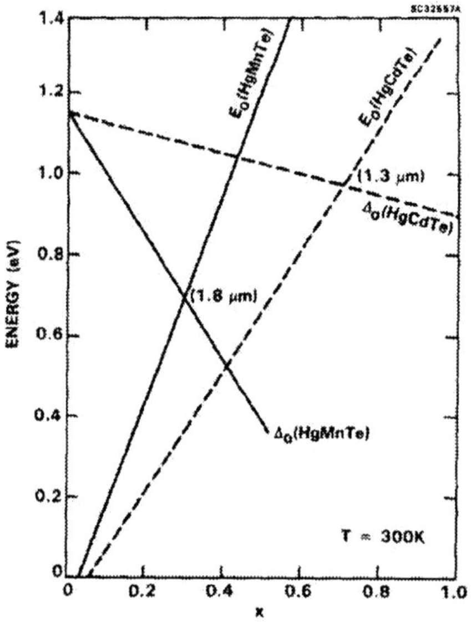

Fig. 1 Bandgap (E_0) and spin-orbit splitting (Δ_0) as a function of Cd and Mn composition for the ternary systems $Hg_{1-x}Cd_xTe$ and $Hg_{1-x}Mn_xTe$.

To calculate the necessary composition of $Hg_{1-x-y}Mn_xCd_yTe$ for the resonance condition, we use analytical expressions obtained by fitting the electroreflectance data obtained by Amirtharaj et al [9] for $Hg_{1-x-y}Mn_xCd_yTe$ as follows:

$$E_0(x,y) = -0.12 + 2.61x + 1.14y + 0.48y^2$$

and

$$\Delta_0(x,y) = 1.17 - 1.5717x - 0.246y \quad .$$

This gives x = 0.19 and y = 0.32 for a gap of about 0.8 eV, which corresponds to 1.55 μm in $Hg_{1-x-y}Mn_xCd_yTe$.

The values of resonance energy $E_0 = \Delta_0$ vs composition x + y were calculated and are shown in Fig. 2. It is evident that E_0 and Δ_0 can be tunable in the range of 1.5-1.6 μm for the $Hg_{1-x-y}Mn_xCd_yTe$ system.

LIQUID PHASE EPITAXY OF $Hg_{1-x-y}Mn_xCd_yTe/CdTe$

Quaternary $Hg_{1-x-y}Mn_xCd_yTe$ alloys have been prepared by two methods: epitaxial layers by the isothermal VPE (ISOVPE) method [10,11] and bulk samples by the modified two-phase mixture (MTPM) crystal growth method [12]. By employing the ISOVPE method, smooth, mirror-like and terrace-free

Fig. 2 Resonance energy of $E_0 = \Delta_0$ vs composition $x + y$ in $Hg_{1-x-y}Mn_xCd_yTe$.

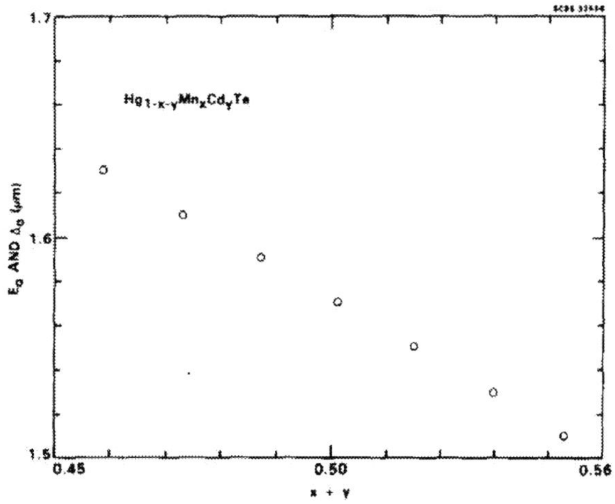

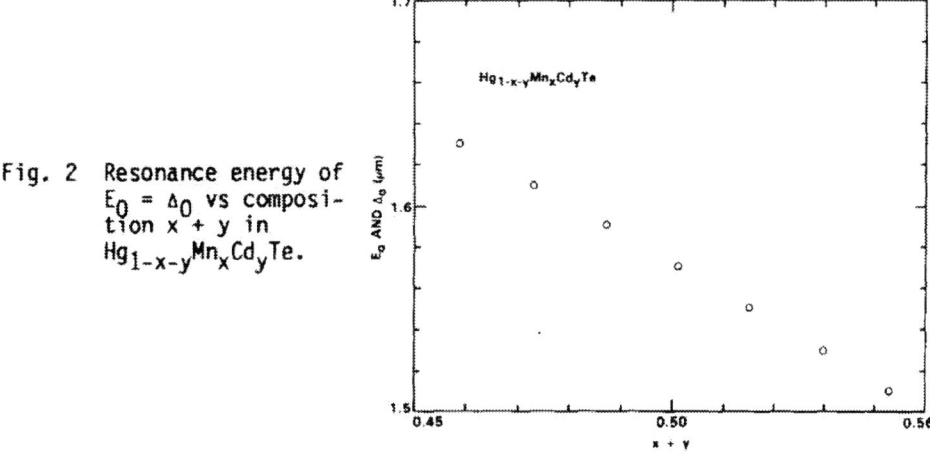

HgMnCdTe films can be grown on CdMnTe substrates. Since the HgMnCdTe epilayers have large compositional gradients across the layer due to the interdiffusion of HgTe and $Cd_{1-x}Mn_xTe$, the ISOVPE method cannot be used to grow homogeneous HgMnCdTe epilayers for fabricating photodiodes. The modified two-phase mixture method [12] was only partially successful in growing a single crystal (16 mm diam) with a few grain boundaries.

We have used LPE for growth of HgMnCdTe, since it has the potential for high quality and uniformity over large areas. LPE is presently the most widely used technique of II-VI epitaxial growth and has been applied successfully to HgCdTe for fabrication of detector arrays [13,14]. Since the lattice mismatch between the CdTe substrates and $Hg_{0.49}Mn_{0.19}Cd_{0.32}Te$ epilayers is less than 0.6%, no lattice mismatch problems are anticipated for LPE growth of 1.55 μm $Hg_{1-x-y}Mn_xCd_yTe$. To avoid mismatch, however, $Hg_{0.49}Mn_{0.19}Cd_{0.32}Te$, having a lattice constant of 6.445Å, can be grown on $Cd_{1-x}Mn_xTe$ ($x = 0.25$) with exact lattice matching.

A vertical furnace containing high-pressure hydrogen (or argon) gas and a quartz tube was used for the growth of HgMnCdTe epitaxial layers on semi-insulating (111)Te-oriented CdTe substrates. A Te-rich solution of HgMnCdTe was employed for the LPE growths. The source materials, Hg, Mn, CdTe and excess Te metal, were loaded into the growth quartz tube in the furnace. After the source materials were reacted at 700°-750°C for 60 min, the melt was cooled to the saturation temperature, approximately 550°C. The growth time was adjusted to produce layers with thicknesses in the range of 10-15 μm. For all compositions, the quantities of HgCdTe and Mn in the melt were experimentally adjusted to achieve the required composition. We have grown $Hg_{1-x-y}Mn_xCd_yTe$ to cover the range from 1.25 to 1.55 μm. The microscopic surface morphologies of the HgMnCdTe epitaxial layers grown to date are characterized by terraced structures similar to the typical surface morphology of HgCdTe layers. A typical IR transmission spectrum for a HgMnTe epitaxial layer is shown in Fig. 3.

Since the epitaxial layers were grown in Te-rich solutions above 500°C, the unintentionally doped epitaxial layers contain the native acceptor presumed to be a cation vacancy. The typical epilayer shows an acceptor carrier concentration on the order of $1.0\text{-}7.0 \times 10^{16}$ cm^{-3} and a hole mobility of 30-50 $cm^2/V\text{-}s$ at room temperature. At 77K, the acceptor concentration drops to 1×10^{15} cm^{-3} and the hole mobility increases to

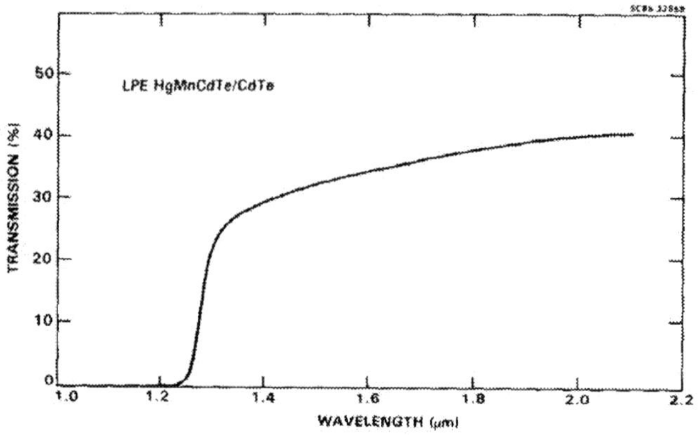

Fig. 3 Transmission spectra (295K) taken on an LPE HgMnCdTe layer grown on a CdTe substrate.

140 cm^2/V-s. Since the device will be operated at room temperature, the electrical parameters at 295K and 77K for typical epilayers with a cutoff wavelength of 1.25 μm are summarized in Table 1.

Table 1

Electrical Properties of As-Grown LPE $Hg_{1-x-y}Mn_xCd_yTe$ Epitaxial Layer[(a)] Grown on CdTe Substrate

	295K	77K
Conduction Type	p	p
Resistivity (Ω-cm)	1.69	24.6
Hall Coefficient (cm^3/Coul)	86.2	3.4×10^3
Hall Mobility (cm^2/V-s)	50.8	139.2
Carrier Concentration (cm^{-3})	7.2×10^{16}	1.8×10^{15}

(a) Cutoff wavelength (295K) = 1.25 μm

One of the shorter wavelength epitaxial layers was analyzed by energy dispersive x-ray analysis (EDAX); a depth profile is shown in Fig. 4. We note that the segregation coefficient of Mn estimated from LPE-grown layers is between 1 and 1.5 in the $Hg_{1-x-y}Mn_xCd_yTe$ system. However, the segregation coefficient of Cd is much higher than that of Mn, and the ratio is at least a factor of 4 for the $Hg_{1-x-y}Mn_xCd_yTe$ system. The growth rate of the $Hg_{1-x-y}Mn_xCd_yTe$ epilayers is 0.3 μm/min and appears to be slower than that of $Hg_{1-x}Cd_xTe$ or $Hg_{1-x}Mn_xTe$ ternary systems.

HIGH-PERFORMANCE $Hg_{1-x-y}Mn_xCd_yTe$ PHOTODIODES

To demonstrate HgMnCdTe photodiode performance, a mesa-type photodiode was fabricated and characterized. The structure consists of a boron-implanted mosaic fabricated on a p-type LPE $Hg_{1-x-y}Mn_xCd_yTe$ layer grown on a CdTe substrate. Conventional lithographic techniques were used during the diode processing. The n^+ region, with an area of 1.6×10^{-4} cm^2 was

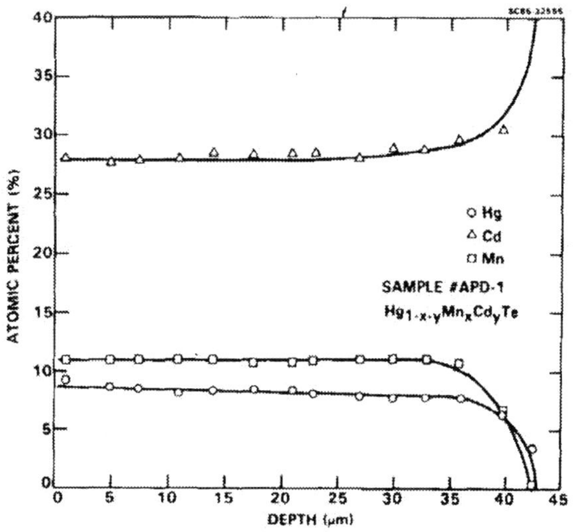

Fig. 4 Depth profiles of Hg, Mn and Cd in $Hg_{1-x-y}Mn_xCd_yTe$ epilayer as a function of distance from surface.

formed by an ion dose of 1×10^{14} cm^{-2} at an energy of 200 KeV. The implanted samples received post-implant annealing at 200°C. Bonded contacts to both implanted and unimplanted regions provided electrical contact for diode and ground connections.

Figure 5a shows a current-voltage (I-V) characteristic of the fabricated 1.54 μm HgMnCdTe photodiode that had a V_B of 50 V defined at I_D = 10 μA. This value is comparable to those of Ge (or InGaAs) avalanche photodiodes [15]. The background carrier concentration of Ge is on the order of 5×10^{16} cm^{-3}; therefore, the value, V_B, of this photodiode should be similar to that of our photodiode, since the V_B is strongly dependent on doping levels for a given energy gap (E_g).

Figure 5b shows an I-V curve at reverse bias. The leakage current at -10 V was 1.5×10^{-4} A/cm², which is relatively low for 1.5 μm photodiode applications and is comparable to some III-V photodiodes [15].

Figure 6 shows the spectral response of the photodiode taken under zero bias and frontside illumination. The cutoff wavelength is 1.54 μm at room temperature. The narrow spectral response might be due to surface recombination or to calibration problems during the measurement. The resistance-area product (R_oA) estimated from the saturation current was 2.62×10^2 Ω-cm². This would be equivalent to a detectivity (D*) value of 1.9×10^{11} cm $Hz^{1/2}$/W at 300K if the noise current is dominated by detector thermal noise at low background and assuming a quantum efficiency of 60%.

Figure 7 shows capacitance measurements as a function of reverse bias on the photodiode. A 0.75 pF capacitance was measured under a 5 V reverse bias at 295K. This value is three times smaller than that of a commercial Ge photodetector measured at 195K (9 pF for 4.6×10^{-4} cm² Ge detectors at 195K) [16]. These results are significant for demonstrating the feasibility of LPE HgMnCdTe for 1.55 μm long-wavelength applications.

Three shorter wavelength HgMnCdTe epitaxial layers have been grown and processed for device fabrication. The V_B of these diodes was over 110 V and the leakage current at V = -10 V was 3×10^{-5} A/cm². This current density is comparable to that obtained for the planar bulk HgCdTe APD reported by Alabedra et al [8]. The main characteristics are summarized in Table 2 and compared with previous published works.

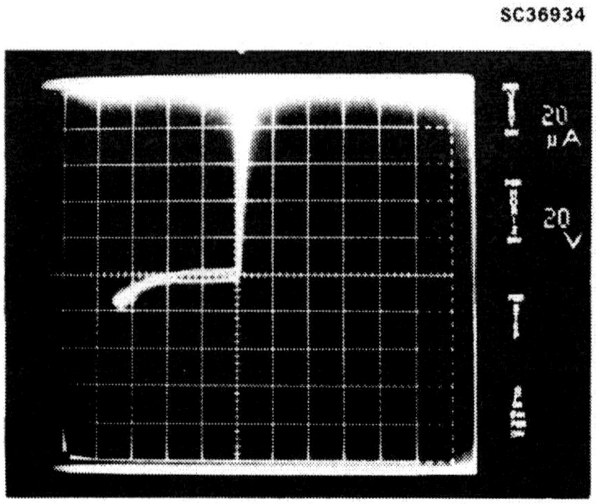

(a)

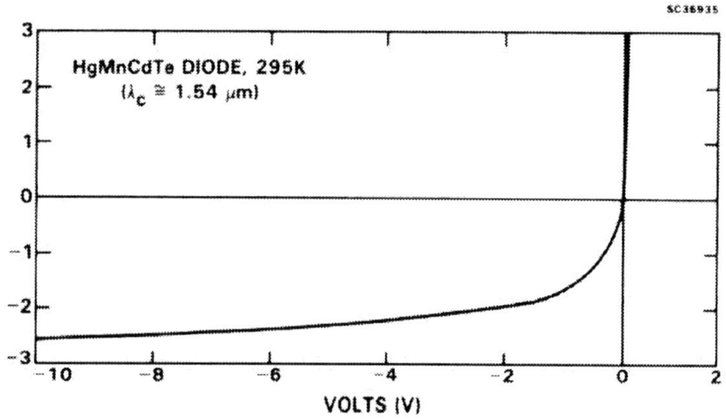

(b)

Fig. 5 (a) and (b) I-V characteristics of a HgMnCdTe photodiode ($\lambda_c \approx$ 1.54 μm at 295K).

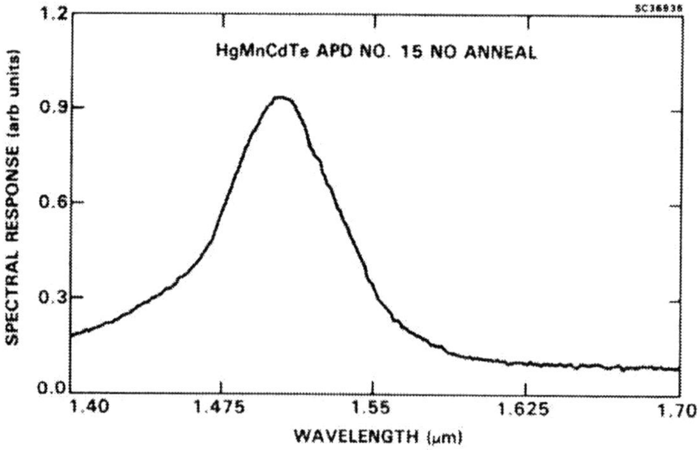

Fig. 6 Spectral response of a HgMnCdTe diode at 295K.

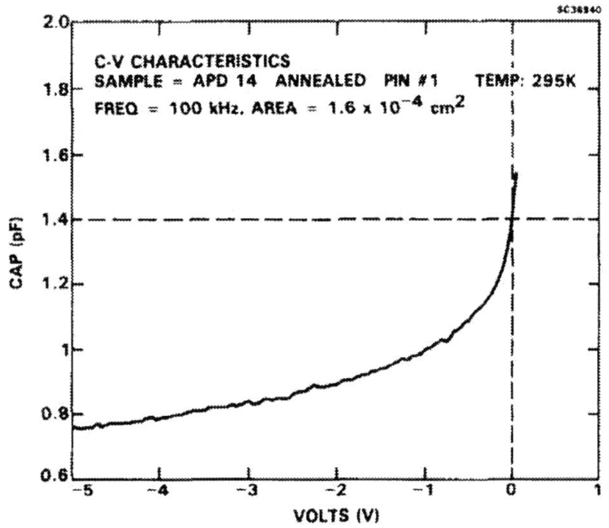

Fig. 7 Capacitance-voltage characteristics of a 1.54 μm HgMnCdTe diode at 295K.

Table 2

Main Characteristics of LPE HgMnCdTe Photodiodes and Comparison of Present Work with Recently Published Work

	HgCdTe Diode[a]	Diode 4	Diode 5	Diode 6	Diode 15
Material	Bulk (HgCdTe)	LPE (HgMnCdTe)	LPE (HgMnCdTe)	LPE (HgMnCdTe)	LPE (HgMnCdTe)
Carrier Concentration (cm^{-3})	$p = 10^{16}$	$p = 10^{16}$	$p = 10^{16}$	$p = 10^{16}$	$p = 10^{16}$
Junction (implanted ion)	Al	Boron	Boron	Boron	Boron
Cutoff Wavelength (μm)	1.3	1.28	1.39	1.46	1.54
Diode Area (cm^{-2})	1.32×10^{-4}	1.6×10^{-4}	1.6×10^{-4}	1.6×10^{-4}	1.6×10^{-4}
Dark Current at -10 V (A/cm^2)	1.5×10^{-5}	3×10^{-5}	6×10^{-5}	3.7×10^{-5}	1.5×10^{-4}
V_B (V)	80	110	120	60	50

(a) R. Alabedra, et al, IEEE Trans. Electron Dev. ED-13, 1302 (1985).

As the cutoff wavelength decreases (bandgap increases), the V_B increases significantly. We note that although the V_B of the 1.46 μm photodiode is only 60 V, the dark current at V = -10 V is about the same as the 1.28 μm photodiode dark current. This dark current for the 1.46 μm photodiode is much lower than that reported for either bulk or LPE HgCdTe avalanche photodiodes. The long-wavelength cutoff, λ_c = 1.45 μm, corresponds to the semiconductor bandgap E_g = 0.84 eV for the $Hg_{1-x}Cd_xTe$ alloy system with composition x = 0.66.

The experimental measurements of the avalanche gain on a planar HgMnCdTe photodiode with ZnS passivation and E_g and spin-orbit splitting (Δ_o) by electroreflectance are in progress.

CONCLUSIONS

The demonstrated performance of HgMnCdTe photovoltaic devices with relatively little development effort compared to III-Vs or IVs has shown equal or superior performance. This technology based on the quaternary HgMnCdTe has both theoretical and practical promise of superior performance and advanced structures for wavelengths in the range 1.3 to 1.8 μm for fiber optic communications.

ACKNOWLEDGEMENT

Research at the University of Maryland was supported in part by DARPA and ARO under Grant No. DAAG29-86-K-0052. The authors gratefully acknowledge the stimulating discussions with Drs. Bill Johnson and Dennis Stone.

REFERENCES

1. T.P. Pearsall and M.A. Pollack, in Semiconductors and Semimetals, Vol. 22, edited by W.T. Tsang (Acdemic Press, 1985), pp. 174-241.
2. R.J. McIntyre, IEEE Trans. Electron Devices ED-13, 164 (1966).
3. P. Becla, J. Vac. Sci. Technol. A 4, 2014 (1986); S. Wong and P. Becla, ibid A4, 2019 (1986).
4. D. Hildebrand, W. Kuebart, K.W. Benz and M.H. Pilkuhn, IEEE J. Quantum Electron. QE-17, 284 (1981).
5. C.S. Verie, F. Raymond, J. Besson and T. Nguyen Duy, J. Cryst. Growth 59, 342 (1982).
6. A. Sher, An-Ben Chen and W.E. Spicer, J. Vac. Sci. Technol. A3, 105 (1985); A. Wall, C. Caprile, A. Francoisi, R. Reifenberger and H. Debska, ibid A4, 818 (1986).
7. S.H. Shin, J.G. Pasko, M.D. Law and D.T. Cheung, Appl. Phys. Lett. 40, 965 (1982); M. Chu, S.H. Shin, M.D. Law and D.T. Cheung, ibid. 37, 318 (1980).
8. R. Alabedra, B. Orsal, G. Lecoy, G. Picard, J. Meslage and P. Fragnon, IEEE Trans. Electron Devices ED-32 (7), 1302 (1985).
9. A.M. Amirtharaj, F.H. Pollak and F.K. Furdyna, Solid State Comm. 39, 35 (1981).
10. P. Becla, P.A. Wolff, R.L. Aggarwal, S.Y. Yuen and R.R. Galazka, J. Vac. Sci. Technol. A3, 119 (1985).
11. U. Debska, M. Dietl, G. Grabeski, E. Janik, E. Kienzek-Pecold and M. Klimkiewicz, Phys. Strat. Sol. (a) 64, 707 (1981).
12. S. Takayama and S. Narita, Jap. J. Appl. Phys. 24 (10), 1270 (1985).
13. C.C. Wang, S.H. Shin, M. Chu, M. Lanir and A.H.V. Vanderwyck, J. Electrochem. Soc. 127, 177 (1980).
14. E.R. Gertner, Ann. Rev. Mater. Sci. 15, 303 (1985).
15. R.D. Dupuis, J.R. Velebir, J.C. Campbell and G.J. Qua, IEEE, Electron. Dev. Lett. EDL-7, 296 (1986).
16. K. Vural (personal communication).

INVESTIGATIONS OF $Cd_{0.9}Mn_{0.1}Te$ DOPED WITH Au, Cu, As AND P ACCEPTORS USING OPTICAL ABSORPTION AND PHOTOLUMINESCENCE

J. MISIEWICZ *, J.M. WROBEL**, P. BECLA AND D. HEIMAN
Francis Bitter National Magnet Laboratory MIT, Cambridge, MA 02139 USA
**Department of Physics, Simon Fraser University, Burnaby, Britisch Columbia V5A 1S6 Canada

ABSTRACT

Using absorption and photoluminescence investigations the energies of shallow acceptors are determined as follows: 0.11 eV for P; 0.12 eV for As; 0.17 eV for Cu and 0.18 eV for Au. Deep level energies are found as 0.28 eV for P; 0.29 eV for As; 0.41 eV and 0.92 eV for Cu; and 0.33 eV, 0.69 eV and 1.25 eV for Au dopants. Acceptor concentrations in the $10^{17}cm^{-3}$ range are achieved for As and P dopants, but for Au and Cu high compensation is found.

1. INTRODUCTION

Cadmium manganese telluride is one of the most intensively studied materials from the group of "diluted magnetic" or "semimagnetic" semiconductors. Considerable attention has been given to these materials due to their interesting semiconducting and magneto-optical properties [1-2]. Because of the large exchange interaction between the valence band and isolated Mn ions, it is important to obtain a high concentration of acceptors or free holes in $Cd_{1-x}Mn_xTe$. $Cd_{0.9}Mn_{0.1}Te$ crystals were doped during melt growth. Room temperature resistivities of crystals were in the range of 1 to $10\Omega\cdot cm$ for As and P dopants, and of 10^5 to $10^6\Omega\cdot cm$ for Au and Cu. For the Au and Cu doping, only highly compensated samples were obtained - the hole concentration, measured by the Hall effect method, was not higher than $10^{12}cm^{-3}$. For As and P doped crystals, hole concentrations were in the range of $10^{17}cm^{-3}$. Resistivity of the samples at 80 K increased typically by factors of 10^5 - 10^6.

2. RESULTS OF MEASUREMENTS

Photoluminescence spectra at 4.2 K indicate a sharp peak at energy 1.75eV very close to the band gap [3], Fig. 1. In some samples there are three less-intense features separated from the main peak by 17-21 meV, 40-47 meV, and 62-66 meV. They are interpreted with the LO CdTe-like phonon in $Cd_{1-x}Mn_xTe$ at 18 meV or LO CdTe-like phonon at 21 meV [4]. The second and third replicas are probably due to two or three LO CdTe-like phonons. In Cu-doped samples, a small additional feature is observed at 1.635 eV. As and P doped crystals exhibited large peaks attributed to donor-acceptor pair recombination; one at 1.594 eV for As and another at 1.635 eV for P dopants. In highly P-doped samples a second high-intensity photoluminescence feature was observed at 1.535 eV.

*present address: Institute of Physics, Technical University of Wroclaw, Wyspianskiego 27, 50-370 Wroclaw, Poland

Mat. Res. Soc. Symp. Proc. Vol. 89. © 1987 Materials Research Society

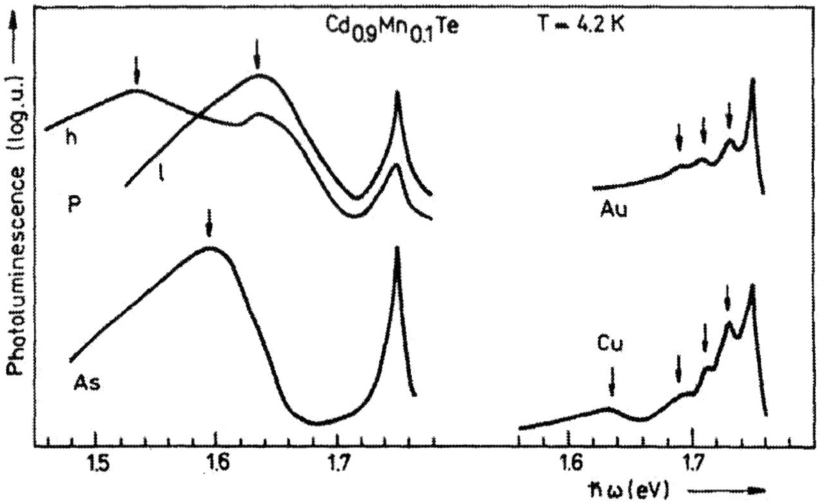

Fig.1. Photoluminescence spectra of doped crystals. Arrows denote characteristic energies; h,l - high and low doping levels.

Absorption spectra for As and P doped crystals indicate the same behavior - Fig.2. At 300 K, there is a distinct absorption increase when the energy is decreased from 0.9 eV to 0.05 eV. A similar dependence is observed for undoped samples but the absorption is at least one order of magnitude smaller.

At low temperatures, absorption is changed drastically - a sharp edge is observed for energies below 0.19 eV where the absorption maximum exists. There is also the second absorption peak at about 0.49 eV. For undoped samples there is only one absorption maximum, at 0.32 eV. Samples with Cu and Au indicate a completely different behavior, see Fig.2. For Cu-doped samples a strong absorption peak is observed at 0.12 eV and two additional features exist on its high energy tail.

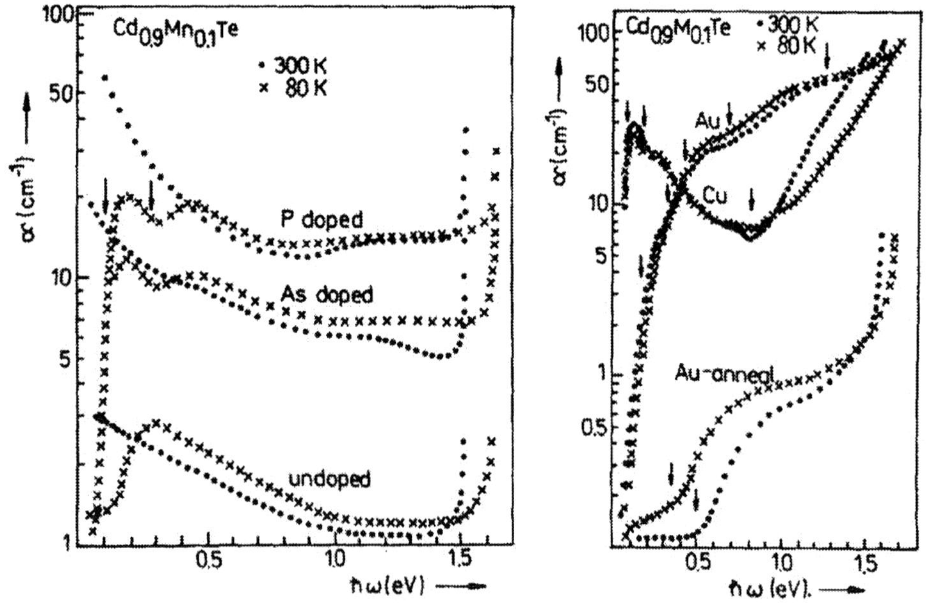

Fig.2. Absorption plots of doped samples, arrows denote characteristic energies.

There are a few apparent changes in the slope of the absorption curves for Au doped samples. In contrast to the previous dopants, there is no temperature dependence of the features observed in Au and Cu doped samples. Fig. 2 shows also plots for Au doped samples annaled in a Cd atmosphere. These plots are similar for Cu, As and P doping annealed with Cd. Only one broad absorption band is observed in the energy range of 0.5-1.1 eV.

3.ANALYSIS OF ABSORPTION RESULTS

The absorption spectra will be discussed separately for As, P and Cu, Au because of their different behavior. The As and P spectra are interpreted in terms of three mechanisms: valence band-to-acceptor level transitions V-A, heavy to light hole interband transitions L-H, as well as intra-heavy hole band transitions H-H. We assumed that at T$\leq$80 K L-H and H-H transitions are neglected, and the absorption is due solely to that photoionization mechanism V-A.

The problem of acceptor photoionization, including the central cell correction, was solved theoretically by Bebb [5] using the quantum defect method (QDM). According to this model, the photoionization cross-section has the form

$$\sigma(\hbar\omega) = \frac{4\pi\alpha_0 a^2}{3n}\,\frac{\nu^3 2^{2\nu} f(y)}{\sqrt{y}(1+y)^{\nu}}, \tag{1}$$

where ν is the effective principal quantum number, $\hbar\omega$ is the photon energy, $\alpha_0 = 1/137$, a the effective Bohr radius, n is the refractive index, $y=(h\nu-E_I)/E_I$ and:

$$f(y) = \left\{\frac{\sin[(\nu+1)\mathrm{arctg}\sqrt{y}]}{\sqrt{y}} - \frac{(\nu+1)\cos[(\nu+1)\mathrm{arctg}\sqrt{y}]}{\sqrt{1+y}}\right\}^2 \tag{2}$$

E_I is the observed ionization energy and $E_1 = E_0/\nu^2$, where E_0 is the hydrogenic (ν = 1) ionization energy. Bebb's formula was fitted to the experimental absorption spectra with two adjustable parameters: ν and the concentration of neutral acceptors N_0 to obtain the absorption coefficient from the cross-section. Because of lack of the data for $Cd_{1-x}Mn_xTe$, values for CdTe were used: n = 2.67 [6] , E_0= 57 meV, and m_{hh}= 0.645 [7].

Figure 3 shows that experimental spectra at 80 K can be well described by Eq.1 except for a low energy tail. The best fit values are: N_0= $2\cdot10^{17}cm^{-3}$ and $1.9\cdot10^{17}cm^{-3}$, and ν = 0.73 and 0.70 for P and As dopants, respectively. It gives the E_I energies equal to 0.11 eV for P and 0.12 eV for As dopants.

To interpret the spectra at 300 K the photoionization components were assumed to be temperature independent so difference between 300 K and low temperature plots is L-H and H-H transitions.

The L-H absorption spectra can be described by the Kahn theory [8]. This theory predicts an absorption coefficient of the form:

$$\alpha_{L-H} = \frac{16\pi^{3/2}e^2\hbar}{ncm_0^2}\cdot\frac{m_{hh}m_{lh}^{5/2}}{(m_{hh}-m_{lh})^{5/2}}\cdot\frac{pA^2(\hbar\omega)^{1/2}}{1+(m_{lh}/m_{hh})^{3/2}}\cdot\frac{\exp\left[-\frac{\hbar\omega}{kT}\cdot\frac{m_{lh}}{m_{hh}-m_{lh}}\right]}{(kT)^{3/2}}, \tag{3}$$

where e, c, k, m_0 constant, p is the total number of free holes, m_{lh} and m_{hh} are the effective masses of holes in light and heavy hole bands, A is the transition matrix element. Using m_{hh}= 0.65 and m_{lh}= 0.115 taken for CdTe from[9] and A = 1.6 calculated in [10] one can determine the α_{L-H} spectral dependence with the free-hole concentration p as a parameter. To get a resonable fit with p-values from electrical measurements, it was necessary to include H-H transitions, having the spectral dependence:

$$\alpha_{H-H} = \alpha_{300K} - \alpha_{V-A} - \alpha_{L-H} = K\lambda^{n} \qquad (4)$$

where K and n are adjustable parameters, and λ is the wavelength. The following values were obtained: K = 0.02, n = 2.2, and p = 1.5 $10^{16}cm^{-3}$ for As doped crystals; and K = 0.2, n = 1.6 and p = 6.1 $10^{16}cm^{-3}$ for P-doping.

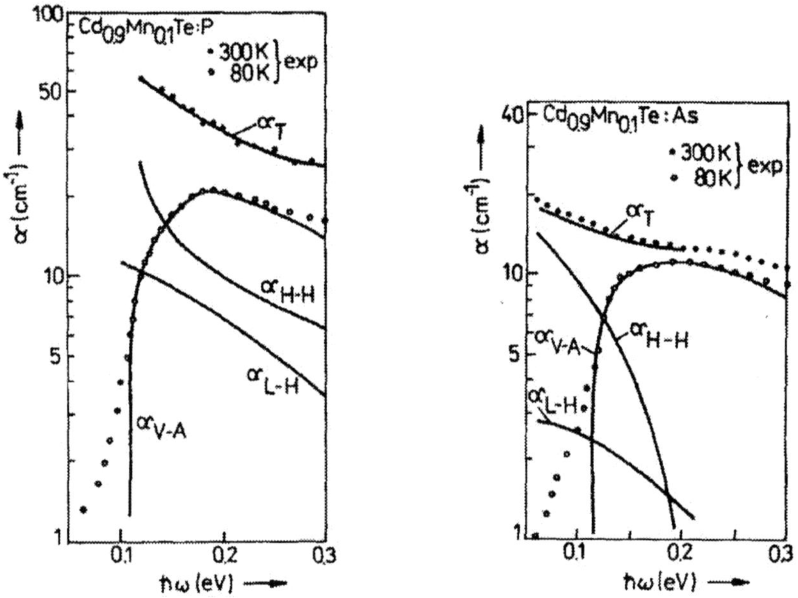

Fig.3. Theoretical description of absorption spectra, $\alpha_T = \alpha_{V-A} + \alpha_{L-H} + \alpha_{H-H}$

Fig.3 shows the contributions to Eq.4, and as we can see the fitting accuracy is quite good.

There is the second acceptor level visible at low temperatures with an activation energy about 0.28 eV and 0.29 eV for P and As doped samples. A small absorption increase in the 0.9 - 1.3 eV energy range can be ascribed to transitions from the spin-orbit split valence band to heavy-hole band. The values of Δ_{SO} determined for CdTe vary between 0.85 - 0.9 eV [9].

Absorption spectra of undoped samples, with hole concentration in the range of $10^{16}cm^{-3}$ indicate one acceptor level at 0.175 eV(when the method of acceptor energy determination is not described, the empirical rule E_I=0.6E_{peak} is used). The same acceptor was identyfied in [10] as 0.17 eV.

For Au and Cu dopants, acceptor level observed at 0.17 - 0.18 eV is compensated with the ratio of N_A-N_D/N_A < 10^{-3} and does not produce carriers. A distinct absorption peak observed at 0.12 eV for Cu-doping cannot be described in terms of V-A transitions. The energies of the other observed transitions are as follows: 0.41 and 0.92 eV for Cu and 0.33, 0.69 and 1.25eV for Au dopants.

Annealing in Cd atmosphere always results in reduction of absorption below fundamental edge - see Fig.2 as an representative of all doped samples. Broad absorption band is characterized by the energy ≈ 0.5 eV at 300 K and ≈ 0.35 eV at 80 K.

4.CONCLUSIONS

Because Au, Cu, As and P elements act as acceptors in CdTe [9], one should expect similar effects in $Cd_{0.9}Mn_{0.1}Te$. But our results indicate that only P and As play the role of effective acceptors in this compound. For the case Au and Cu dopants the deep level creation completely destroys the occupation of level at 0.17 eV. This level in undoped $Cd_{0.9}Mn_{0.1}Te$ (see Fig.2) gives hole concentration in the range of $10^{16}cm^{-3}$. A feature in the energy range of 0.1 eV for Cu doping, observed in PL and α spectra, must be connected with some transition between levels within the bandgap. To interpret the donor-acceptor pair recombination spectra we take into consideration the donor energies as 50 meV e.g. residual In or Cl deep complex donor and 23 - 27 meV one,e.g. Br, Cd_i^+,V_{Te}^+ in CdTe [9]. Then we get E_I = 0.119 eV for P and 0.120 eV for As dopants. These values are in very good accordance with those obtained from absorption.

The second row acceptor levels observed in P and As doped crystals with the ionization energy in the range of 0.25-0.29 eV is strongest in the high doped samples. We cannot see these transitions in photoluminescence for the low doped P and As samples. In the absorption spectra for As-doped crystals, which have lower doping level in comparison with the P-doped, the higher energy transition is a few times weaker than for the high P-doped crystal - see Fig.2. This level was not observed in CdTe doped with P or As. The lowest acceptor ionization energies for As and P dopants: 0.11 and 0.12 eV respectively, found in $Cd_{0.9}Mn_{0.1}Te$ crystals are higher than in CdTe: 0.068 eV and 0.092 eV respectively - see [11]. This ionization energy increase, is similar to that found in [12], where the native acceptor energies versus manganese concentration were studied. This ionization energy increase can be explain in terms of BMP effect [13].

ACKNOWLEDGMENTS

This work was supported by DARPA contract N00014-83-K-0454. Francis Bitter National Magnet Lab is supported by the NSF. One of us (J.M.W.) is financed by The Natural Sciences and Engineering Research Council of Canada.

REFERENCES

1. R.R. Galazka, Proc. 14 Int. Con. Phys. Semicond. Edinburgh, 1978, Inst. Phys. Conf. Ser. 43, 133 (1978)
2. J.K. Furdyna, J. Appl. Phys. 53, 7637 (1982)
3. A. Golnik, J. Ginter, and J.A. Gaj, J. Phys.C. 16, 6073, (1983)
4. S. Venugopalan, A. Petrou, R.R. Galazka and A.K. Ramdas, Solid State Commun. 35, 401 (1980)
5. H.B. Bebb, Phys. Rev. 185, 1116 (1969)
6. E.J. Danielewicz and P.D. Coleman, Appl. Optics 13, 1164 (1975)
7. Le Si Dang, G. Neu and R. Romestain, Solid State Commun. 44, 1187 (1982)
8. A.H. Kahn, Phys. Rev. 97, 1647 (1955)
9. K. Zanio in Semiconductors and Semimetals vol. 13 "Cadmium Telluride", (Academic Press N.York 1978 p.p. 129-163)
10. P. Wojtal, A. Golnik and J.A. Gaj, phys. stat. solidi b 92, 241 (1979)

11. J.L. Pautrat, J.M. Francou, N. Magnea, E. Molva, and K. Saminadayar, J. Crystal Growth 72, 194 (1985)
12. J. Jaroszynski, T. Dietl, M. Sawicki, E. Janik, Physica 117B/118B, 473 (1983)
13. J. Misiewicz, E.D. Isaacs, J. Wrobel, P. Becla, and D. Heiman - to be published.

OPTICAL PROPERTIES OF DOPED $Cd_{1-x}Mn_xTe$

Y. LANSARI, N. C. GILES, AND J. F. SCHETZINA
Department of Physics, North Carolina State University, Raleigh, North Carolina 27695-8202

P. BECLA, D. KAISER[a)]
Massachusetts Institute of Technology, Francis Bitter National Magnet Laboratory, Cambridge, Massachusetts 02139

[a)] Present address: IBM Research, Yorktown Heights, New York 10598

ABSTRACT

The introduction of phosphorus and arsenic dopants into bulk $Cd_{1-x}Mn_xTe$ crystals grown by the Bridgman-Stockbarger technique has been studied with respect to the resulting optical properties. Samples with a Mn composition in the range $0.10 < x < 0.30$, both as-grown and annealed, were investigated. A combination of room temperature transmittance and reflectance measurements over the spectral range from the ultraviolet to the far infrared has been used to gain information concerning the structural quality of the samples. Low temperature photoluminescence measurements (1.6-5 K) were used to determine optical quality and excitonic energies.

I. INTRODUCTION

$Cd_{1-x}Mn_xTe$ belongs to the class of materials referred to as diluted magnetic semiconductors (DMS). These DMS alloys, which contain a fraction x of magnetic ions (such as Mn^{++}) substitutionally placed on the group II sites, have received a great deal of interest in recent years because of their unique magnetic and magneto-optic properties [1-4]. Bulk $Cd_{1-x}Mn_xTe$ crystals can be grown in the zinc-blende (cubic) structure with Mn fractions of up to $x = 0.7$. The present paper deals with the intentional doping of CdMnTe with arsenic and phosphorus atoms. These dopants, which create shallow acceptor levels in CdTe, are expected to replace the Te atoms in the DMS lattice as well. We present a study of the optical properties to characterize the quality of the samples. The experimental methods used here include transmittance and reflectance measurements, and low temperature photoluminescence (PL) measurements. Undoped, As-doped, and P-doped bulk samples of CdMnTe were studied both in as-grown form, and after annealing in an overpressure of pure Cd.

II. EXPERIMENTAL DETAILS

A. Sample Preparation

Undoped $Cd_{1-x}Mn_xTe$ crystals were grown by the vertical Bridgman-Stockbarger technique. As- and P-doped $Cd_{1-x}Mn_xTe$ crystals were obtained by incorporation of the dopant impurities directly to the melt. Initial concentration of the dopant impurities was 3×10^{18} cm^{-3}. Electron microprobe analysis was used to determine the absolute value of the composition x, and the variation of x in the boule. The typical dislocation density obtained from etch pit density measurements was in the range of 10^4 to 10^5 cm^{-2}. Additional details on the growth method and parameters are given elsewhere [5].

Mat. Res. Soc. Symp. Proc. Vol. 89. ©1987 Materials Research Society

B. Optical Characterization

Room temperature optical transmission and reflectance spectra, and liquid helium PL measurements were used to determine the optical properties of the undoped and doped samples. Prior to this investigation of the samples, both front and back surfaces were chemi-mechanically polished and etched in a bromine-methanol solution to remove near-surface damage.

The transmittance measurements in the far infrared (IR) range (5000 cm^{-1} to 180 cm^{-1}) were performed using a double-beam, ratio-recording Perkin-Elmer 983-G spectrophotometer. In the ultraviolet (UV)-visible (VIS)-near IR range (185 nm to 3200 nm), a double-beam, double-monochromator, ratio-recording Perkin-Elmer LAMBDA-9 spectrophotometer was used for transmittance and reflectance measurements. The UV reflectance measurements give the positions of the E_1 and $E_1 + \Delta_1$ peaks, which change with alloy concentration. The optical absorption edge of each sample was determined from transmittance data in the region from 600-3200 nm. Far IR transmittance spectra were used to obtain qualitative information concerning structural defects which produce extrinsic optical absorption.

Low temperature PL measurements were performed to investigate sample quality and determine excitonic energies associated with the introduction of phosphorus and arsenic atoms into the host II-VI lattice. The samples were mounted on a copper sample holder and cooled to liquid helium temepratures in a Janis Research Products SuperVaritemp dewar. A Spectra-Physics He-Ne model 125A laser and an argon ion model 171 laser with single-line output optics were used as excitation sources for the luminescence measurements. The input excitation light and output luminescence were directed to and from the sample in a near-backscattering geometry. The luminescence obtained from the focused laser beam was directed by a spherical mirror onto the entrance slit of an ISA model HR-640 grating spectrometer and detected by a photomultiplier tube (PMT) with S-1 response. The output from the PMT was sent to an INSACO lock-in amplifier and then analyzed with a Digital Equipment Corporation Micro PDP-11 computer which also controlled the wavelength setting of the spectrometer.

III. Results and Discussion

Room temperature transmittance spectra of the undoped samples in the UV/VIS range show a shift in the absorption edge for the different x values, or Mn concentrations. As expected, for $x < 0.3$, the absorption edges shift linearly toward higher energies with increasing Mn concentration [6,7]. We obtained an empirical relation for the shift in energy position of the absorption edge as a function of Mn concentration by performing a linear fit to the room temperature optical absorption data of Abreu et al. [7] over the range investigated, $0.10 < x < 0.3$, allowing a determination of the x value for the undoped samples studied in this report:

$$x = (E_{co}(eV) - 1.462)/(1.435) \qquad (1)$$

where x is the Mn concentration and E_{co} is the energy position of the edge for the maximum absorption recorded. The values obtained using Eq. 1 are in close agreement with those obtained by electron microprobe measurements.

The transmittance spectra for the undoped samples in the energy range below the band gap are reasonably flat, although some tailing is observed in the near-edge region for the undoped, as-grown samples. As in the case of CdTe, this tailing may be related to free carrier absorption and/or impurity levels within the band gap [8].

In the far IR spectral region, the undoped samples are transmitting out to 500 cm^{-1} (20 um), where the phonon absorption peaks occur. Typically, the measured % transmittance at 5000 cm^{-1} (2 um) is close to 60% for the samples studied.

UV reflectance measurements recorded in the range 300 to 400 nm are used to determine the positions of the E_1 and the $E_1 + \Delta_1$ peaks [5]. Broadening of the peaks with increasing Mn concentration is observed, in agreement with the observations of Bucker et al. [9]. Interestingly enough, the energy position of the $E_1 + \Delta_1$ feature does not depend on the Mn concentration as strongly as the energy of the E_1 feature does.

The measured transmittance amplitudes near the absoprtion edge for the p-type samples are, in general, lower in magnitude than those observed for the undoped samples. Significant tailing of the transmittance curves in the near edge regions are noted. As in the case for the undoped samples, this may be related to the concentration of impurities in the material and to carrier absorption. More striking to note is the shift in the absorption edges from the undoped to the doped samples. Energy shifts in the absorption edges due to the annealing process are also observed. In addition to shifting E_{co}, the anneal introduces considerable absorption in the near-edge region, making the tailing even more noticeable in the case of the doped samples. This may indicate that the annealing process introduces new defects in the material, as was suggested by D.J. As et al. for thermally annealed epitaxial CdTe layers [10]. Figure 1 shows the transmittances of CdMnTe:As (x = 0.109) as-grown and annealed samples, respectively.

All of the intentionally doped samples are transmitting over the range from 2 um to approximately 20 um, where phonon absorption peaks occur. Maximum transmittances for as-grown, doped samples range from 39% to 62% at 2 um, with the highest values corresponding to the As-doped material. It should be pointed out that the IR transmittances for As-doped samples are comparable to those obtained for the undoped samples. The IR transmittance spectra for As-doped (x = 0.109) samples are shown in Fig. 2. Again, the annealing treatment changes the shape of the transmittance spectrum, lowering the transmittance amplitude in the 5000-3000 cm^{-1} region.

The E_1 peaks measured for the As-doped material agree well with the positions measured for the undoped samples. Fig. 3 show the UV reflectance peaks for As-doped (x = 0.109) samples. The annealing treatment shifts the $E_1 + \Delta_1$ peak to higher energy and the E_1 peak to lower energy in the As-doped sample. The opposite trend is observed in the P-doped material. The reflectance peaks for the P-doped samples are at lower energies than those measured for either the undoped or As-doped materials.

Low temperature PL spectra (1.6-5 K) for the undoped, P-doped and As-doped samples of different Mn concentrations were measured and reported in detail in an extended paper [5]. In this work, we discuss the spectra obtained for the CdMnTe:As samples, both as-grown and annealed, with 10.9% Mn. As can be seen in Fig. 4, the edge emission peaks which are believed to be (A^0,X) recombination, dominate the spectra. These peaks give a FWHM (full-width-at-half-maximum) of 8.7 meV and 11.0 meV, for the as-grown and annealed samples, respectively. These FWHM's correspond to fairly narrow peaks when compared to 11.0 meV, which is the value predicted for $Cd_{1-x}Mn_xTe$, x = 0.109, using the model of alloy broadening given by Schubert et al. [11]. The energy position of the free exciton recombination for this Mn concentration should occur at 1.7681 eV [12]; therefore, the observed PL edge emission peaks at 1.7504 eV (as-grown) and 1.7521 eV (annealed) give (A^0,X) binding energies of 17.7 meV and 16.0 meV, respectively. The asymmetric lineshape of the (A^0,X) peaks is probably due to the presence of phonon replicas, starting at about 21 meV below the main line. Low amplitude defect emission bands are observed at 1.5963 eV for the as-grown material, and 1.5933 eV (FWHM = 69 meV) for the annealed sample.

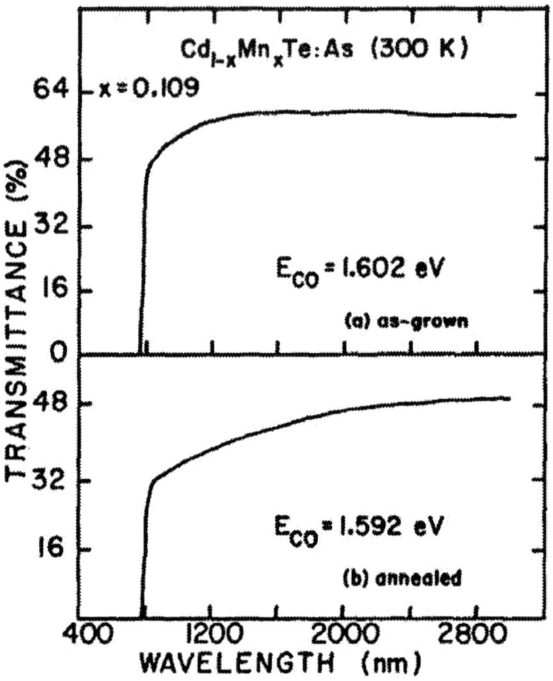

Fig. 1. Transmittance spectra (300 K) for $Cd_{1-x}Mn_xTe$:As, x = 0.109, (a) as-grown, (b) annealed

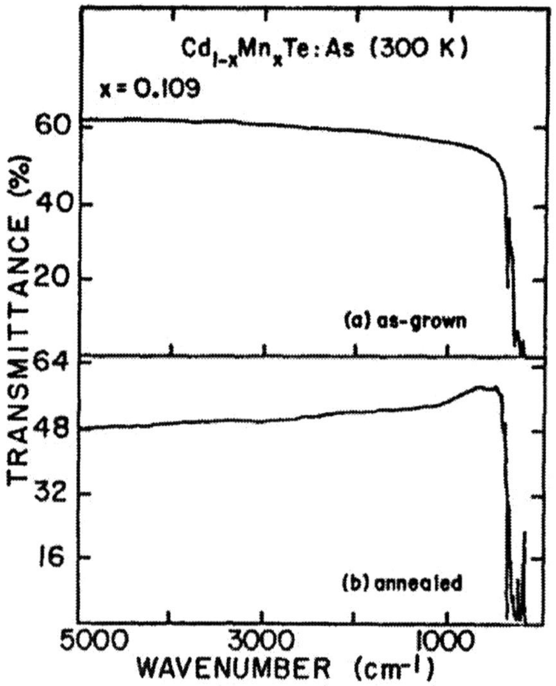

Fig. 2. IR transmittance spectra (300 K) for $Cd_{1-x}Mn_xTe$:As, x = 0.109, (a) as-grown, (b) annealed

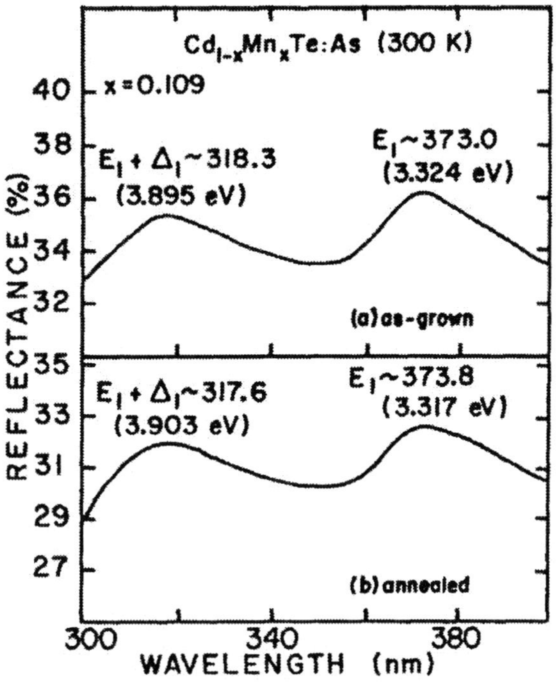

Fig. 3. Reflectance spectra (300 K) for $Cd_{1-x}Mn_xTe$:As, x = 0.109, (a) as-grown, (b) annealed

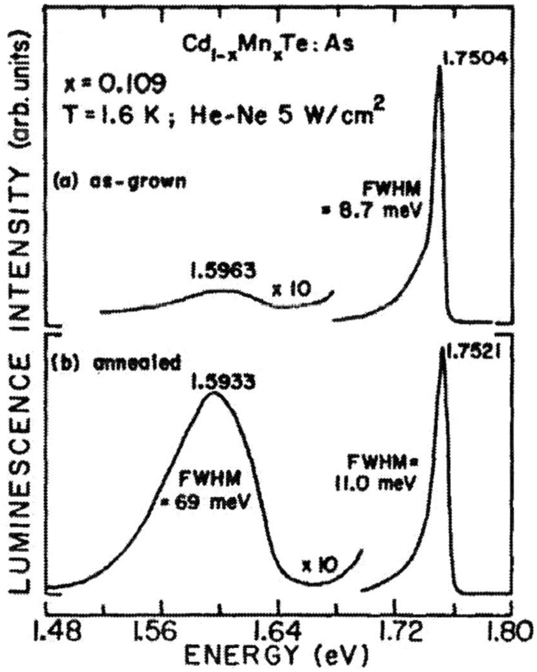

Fig. 4. PL spectra (1.6 K) for $Cd_{1-x}Mn_xTe$:As, x = 0.109, (a) as-grown, (b) annealed

IV. CONCLUSIONS

In summary, the optical properties of bulk undoped and p-type Bridgman-grown $Cd_{1-x}Mn_xTe$ boules in the composition range $0.10 < x < 0.30$ have been studied.

Room temperature transmittance spectra in the UV/VIS range were used to determine the absorption edges E_{co} of the undoped samples. The dependence on Mn concentration of the measured values of E_{co} was found to agree closely with that reported earlier [7] in the range $0.10 < x < 0.3$. The IR transmittance spectra for the undoped and doped samples were also measured.

Reflectance peaks, the E_1 and $E_1 + \Delta_1$ transitions, are broadened with increasing Mn concentration. The energy position of the E_1 peak shows a dependence on Mn concentration, while the energy position of the $E_1 + \Delta_1$ peak shows little or no dependence on x-value.

Low temperature PL measurements at liquid helium temperatures were used to determine excitonic binding energies for As-doped samples, both as-grown and annealed, with 10.9% Mn concentration. Edge emission peaks, believed to be (A^0,X) transitions, were observed to dominate the spectra in these As-doped samples.

V. ACKNOWLEDGEMENTS

The authors are grateful to R. Burns and M. Gildner for assistance in polishing the samples prior to the optical measurements. The work at NCSU was supported by DARPA/ARO contract DAAG29-83-K-0102 and NSF grant DMR83-13036. The work at MIT was supported by DARPA through contract N00014-83-K-0454.

REFERENCES

1. J. K. Furdyna, J. Vac. Sci. Technol. 21, 220 (1982).
2. D. L. Peterson, A. Petrou, M. Datta, A. K. Ramdas, and S. Rodriguez, Solid State Commun. 43, 667 (1982).
3. J. A. Gaj, R. R. Galazka, and M. Nawrocki, Solid State Commun. 25, 193 (1978).
4. S. M. Ryabchenko, O. V. Terletskii, I. B. Mizetskaya, and G. S. Olenik, Fiz. Tekh. Poluprovodn. 15, 2314 (1981) [Sov. Phys.-Semicond. 15, 1345 (1981)].
5. P. Becla, D. Kaiser, N. C. Giles, Y. Lansari, and J. F. Schetzina, submitted to J. Appl. Phys.
6. N. T. Khoi and J. A. Gaj, Phys. Stat. Sol. B83, K133 (1977).
7. R. A. Abreu, W. Giriat, and M. P. Vecchi, Phys. Lett. 85A, 399 (1981).
8. K. Zanio, Semiconductors and Semimetals, Vol. 13, ed. R. K. Willardson and A. C. Beer, (Academic Press, 1978).
9. R. Bucker, H. E. Gumlich, and M. Krause, J. Phys. C: Sol. State Phys. 18, 661 (1985).
10. D. J. As and L. Palmetshofer, J. Cryst. Growth 72, 246 (1985).
11. E. F. Schubert, E. O. Gobel, Y. Horikoshi, K. Ploog, and H. J. Queisser, Phys. Rev. B30, 813 (1984).
12. D. Heimer, P. Becla, R. Kershaw, D. Ridgley, K. Dwight, A. Wold and R. R. Galazka, Phys. Rev. B34, 3961 (1986).

SURFACE AND BULK PHOTOCONDUCTIVITY OF $Cd_{1-x}Mn_xTe$

H. Neff[a], K. Y. Lay[b], K. Park[c] and K. J. Bachmann[a,b]
Departments of Chemistry[a], Materials Engineering[b] and Physics[c]
North Carolina State University
Raleigh, North Carolina 27695-8204

ABSTRACT

Double beam photoconductivity experiments are reported for the system $Cd_{1-x}Mn_xTe$. The technique allows a separation of surface and bulk contributions, respectively. Bulk effects dominate for manganese rich material and reveal a sharp peak at the band gap energy while surface conductivity reveals a step function type spectral behavior. The growth of a native oxide on the surface causes an increase in the surface recombination velocity and a change from surface to bulk conduction. An oxide related trap state was discovered that is located at approximately 400 meV above the valence band edge.

INTRODUCTION

The photoconductivity (PC) effect in semiconductors is used as a tool for optical materials characterization [1] and sensor applications as well. As demonstrated previously [2], both, the bulk and the surface contribute to the PC. The relative extent of these contributions depends primarily on the geometric arrangement of the electrodes. In defect free material, the spectral behavior of a prominent bulk photoconductive effect is characterized by a pronounced feature close to the band gap energy. In contrast, a prominent surface effect typically reveals a step-function type spectral distribution and a low energy cut-off at the same energy [2]. Subband gap states induced by defects, adsorbed gases [3,4] or the formation of an oxide layer may affect the spectral distribution and yield as well. In this communication we report on double-beam photoconductivity measurements on single crystals of $Cd_{1-x}Mn_xTe$ ($0.3 < X < 0.6$) using the planar-surface sensitive-two electrode configuration with a spacing of 0.5 mm between the contacts. Double beam spectroscopy has been developed primarily for the assessment of minority-carrier traps in defect rich material using a long wavelength probe (below band gap excitation) to empty/fill sensitizing trap levels [5]. In this study we use above band gap laser illumination as a pump beam in nearly defect free material. The technique allows a clear cut identification of bulk and surface contributions to the photoconductivity and is sensitive to the alterations of the surface properties due to native oxide formation.

EXPERIMENTAL

Bulk crystals of nominally undoped p-type $Cd_{1-x}Mn_xTe$ at a net carrier concentration of 10^{15}-10^{16} were prepared using the modified Bridgman technique [6]. For a discussion of the low temperature PC and PL spectra of $Cd_{1-x}Mn_xTe$ we refer the reader to a separate publication [7]. The experiments were performed at room temperature on freshly cleaved (110) surfaces. Figure 1 shows a schematic representation of the experimental configuration. The sample is illuminated by chopped (ac) light from a monochromator (probe beam). Simultaneously a second (dc) pump beam is focussed on the crystal surface employing an Ar^+-ion laser $h\nu$(488 nm) > E_g and an intensity of approx. 100 mW/cm^2. A detailed description of the optical set-up is given elsewhere [8]. The measurements were performed in air using phase sensitive lock-in detection in combination with a current-voltage converter and a Tektronix 4052 microcomputer for data processing.

Mat. Res. Soc. Symp. Proc. Vol. 89. ©1987 Materials Research Society

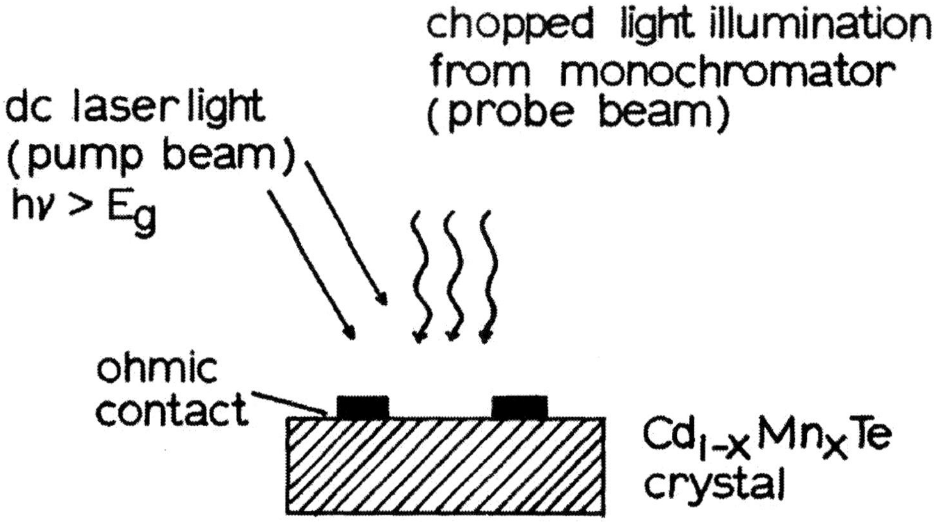

Figure 1. Sketch of the experimental set-up for double beam photoconduction spectroscopy.

RESULTS AND DISCUSSION

Figure 2 shows PC spectra obtained under chopped (ac) monochromatic light (solid line) and in the presence of an additional laser illumination (dashed line) using the 488 nm line of an Ar^{+}-ion laser at an incident power of approximately 100 mW/cm^2 for two $Cd_{1-x}Mn_xTe$ crystals (a) x = 0.3 and (b) x = 0.45. Without laser illumination the spectra show sharp low-energy cut offs of the PC-signals at photon energies of (a) 1.88 eV and (b) 2.09 eV, respectively. For photon energies $h\nu > E_g$ the spectrum (a) remains nearly constant while the shape of spectrum (b) is strongly affected by surface recombination. In spectrum (a) a distinct feature is resolved at a photon energy of 1.48 eV and is attributed to a surface defect. This structure does not appear in the presence of the additional laser illumination that transforms the PC-spectrum into a sharp single peak. The energetic position of this peak agrees well with the location of the fundamental gap E_0 as confirmed independently by means of photoreflectance spectroscopy [9] of the same samples. The photoreflectance spectrum (PR) for the sample with x = 0.3 is displayed as an insert in fig. 1. Note that in the presence of mixed surface and bulk conduction the peak position is shifted off the band gap energy explaining the difficulties in achieving reproducible band gap data from PC spectra without attention to this point [10]. Also, we would like to mention that the bulk acceptor states contribute to the PC as demonstrated for example by the shallow feature at 1.71 eV in Fig. 2a that does not vanish upon laser illumination. The position of the bulk acceptor state is in agreement with previous reports [11,12]. Furthermore, it should be noted that the effects reported here for $Cd_{1-x}Mn_xTe$, also, were observed on (110) InP-crystals [13].

Figure 3 shows the effect of aging in air on the PC spectra of $Cd_{1-x}Mn_xTe$ recorded without laser illumination. The solid line is obtained for the freshly prepared sample. The dashed-dotted line shows the photoconductive behavior of the same crystal recorded after exposure to air for 6 hr.

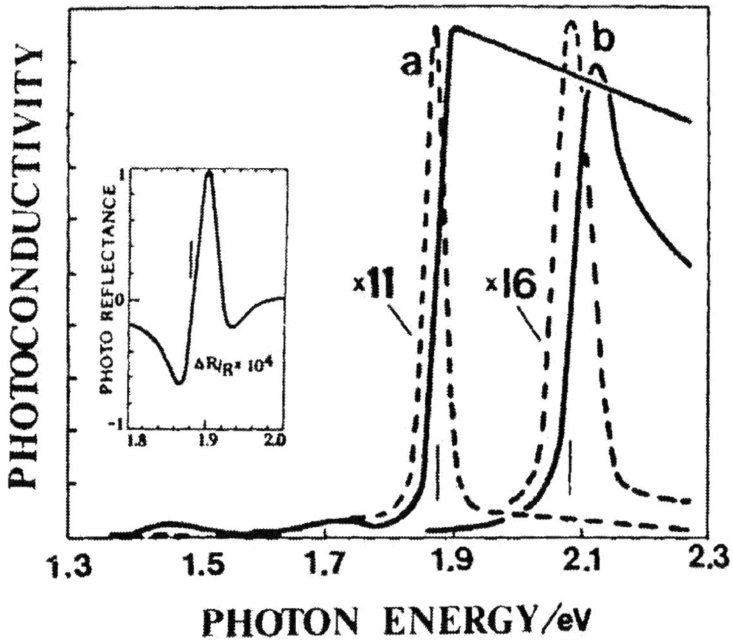

Figure 2. PC spectra of $Cd_{1-x}Mn_xTe$ obtained with the probe beam (solid line) and in the presence of the addition of laser illumination (dashed line), for (a) x = 0.3 and (b) x = 0.45. Insert: PR spectra for x = 0.3. The bars denote the position of the fundamental E_0-transition. Note the different plot factors for spectrum a and b in the presence of the dc laser illumination.

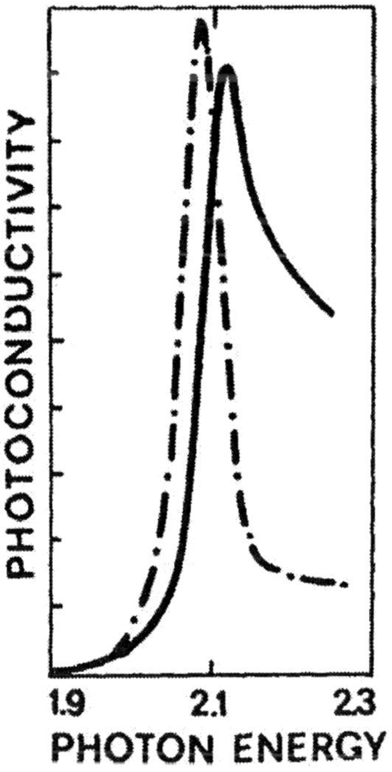

Figure 3. PC spectra of $Cd_{0.55}Mn_{0.45}Te$ obtained from the freshly cleaved sample (solid line) and after an exposure of 6 hrs to air (dashed line). Note the shift of the PC-maximum to lower energies.

Both, the photoyield and the spectral distribution exhibit a drastic change. A pronounced peak occurs shifted approximately 50 meV to lower values. It should be noted that this spectrum is almost identical with the data obtained for the freshly prepared sample recorded in the presence of the laser illumination (see spectrum (b) in fig. 2). Aging effects were observed for all samples but are more pronounced for crystals with a high manganese content.

Figure 4 shows a typical photoconduction transient signal recorded for the same crystal as presented in spectrum (a) in fig. 2. Upon switching on the laser beam a nearly complete suppression of the (ac) photoconductive signal is observed, followed by a rather slow recovery of the photocurrent when the laser is switched off reaching steady state conditions after approximately 2 sec. The increase of the PC-transient upon switching off the laser beam clearly exhibits a two-step-process.

We explain the observed phenomena on the basis of the absorption behavior of the incident light and the effect of the formation of an oxide layer. As described before, the photoconductivity effect in the planar configuration employed in this study accounts for both, bulk and surface transport. We attribute the quenching of the surface conductivity at photon energies larger than the band gap to a "flooding" in the near surface regime where the laser light is absorbed that drives the "dc" photocurrent into the saturation regime (penetration depth of the incident laser light approx. 1000 A). Hence, electron-hole pairs additionally excited through the primary (chopped) light beam do not contribute to the photocurrent in the near surface regime for photon energies $h\nu > E_g$. However, photons of energy close to E_g that penetrate to greater depth than the laser beam are absorbed in the bulk, still contribute to the PC gain. Therefore, the corresponding pronounced peak in the PC-spectrum in the presence of additional laser illumination represents the bulk response. As shown in several studies the surface conductivity is strongly affected by gas adsorption and defects on semiconductor surfaces [3,4]. We attribute the change from a dominant surface conductivity effect to bulk conduction behavior shown in fig. 3 primarily to a chemical alteration of the surface due to the formation of a native oxide. Since the oxide consists primarily of MnO and TeO_2 [14], the oxidation process leads to changes in the stoichiometry of the semiconductor surface layer beneath the native oxide film. Therefore, we explain the decrease of the surface photoconductivity by the formation of deep interface states as a consequence of the oxidation process that enhances the surface recombination rate, while the bulk properties remain almost unchanged.

Further evidence for the operation of slow trapping centers at the oxidized $Cd_{1-x}Mn_xTe$ surface is given by the PC transients (fig. 4) that display two distinct time intervals in the overall decay process. The role of the 400 meV deep surface state observed in the PC spectrum (a) of fig. 2 without laser illumination in the discharge mechanism and the chemical nature of this state are presently unknown and merit further investigation.

In summary, double beam PC spectroscopy has been employed for the investigation of bulk and surface contributions to the photoconductivity of $Cd_{1-x}Mn_xTe$. The experimental data show that native oxide growth on $Cd_{1-x}Mn_xTe$ is accompanied by a strong degradation of the surface contribution to the photocurrent. We interpret this result as a consequence of changes in the stoichiometry of the material at the semiconductor/oxide interface. Furthermore, we have demonstrated that the peak position in the PC spectra depends on the relative magnitudes of the surface and bulk contributions, respectively, to the PC. This explains the inconsistency of previous attempts at deducing bandgap information from the PC spectra in the literature. Also a deep surface state located 400 meV above the valence band edge and two distinct contributions to the slow discharge of traps at the oxidized $Cd_{1-x}Mn_xTe$ (110) surface have been observed, but the chemical nature of these states is presently unknown.

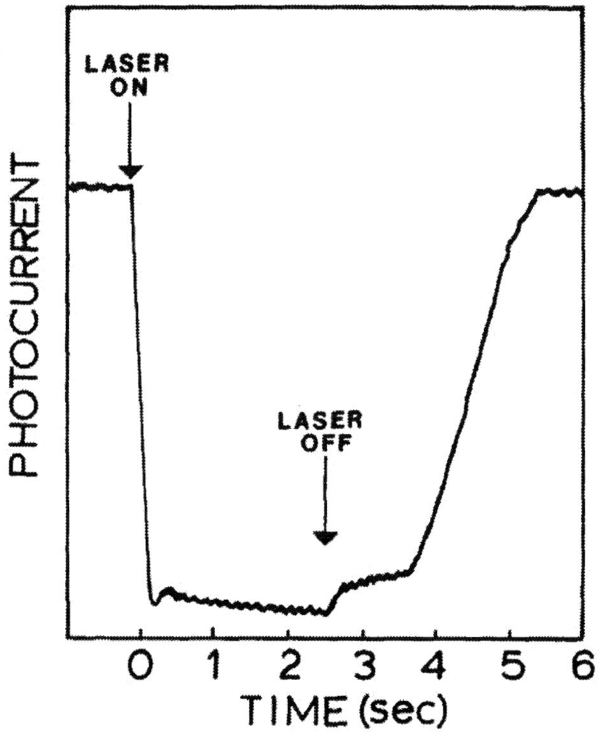

Figure 4. Photoconduction transient signal for $Cd_{0.7}Mn_{0.3}Te$ recorded at a photon energy of 2.2 eV.

ACKNOWLEDGMENTS

We would like to acknowledge the support of this work by NSF Grant DMR-8414580.

REFERENCES

1. See, for example: R. H. Bube, Photoconductivity in Solids, Wiley, New York, 1960.

2. L. J. Brillson, Surface Sci. 69, 62 (1977).

3. I. E. Ture, G. J. Russell, and J. Woods, J. Cryst. Growth 59, 223 (1982).

4. M. Tanielian, H. Fritzsche, C. C. Tsai, and E. Symbalisty, Appl. Phys. Lett. 33, 353 (1978).

5. R. G. Humphreys, D. C. Herbert, B. R. Holeman, P. Tapster, and W. P. Bickley, J. Phys. C: Solid State Phys., 16, 1469 (1983).

6. K. Y. Lay, N. C. Giles-Taylor, J. F. Schetzina, and K. J. Bachmann, J. Electrochem. Soc. 13, 1049 (1986).

7. H. Neff, K. Y. Lay, K. J. Bachmann, and R. Kotz, J. Luminescence, In press.

8. P. Lange, H. Neff, M. Fearheiley, and K. J. Bachmann, J. Electron. Mat. 14, 667 (1985).

9. K. Y. Lay, H. Neff, and K. J. Bachmann, Phys. Stat. Sol. (a) 92, 567 (1985).

10. N. V. Yoshi, J. Martin, and P. Quintero, Appl. Phys. Lett. 39, 79 (1981) and references cited in this paper.

11. K. Yamada, M. Lindström, J. Heleskivi and R. R. Galazka, Jap. J. Appl. Phys. 19, Supplement 19-3, 361 (1980).

12. J. Stankiewicz and A. Aray, J. Appl. Phys. 53, 3117 (1982).

13. M. S. Su and H. Neff, unpublished results.

14. H. Neff, K. Y. Lay, P. Lange, G. Lucovsky, and K. J. Bachmann, J. Appl. Phys. 60, 151 (1986).

INFLUENCE OF ELECTRIC FIELDS ON EXCITON LUMINESCENCE IN ZnSe/(Zn,Mn)Se SUPERLATTICES

QIANG FU AND A. V. NURMIKKO
Division of Engineering and Department of Physics
Brown University, Providence, RI 02912

L. A. KOLODZIEJSKI AND R. L. GUNSHOR
School of Electrical Engineering
Purdue University, West Lafayette, IN 47907

ABSTRACT

Application of moderate external electric fields to ZnSe/(Zn,Mn)Se superlattices is shown to yield readily measurable spectral shifts of the exciton ground state resonance in the quantum well limit. This shows that confinement effects increase the classical field ionization threshold. At high applied fields and low temperatures, externally injected hot electron excited luminescence from internal excitations of the Mn-ion is observed.

INTRODUCTION

In this paper we outline results of experiments in which the influence of applied electric fields on the exciton ground state in ZnSe/(Zn,Mn)Se superlattices [1] is studied through their luminescence, with emphasis on the quantum well limit. The motivation was in part to test carrier confinement effects in this heterostructure against field induced ionization. This occurs against the backdrop of recent magneto-optical studies [2] which have suggested that the valence band offset, in particular, of this II-VI wide bandgap system may be quite small and in actual practice deermined by the lattice mismatch strains in a given structure. The uniaxial strain components also define the light-hole exciton as the lowest energy interband excitation. Furthermore, electric field induced effects on electron and hole states in DMS quantum wells can be expected to provide a counterpoint to analogous effects in external magnetic fields which also can induce spatial shifts in the confined particle wavefunctions towards DMS barrier layers [3,4]. The work discussed here has, from the non-DMS point of view, a direct connection to the 'quantum Stark effect' discovered and formulated recently for the GaAs/(Ga,Al)As quantum well case [5].

EXPERIMENTAL

One principal ZnSe/(Zn,Mn)Se structure formed the basis of this study: a multiple quantum well structure which contained a single ZnSe well significantly wider than all the others (of equal thickness) so as to approximate a single quantum well structure (SQW). The structure was grown on a n+GaAs substrate following the deposition of an approximately 1 micron thick buffer layer of ZnSe. The dimensions of the SQW section were L_w = 70 Å for the ZnSe well and L_b = 185 Å for the (Zn,Mn)Se barrier of x=0.45 Mn concentration. Another structure tested was closer to the superlattice (electronic) limit with approximately 100 layer pairs of ZnSe/(Zn,Mn)Se with L_w = 45 Å and thin barrier layers of x=0.45 and L_b = 9 Å (approximately three monolayers of ZnMnSe). Because of finite lattice mismatch (up to 2%) these superlattices are under considerable strain. However, previous work has shown that the strain can be elastically accommodated. A Schottky barrier electrical contact was evaporated on the topmost (Zn,Mn)Se layers; reproducible results were obtained with a thin transparent layer of Cr defining an optical aperture of 100 micron diameter and a thicker Au layer on its perimetry allowing electrical contacting to external circuitry. The back contact to

Mat. Res. Soc. Symp. Proc. Vol. 89. ©1987 Materials Research Society

the n+GaAs substrate was assumed to be nearly an ohmic one. All the data discussed below was taken in this 'perpendicular' field geometry.

Figure 1 shows the photoluminescence (PL) emission from the SQW portion of the quantum well sample at T=2K with an without an applied electric field. The PL was excited either by the 3250 Å or 4416 Å lines from an HeCd laser. The SQW intensity was comparable to that of the MQW segment indicating good collection efficiency of carriers to the 70 Å wide ZnSe well. A spectral shift is clearly induced by the field; for the particular polarity used (with respect to the top electrode) the shift is towards higher photon energies. This, however, also includes the effects of the built-in Schottky barrier field as can be seen in Fig. 2 which summarizes the spectral peak positions of PL as a function of the applied voltage. We observed that the application of approximately +9 V of external bias reproduced the PL spectrum from an identical SQW control structure without a Schottky barrier electrode. The estimated total electric field in the SQW region is shown on the top edge in the figure. The presence of a finite field thus produces a spectral redshift and some broadening in the moderate range of fields considered here.

In the region of biases shown in Fig. 2 the currents measured in the external circuit were negligible or small and the PL quantum efficiency was relatively constant as shown. Beyond approximately 14 V, however, a strong 'forward bias' conduction was observed together with rapid quenching of the SQW PL emission as also indicated. This occurs as a consequence of efficient injection of electrons across the n+GaAs/ZnSe heterojunction as also verified by the onset of bright yellow Mn-ion internal luminescence in the structure. This emission well known as the basis for thin film electroluminescent devices, is generally though to be induced by hot electron impact excitation of the Mn-ion d-electron states. Its presence here shows the high interfacial quality of the n+GaAs/ZnSe heterojunction.

In strong contrast to the distinct field induced spectral shifts observed in the quantum well structure with nearly constant PL efficiency, the corresponding exciton luminescence from the thin barrier ZnSe/(Zn,Mn)Se superlattice decreased rather continuously with field with accompanying spectral broadening and very little evidence for shifts. The good structural quality of this structure was indicated by the bright low temperature PL at photon energies expected from a simple superlattice band calculation and a linewidth comparable to that of the SQW emission. Therefore, we assign the difference between the two cases to effects of quantum confinement of the exciton which are largely absent in the superlattice.

DISCUSSION

Exciton states in semiconductor quantum wells and superlattices have been subject to many previous investigations, particularly in the GaAs/(Ga,Al)As heterostructure, including an elegant treatment of the problem in connection with room temperature electroabsorption by D. Miller et al. [6]. The influence of applied electric fields on PL spectra presents additional complications due to possible relaxation and localization effects on the final state section, factors which are present in the wide gap strained II-VI heterostructures. Nevertheless, we consider only the following three effects as being of importance in the ZnSe/(ZnMn)Se SQW case described above: (i) Shift of the confined particle energy levels in a quantum well (Stark effect); (ii) Reduction of the exciton binding energy due to the spatial separation of the electron and hole; and (iii) Possible spin exchange effects arising form the enhanced penetration of the wavefunctions into the DMS barriers. The last factor is specific to DMS quantum well structures where increasing the penetration, e.g. of the hole wavefunction (by electric or magnetic fields) may give rise to magnetic polaron effects (particularly if the exciton lifetime is long enough) [3].

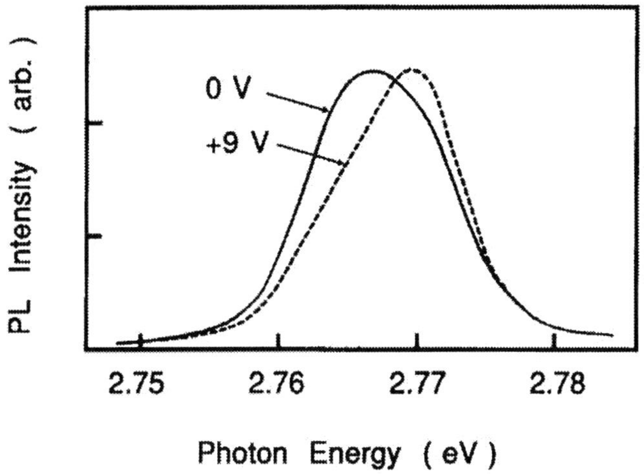

Figure 1: Photoluminescence at the light-hole exciton transition from a single ZnSe/(Zn,Mn)Se quantum well at T = 2K.

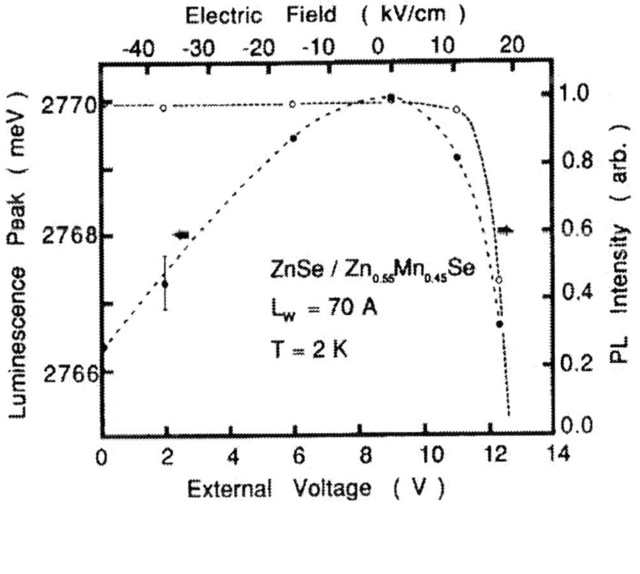

Figure 2: Summary of the field induced shifts and the changes in luminescence amplitude as a function of voltage. The real electric field (top scale) in the Schottky barrier structure is obtained by comparing exciton luminescence from an unmetallized reference sample.

In considering the implications of data in Figs. 1 an 2, we note the following key parameters in the ZnSe/(Zn,Mn)Se system: the 3D exciton binding energy in ZnSe is approximately 20 meV, its Bohr radius some 30 Å, and the classical ionization field is estimated as E_i=9.5×10^5 V/cm. As noted above, the uniaxial component of the lattice mismatch strain splits the valence band degeneracy so that the light hole exciton dominates the PL emission [2]. For x=0.45 this split is approximately 45 meV. While the effective masses in ZnSe are approximately known (m_e=0.17×m_o, m_{lh}=0.15×m_o), those of (Zn,Mn)Se have not been determined. With an Mn-ion concentration of x=0.45 in the barrier regions, the overall bandgap difference is approximately 277 eV. As already noted, the details of the conduction and valence bandoffsets remain unknown although the latter as a smaller quantity may be strongly influenced by the lattice mismatch strain.

The rapid quenching of PL emission under large positive bias (Fig. 2) is associated with large currents (even in the absence of illumination) through the structure. The disappearance of the exciton in this regime is not due to exciton dissociation by field effects but, most likely, from impact ionization by energetic electrons and holes injected into the SQW regions (the concomitent increase in the Mn-ion yellow electroluminescence is consistent with that). A useful region for analysis in our SQW structure is one smaller positive biases, where internal electric fields up to about 5×10^4 V/cm are present (negative external bias also led to increase PL quenching through injection from the Schottky contact). In this region it is clear that quantum confinement is playing a major role in keeping the exciton from dissociating in spite of fields which are comparable to the 3D exciton dissociation case.

We have made calculations for the "light hole" n=1 exciton energy of the ZnSe/(Zn,Mn)Se quantum well system by using a variational approach which properly takes into account the total net internal field acting on the exciton. Our treatment, which will be detailed elsewhere [7], self-consistently includes both the "external" field (applied plus Schottky) and the exciton Coulomb fields. Such an approach is relevant under conditions considered here, namely of a relatively large exciton binding energy and a small valence bandoffset. As already noted, the details of the conduction and valence bandoffsets remain unknown, although evidence shows that the latter is small [2]. In the calculation we take the valence bandoffset to be 60 meV, keeping with $\Delta E_v << \Delta E_c$. While this choice is somewhat arbitrary, the results are not altogether that sensitive to the actual choice of ΔE_v and the calculation provides good agreement for the experimental energy shift for the n-1 light hole exciton (in spite of the fact that we have ignored any light-heavy hole mixing. The calculation also shows the particular feature of the exciton with a large binding energy in wide gap II-VI quantum wells; namely, that the internal Coulomb field is quite effective in opposing the tendency of electron-hole separation by the external fields. Nonetheless, a form of "quantum confined Stark effect" [5] does occur and the spectral redshifts observed by us are rather large (several meV) considering the moderate fields employed in these experiments. This net shift is, of course, a result of two opposing effects, namely, a reduction in the exciton binding energy (blueshift) and changes in the one particle energies (redshift). Finally, we have estimated that the magnetic polaron contribution in these circumstances remains in fact small, contributing less than 1 meV to the exciton energy.

In conclusion, we have observed electric field induced spectral shifts of the straon split n=1 light-hole exciton in ZnSe/(Zn,Mn)Se quantum wells in a moderate field regime. Our results suggest that, in spite of the relatively small valence bandoffset, it should be possible to exploit the such effects in electro-optical modulator devices as the evolution of advanced epitaxy of this and related wide gap II-VI materials permits the design of structures where higher applied fields are possible. Furthermore, the generation of the yellow Mn-ion electroluminescence indicates additional possibilities for superlattice based (Zn,Mn)SE light emitters.

The authors wish to acknowledge M. Yamanishi for useful discussions. This work was supported by National Science Foundation grant ECS-861101 (Brown) and Office of Naval Research Contract N00014-82-K0563 (Purdue).

REFERENCES

[1] for review see L. A. Kolodziejski, R. L. Gunshor, N. Otsuka, S. Datta, W. M. Becker, and A. V. Nurmikko, *IEEE J. Quant. Electron.* QE-22, 1666 (1986).

[2] Y. Hefetz, J. Nakahara, A. V. Nurmikko. L. A. Kolodziejski, R. L. Gunshor, and S. Datta, *Appl. Phys. Lett.* 47, 989 (1985).

[3] Ji-Wei Wu, A. V. Nurmikko, and J. J. Quinn, *Phys. Rev.* B34, 1084 (1986).

[4] C. E. T. Goncalves da Silva, *Phys. Rev.* B33, 2923 (1986).

[5] D. A. B. Miller, D. S. Chemla, T. C. Damen, A. C. Gossard, W. Wiegmann, T. H. Wood, and C. A. Burrus, *Phys. Rev. Lett.* 53, 2173 (1984); E. E. Mendez. G. Bastard, L. L. Chang, L. Esaki, H. Morkoc, and R. Risher, *Phys. Rev.* B26, 7101 (1982); M. Yamanishi and I. Suemune, *Jap. J. Appl. Phys.* 22, 622 (1983).

[6] D. A. B. Miller, D. S. Chemla, T. C. Damen, A. C. Gossard, W. Wiegmann, T. H. Wood, and C. A. Burrus, *Phys. Rev.* B32, 1043 (1985).

[7] Q. Fu, A. V. Nurmikko, and J.-W. Wu, *Appl. Phys. Lett.* (in press).

OPTICAL PROPERTIES OF (Pb,Eu)Te THIN FILMS AND SUPERLATTICES

W. C. Goltsos*, A. V. Nurmikko*, and D. L. Partin**
*Division of Engineering and Department of Physics, Brown University, Providence, RI 02912
**General Motors Research Laboratories, Warren, MI 48090

ABSTRACT

Photoluminescence, transmission, and reflectance measurements have yielded information about the states defining an optical gap in thin films and superlattices based on the (Pb,Eu)Te system, including the limit of high Eu concentration. Magneto-optical measurements show the presence of finite spin exchange processes at low Eu-concentrations.

INTRODUCTION

The developments in molecular beam epitaxial growth of (Pb,Eu)Te have recently amply demonstrated the usefulness of this material in infrared optoelectronic applications, especially as low threshold injection lasers in quantum well heterostructures [1,2]. In the devices to date, the Eu-concentrations have been relatively low ($x<0.10$). Initial characterization of e.g. PbTe/(Pb,Eu)Te quantum wells in terms of basic electronic properties has in this case been fairly successful [3,4] as the relevant electronic states of (Pb,Eu)Te can be assumed to be PbTe-like (p-like at L-point). However, the bandgap of (Pb,Eu)Te is seen to vary anomalously beyond $x>0.10$ as reported recently [5]. In this paper we extend this work to higher Eu concentrations (up to $x=1$) and discuss some early results from related superlattice structures.

EXPERIMENTAL RESULTS AND DISCUSSION

(a) (Pb,Eu)Te single crystal films

The samples used in these studies were MBE grown single crystal films of (Pb,Eu)Te on (111) oriented BaF_2 substrates. Substantial lattice mismatch effects occur which are presumably removed in part by misfit dislocations nucleating at the film/buffer interface (lattice constants for PbTe, EuTe, and BaF_2 at room temperature are 6.460 Å, 6.585 Å, and 6.200 Å, respectively). In analyzing results of optical measurements, however, we have not made attempts to correct for strain induced effects on the relevant electronic energies. This is in part due to unknown deformation potential constants in (Pb,Eu)Te and also the large energy range which we investigate. The thin films varied in thickness from 0.3 to 30 microns and were generally undoped. However, nonstochiometric growth in excess Te vapor usually leads to p-type material with hole density on the order of $1\text{-}2\times10^{17}cm^{-3}$.

The low temperature photoluminescence experiments were performed with a continuous wave Nd:YAG laser as the direct excitation source except for samples of higher Eu concentration where second harmonic excitation at 0.53 micrometers was employed. Transmission and reflection measurements, as needed, were carried out in standard spectrophotometers. Reference 5 describes the method of extracting optical bandgap information and its precision. A summary of the data is given in Figure 1 which displays the effective bandgap in (Pb,Eu)Te at $T=10K$ (onset of strong absorption with a well-defined edge) as a function of the Eu-composition.

Mat. Res. Soc. Symp. Proc. Vol. 89. © 1987 Materials Research Society

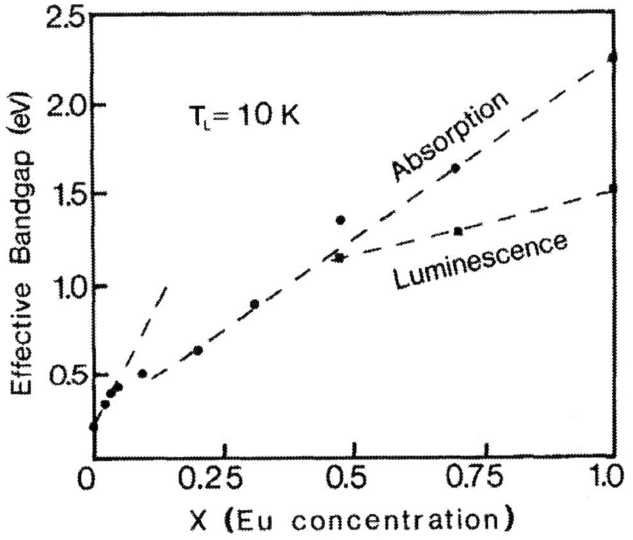

Figure 1: Optical gap of thin single crystal films of (Pb,Eu)Te at T=10K

Beyond a composition range of approximately x=0.30, an increasing Stokes shift is seen to occur between the absorption edge and the associated luminescence emission. The dashed lines are to guide the eye and correspond to the following approximate slopes: in the low energy region ($x<0.05$) $dE_g/dx=5.8$ eV, and in the high energy region $dE_g/dx=2.1$ eV. Measurements of Zeeman shifts in photoluminescence have been recently made by us to show that relatively small effective interband g-factors resulting from p-f exchange are typical in (Pb,Eu)Te at moderate concentrations ($x<0.20$); these are approximately one order of magnitude smaller than in II-VI and IV-VI DMS materials with Mn as the magnetic ion (the f-f exchange is also substantially weaker. For example, for a sample with x=0.032, we obtained $A=\alpha-\beta=49$ meV whereas for (Cd,Mn)Te $A\gg1$ eV. Note that an interband measurement is sensitive to the different in the conduction and valence band exchange coefficients; it is possible (although here unlikely) that cancellation effects occur for exchange of the same sign for the two bands.

The absorption edge in EuTe (which shows several distinct features) is generally considered to be due to the 4f to 5d (t2g Eu-ion transition (the corresponding luminescence Stokes shifts include a configurational relaxation which might contain magnetic polaron effects) [6]. Therefore, as the Eu-ion concentration is increased in (Pb,Eu)Te, on expects the eventual appearance of the 4f state (S=7/2) within the p-p bandgap, the latter defining the absorption edge in PbTe. We assign the observation of the incipient Stokes shift at approximately x=0.30 to such a 'crossover' behavior.

(b) (Pb,Eu)Te Superlattices

The basic electronic properties of PbTe/(Pb,Eu)Te superlattices and multiple quantum wells have been studied in connection with laser development [1,2] and in other spectroscopic studies of these structures [3,8] at low concentrations of the Eu-ion in the barrier layers. The structures are of high quality as evidenced by their excellent luminescent efficiencies and spectral signatures typical of a quasi-2D electron-hole plasma. This is also clear from well behaved magnetic field effects such as in Figure 2 which shows the cyclotron anisotropy of the lowest spin-Landau level interband transition at a measured temperature T_L=2K for a structure with a PbTe well width of L=51 Å and x=0.04 (lower panel) [8]. The next interband transition is also visible on this sample at an elevated temperature

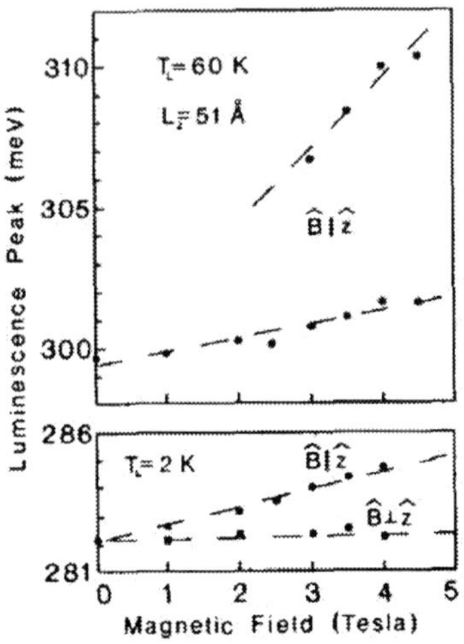

Figure 2: Lowest interband resonance as influenced by an external magnetic field for a PbTe/(Pb,Eu)Te structure of 51 Å well and 355 Å barrier width. Lower trace: Dependence of field orientation versus superlattice axis at T=2K. Upper trace: Lowest resonance and the appearance of the next spin/Landau transition at T=60K.

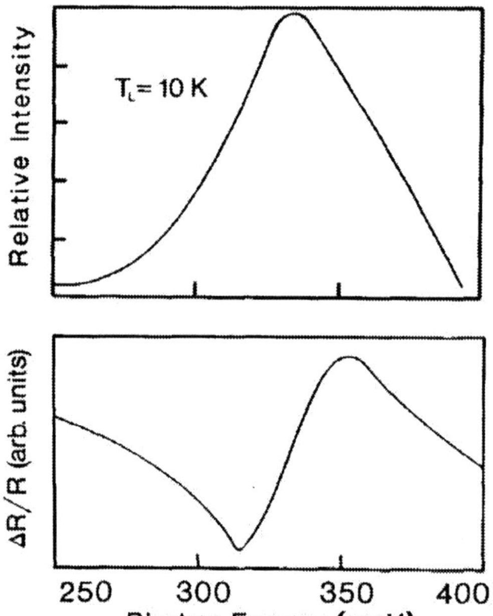

Figure 3: Luminescence from a PbTe/EuTe superlattice at T=10K with a 50 Å well and a 235 Å barrier (top panel); corresponding modulated reflectance spectrum at the superlattice bandgap (lower panel).

(upper panel). The field anisotropy evident in the lower panel is consistent with a quasi-2D transition involving free electrons and holes for which cyclotron motion is inhibited for $B \perp \hat{z}$. Additional details of this work are given in Ref. 8.

We have recently also studied some optical properties of PbTe/EuTe superlattices grown either on PbTe or BaF_2 substrates. Unlike the structures with lower Eu-concentration in the barrier layers, where the addition of a Se constituent (up to 5%) facilitates a nearly strain free superlattice, the PbTe/EuTe structures are highly strained. This is clear, among other indicators, from the appearance of a relative robust signature in Raman scattering from optical phonons [9] which is symmetry forbidden in bulk PbTe and EuTe (in the latter, spin dependent interactions give rise to a finite but weak Raman process). We have investigated the signature of the superlattice bandgap through photoluminescence and modulated reflectance techniques as summarized in Figure 3 for a PbTe/EuTe structure with 50 Å wells and 235 Å barriers (30 layer pairs on a PbTe substrate). The top panel in the figure shows the luminescence at $T \approx 10K$; the emission is characterized by a broad transition with a peak at approximately 330 meV (in contrast, spontaneous emission from PbTe/(Pb,Eu)Te structures at low Eu-concentrations had linewidths less than 5 meV at low temperatures). To verify that the luminescence is an indicator of the superlattice bandgap, a photomodulated reflectance spectrum was also measured (lower panel in the figure). Its lineshape is qualitatively in agreement with expectations for a simple interband transition; however, significant damping is also implied by the broad linewidth. A Kronig-Penney calculation is in reasonable agreement with the observed superlattice bandgap which has nearly doubled form the value of bulk PbTe. The origin of the spectral broadening may involve mixing of the superlattice state with the Eu-ion f-electron states in the barrier layers; early magnetic field experiments have not yet given sufficiently clear indicators about the role of such complications. Nonetheless, the PbTe/EuTe superlattices presents intriguing prospects for further research as a rare combination of a narrow bandgap and magnetic semiconductor.

The work at Brown was supported by an Office of Naval Research contract N00014-83-K0638.

References:

[1] D. L. Partin, *Appl.* Phys. Lett. **45**, 487 (1984).

[2] D. L. Partin and C. M. Thrush, *Appl. Phys. Lett.* **45**, 193 (1984).

[3] W. Goltsos, J. Nakahara, A. V. Nurmikko, and D. L. Partin, *Appl. Phys. Lett.* **46**, 1173 (1985).

[4] J. Heremans, D. L. Partin, P. D. Dresselhaus, M. Shayegan, and H. D. Drew, *Appl. Phys. Lett.* **48**, 928 (1986).

[5] W. C. Goltsos, A. V. Nurmikko, and D. L. Partin, *Solid State Comm.* **59**, 183 (1986).

[6] e.g. P. Wachter, *Crit. Rev. Solid State Sci.* **3**, 189 (1972).

[7] A. Krost, B. Harbecke, R. Raymonville, H. Schlegel, E. J. Fantner, K. E. Ambrosch, and G. Bauer, *J. Phys.* **C18**, 2119 (1985).

[8] W. Goltsos, J. Nakahara, A. V. Nurmikko, and D. L. Partin, *Surface Sci.* **174**, 288 (1986).

[9] S.-K. Chang, H. Nakata, and A. V. Nurmikko, unpublished.

ATOMIC LAYER EPITAXY OF DILUTED MAGNETIC SEMICONDUCTORS

M. Pessa, J. Lilja, O. Jylhä, M. Ishiko* and H. Asonen
Department of Physics, Tampere University of Technology
P.O. Box 527, SF-33101 Tampere, Finland

ABSTRACT

At the time being, there is evidently a sharp increase of interest in growth of low-dimensional thin-film structures of semiconducting compounds by the Atomic Layer Epitaxy (ALE) method. Many of the wide-gap II-VI compounds can be grown from elemental source materials by an ALE mode related to Molecular Beam Epitaxy (MBE). This paper concentrates on recent ALE work involving ZnSe- and CdTe- derived diluted magnetic semiconductors (DMS) where a fraction of Zn or Cd is substituted by Mn atoms. It shows that $Zn_{1-x}Mn_xSe/ZnSe$ and $Cd_{1-x}Mn_xTe/CdTe$ DMS structures with abrupt interfaces and high structural perfection can be produced at low growth temperatures by the ALE method.

INTRODUCTION

Atomic layer epitxy (ALE) is a relatively new method of making low-dimensional thin-film structures of semiconductor and insulator compounds [1]. ALE makes use of the difference between chemical and physical adsorption of molecular species brought onto the substrate surface as alternate molecular beam or vapor pulses. The growth proceeds stepwise in a layer-by-layer fashion resulting in "digital epitaxy" where film thickness is precisely controlled. It also provides advantages for coating several large-area substrates simultaneously in an industrial environment [2].

The characteristic feature of the simplified layer-by-layer model of ALE [3] is that the number of layers deposited is determined solely by the number of operational cycles performed. Since all the beams are turned off for a short time between two successive pulses, a thermodynamical equilibrium is approached at the end of each reaction step. This is not normally true in other vapor phase deposition methods, where the deviation from equilibrium is actually the driving force for the film growth.

Although ALE has the attractive feature of producing one complete monolayer of a compound film per operational cycle in a self-limiting manner, achieving such a growth strictly in practice requires an optimization of growth conditions[4–6]. Growth is affected by the stability of the surface layer after different pulses, surface defects, non-unity sticking coefficients and details of the reaction mechanism. All these factors evidently depend upon the material to be grown, its crystal structure and the reactants used, and make the choice of growth parameters an important, non-trivial procedure.

Mat. Res. Soc. Symp. Proc. Vol. 89. ©1987 Materials Research Society

Epitaxy of II-VI compounds, viz., CdTe [3-5], CdMnTe [7,8], ZnTe [9,10], ZnSe [9,10] and ZnMnSe [11], has been achieved using the MBE- variant (molecular beam epitaxy) of ALE where the beams are produced from Knudsen-like effusion cells in ultra-high vacuum. On the other hand, III-V compounds, such as GaAs [12-18], AlAs [13], InAs [17,18], GaInAs [18], InP [19], GaP [19] and GaInP [19] have been grown by applying metallo-organic chemical vapor deposition (MOCVD) ALE and chloride ALE.

ALE has begun to attract considerable attention as a method for producing thin films of high structural quality. Fabrication of ternary compounds and low-dimensional (semimagnetic) structures is of particular interest. ALE may also prove a vechile for investigating fundamental aspects of compound semiconductor growth.

ALE OF DMS MATERIALS

Growth of ZnSe, ZnTe and CdTe

Growth of zinc chalcogenides (ZnSe, ZnTe) and CdTe on GaAs (100) and CdTe (111) substrates, respectively, has been carried out by MBE-like ALE [3-5,9-11]. ZnSe is an "ideal" compound for alternate deposition; one molecular layer per operational cycle can easily be achieved. This is illustrated in Fig. 1 where the total thickness of ZnSe (and ZnTe) is shown as a function of number of opening and closing cycles of the effusion cells [9]. The measured thickness is in complete agreement with the calculated one on assumption that one molecular layer per operational cycle is obtained. For ZnTe, the net growth rate appears to be slightly enhanced; a better agreement with the theoretical growth rate would readily be obtained if the doses of Zn and Te_2 were somewhat reduced.

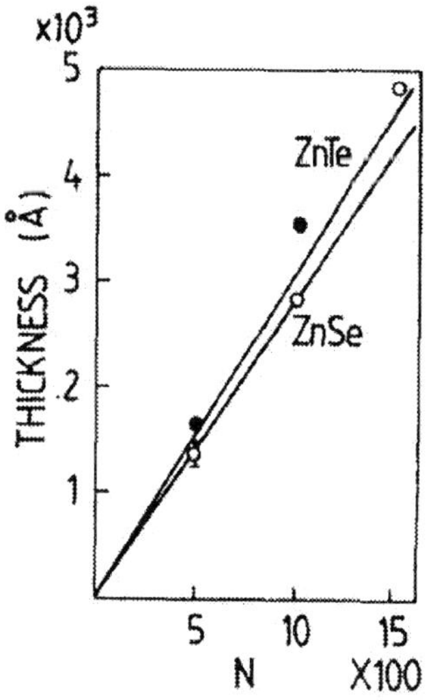

Fig. 1. Thickness of ZnSe and ZnTe films plotted against the total number N of operational cycles. The solid lines give thickness of one molecular layer multiplied by N.

The length of duration of the pulse, Δt, and the recovery time ("dead time") between two successive pulses, $\Delta\tau$, are typically 2 - 10 sec and 1 sec, respectively. A net growth rate in ALE is seldom greater than 0.3 μm/h.

An appropriate substrate temperature for growth of ZnSe is $T_s = 545 - 560$ K, about 100 K lower than that required in the MBE method. ZnTe grows in a temperature range from 530 to 670 K; $T_s \approx 620$ K is normally used. Growth of CdTe has been carried out at $490 \preceq T_s \preceq 570$ K, and even lower T_s may be possible.

Reflection high-energy electron diffraction (RHEED) patterns of ZnSe and ZnTe [9,10] and low-energy electron diffraction (LEED) patterns of CdTe [3] exhibit the presence of reconstructed surfaces which depend on the pulse to which the substrate was last exposed. The Se- or Te-covered surface show a streaky (2x1) RHEED structure, while a mixture of (1x2) and c(2x2) patterns appears on the Zn- covered surface [9,10]. This behavior of reconstruction is periodic during deposition indicating that ZnSe and ZnTe grow in a layer-by-layer fashion. Two-dimensional layer growth was further evidenced by observing a change in RHEED intensity after each pulse [10].

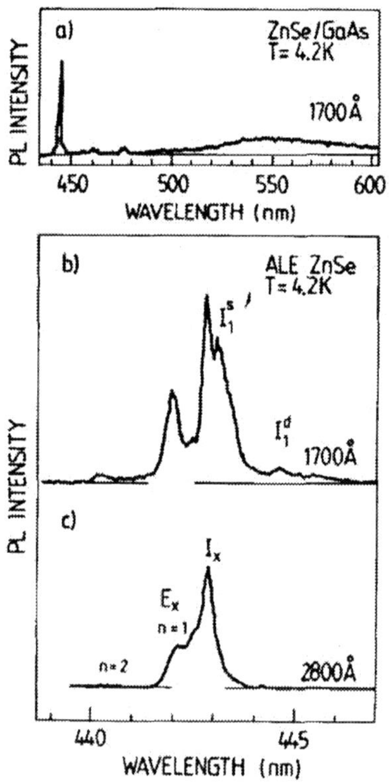

Fig. 2. 4.2-K PL spectra from a 0.17-μm ZnSe showing a) full visible range, b) excitonic emission band, and c) PL spectrum from a 0.28-μm ZnSe (Refs. 9 and 10).

For CdTe overlayers on CdTe (111)B substrates, termination at a Cd pulse gives rise to a clearly resolved p(2x2) LEED symmetry, while termination at a Te_2 pulse results in a p(1x1) pattern with very weak traces of p(2x2) structure. This dependence of the surface geometry on the polarity of the last pulse is observed repeatedly in successive tests. The effect of termination, although less pronounced, was evidenced with the (111)A substrates, too. These results support the idea of layer-by-layer growth of CdTe (111). Moreover, it is evident from LEED patterns that the quality of the surface after deposition is remarkably improved in comparison to the initial substrate surface.

ALE ZnSe films are of high quality in terms of photoluminescence (PL) properties [10]. Figs. 2a and 2b show 4.2-K PL obtained for 0.17 -μm ZnSe layer grown on GaAs (100). The distinctive feature is the presence of dominant excitonic emission lines around $\lambda = 440$ nm, in contrast to a weak deep emission band centered at $\lambda = 550$ nm. This observation provides a clear-cut evidence for good optical properties of such a thin heterostructure where the misfit strain is still present. In an attempt to study the evolution of PL for thicker films Yao et al. [10] grew a 0.28-μm ZnSe overlayer. The spectrum of this sample (Fig. 2c) shows an increase in intensity of the donor bound exciton emission and a comparatively weak free exciton line. This behavior of photoluminescence as a function of film thickness was related to impurities originating from the molecular beam sources [10]. Apparently, clean ultra-high vacuum conditions are of great importance for the MBE-like ALE method.

Growth of DMS's containing Mn

Studies of Zn chalcogenides and CdTe provide evidence that MBE-like ALE is capable of producing these materials with high structural perfection. Therefore, a preparation of diluted magnetic semiconductors (DMS) [20-26] by alloying Mn with the II-VI's is a new and interesting application of ALE.

Manganese is an incongruently evaporating element whose vapor pressure is about 2×10^{-15} Pa, eleven orders of magnitude smaller than vapor pressures of a group II element A (A = Zn, Cd), or a group VI element B (B = Se, Te), at growth temperatures used in ALE. This large difference in vapor pressures means that although the excess of A and B still re-evaporates, the amount of Mn brought onto the substrate within the pulse remains on the surface. Thus the mole fraction x is determined by the intensity and length of the Mn pulse. As a consequence, a special mode of deposition is needed for the incorporation of Mn.

Fig. 3 illustrates the principle of growing an $A_{1-x}Mn_xB$ DMS structure. Alloying Mn with element A produces a two-dimensional, substitutionally disordered alloy layer of $A_{1-x}Mn_x$ per pulse of A and Mn. The x value may be determined precisely in each layer by varying the length of the Mn pulse while evaporating A (cell B being shut). This procedure allows growth of DMS's with abrupt interfaces and desired composition profiles at the same low temperatures as those used for growth of the binary compounds. Another interesting feature is that because the Mn pulse is adjusted to produce a coverage of at most one full atomic layer, a fixed $x \lesssim 1$

is always obtained; excessively arriving A or B or slight variations in these fluxes do not alter x.

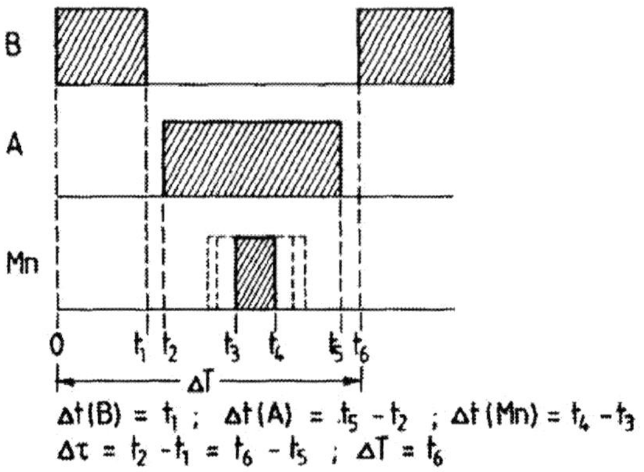

Fig. 3. Deposition sequence in the growth of a DMS structure by ALE.

An example of actual sequences of growth of ZnSe with $\Delta T(ZnSe) = 17$ sec, $\Delta t(Zn) = 10$ sec, $\Delta t(Se) = 5$ sec, $2 \times \Delta\tau(Zn,Se) = 2$ sec and $T_s = 548$ K, is given in Fig. 4a, showing beam pressure variations during growth. To grow $Zn_{1-x}Mn_xSe$, the Mn cell was opened for 1 sec during evaporation of Zn (Fig. 4b). The corrected beam pressures were 3.5×10^{-5} for Se, 2×10^{-5} for Zn and 0.3×10^{-5} Pa for Mn. This set of the growth parameters yielded $x \approx 0.15$.

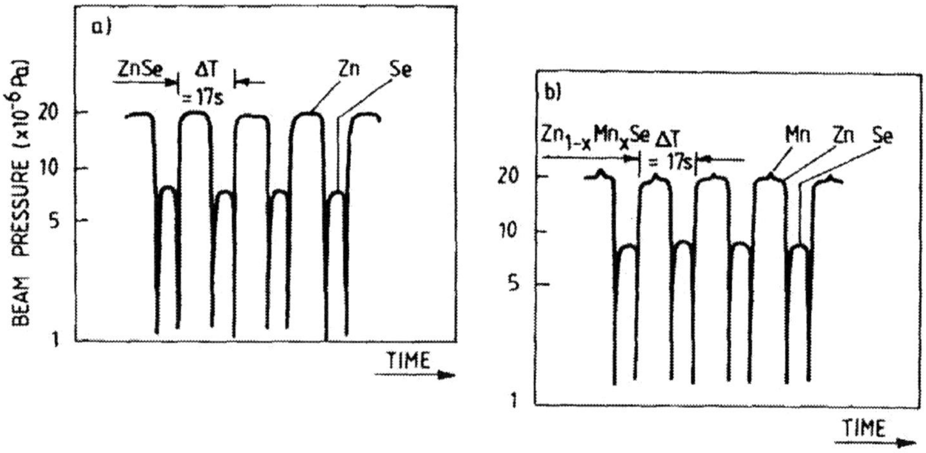

Fig. 4. Beam pressure variations when growing a) ZnSe and b) $Zn_{1-x}Mn_xSe$ ($\Delta t(Mn) = 1$ sec).

The depth profile of an ALE- grown 5-layer $Zn_{1-x}Mn_xSe$ structure is shown in Fig. 5, as obtained by using Auger electron spectroscopy in conjuntion with Ar^+-ion sputtering. Each layer is 50 nm in thickness, grown on a 2-μm ZnSe buffer layer which was deposited onto a GaAs (100) substrate by MBE. For the first, second and third layers of $Zn_{1-x}Mn_xSe$ the Mn cell was opened for 3, 2 and 1 sec, respectively, in the middle of the 10-sec Zn pulse. When growing the fourth layer the Mn cell was open for 2 sec in the beginning of the Zn pulse, while for the topmost layer the Mn pulse was given in the end of the Zn pulse.

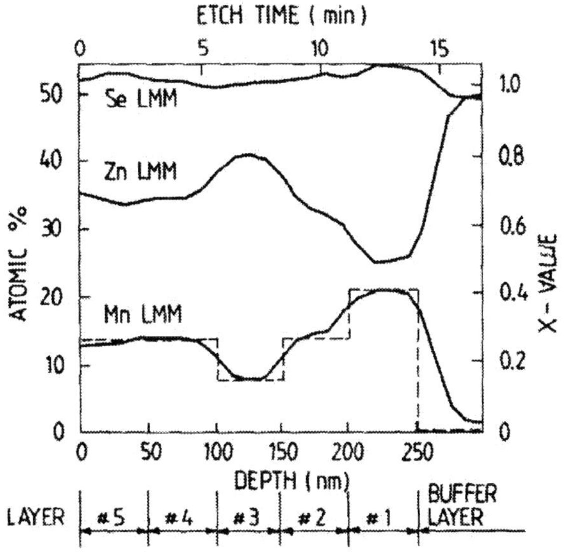

Fig. 5. Compositional depth profile of a multilayer $Zn_{1-x}Mn_xSe/ZnSe$ structure grown on a GaAs (100) substrate.

It can be seen from Fig. 5 that Mn replaces Cd, as expected on the basis of the growth mode, resulting in alternate monolayers of $Zn_{1-x}Mn_x$ and Se. The composition is determined by the length of duration of the Mn pulse. Judging from the projected depth profile the mole fraction is approximately 0.14 (layer #3), 0.26 (layers #2, 4 and 5) and 0.41 (layer #1). This is in close agreement with the expected ratios of x of 1:2:3. The depth profile also shows, to within an accuracy of the measurement, that x is independent of the moment of opening the Mn cell during evaporation of Zn.

A limited number of $Zn_{1-x}Mn_xSe$ and $Cd_{1-x}Mn_xTe$ samples grown so far by ALE in the composition range $0 \precsim x \precsim 1$ have been characterized using X-ray diffraction, low-energy electron diffraction and optical (Nomarsky) microscopy.

X-ray diffraction from multilayer $Zn_{1-x}Mn_xSe/ZnSe/GaAs$ show that growth occurs in the (100) direction. $Zn_{1-x}Mn_xSe$ on ZnSe undergoes a hydrostatic compressive and uniaxial tensile strain because the lattice constant increases significantly with increasing x; for zinc-blende structure $a_o(x) = 0.56670 + 0.0267x$ (in nm). The large built-in strain in $Zn_{1-x}Mn_xSe$ causes lattice distortions and misfit dislocations, as growth continues. In the present experiments on $Zn_{1-x}Mn_xSe$ of thicknesses ranging from 50 to 150 nm, and $x \prec 0.5$, we have observed the full width at half maximum (FWHM) breadths of 11 min of arc for (400) diffraction. This is to be compared with FWHM of 3.1 min of arc of the GaAs substrate, the limit of intrumential resolution.

$Cd_{1-x}Mn_xTe$ films of less than 30 nm in thickness have been grown on CdTe (111) substrates. The (111) surfaces of these films are well ordered, exhibiting sharp p(1x1) LEED spots with low background intensity when $0 \preceq x \prec 0.7$. For $x \succeq 0.7$, LEED becomes relatively poor indicative of a presence of many imperfections; although these overlayers still retain essentially their zinc-blende structure they probably contain other phases, too [7,8].

ZnSe and $Zn_{1-x}Mn_xSe$ possess extremely smooth, specular surfaces. Fig. 6 shows a Nomarsky micrograph of surface morphology typical of the layer structures produced by ALE in a routine run. Surfaces of CdTe and $Cd_{1-x}Mn_xTe$ are of similar quality. Such good surface morphology will be of great advantage when growing superlattices of II-VI compounds with abrupt interfaces.

Measurements of photoluminescence and Hall mobilities would now be desired for further evaluating the quality of these materials in terms of optical and transport properties.

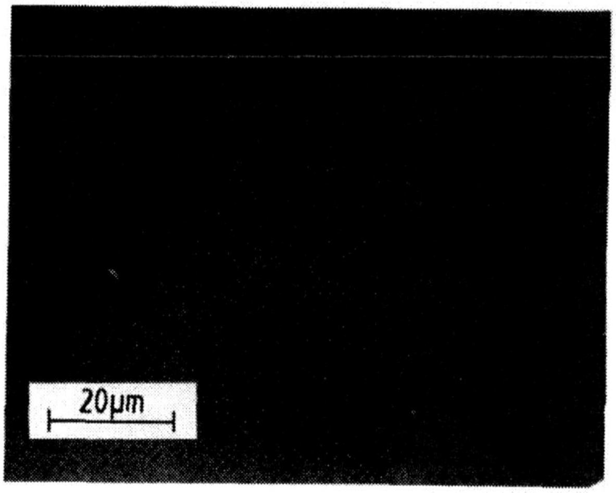

Fig. 6. Nomarsky micrograph of a 35-nm ZnSe/115-nm $Zn_{0.02}Mn_{0.98}Se$/ZnSe/GaAs (100) structure.

This work is supported in part by the Center of Technological Development (TEKES, Finland) and The Academy of Finland.

References

1. C.H.L Goodman and M. Pessa, J. Appl. Phys. **60**, R65 (1986)
2. T. Suntola and J. Hyvärinen, Ann. Rev. Mater. Sci. **15**, 177 (1985)
3. M. Pessa, O. Jylhä and M. A. Herman, J. Crystal Growth **67**, 255 (1984)
4. M. A. Herman, M. Vulli and M. Pessa, J. Crystal Growth **73**, 403 (1985)
5. M. A. Herman, O. Jylhä and M. Pessa, Cryst. Res. Technol. **21**, 841 (1986) and Cryst. Res. Technol. **21**, 969 (1986)
6. T. Pakkanen, M. Lindblad and V. Nevalainen, *in The Extended Abstracts of The First Symposium on Atomic Layer Epitaxy,* Espoo, Finland, 1984, p. 14
7. M. A. Herman, O. Jylhä and M. Pessa, J. Crystal Growth **66**, 480 (1984)
8. M. Pessa and O. Jylhä, Appl. Phys. Lett. **45**, 646 (1984)
9. T. Yao and T. Takeda, Appl. Phys. Lett. **48**, 160 (1986)
10. T. Yao, T. Takeda and R. Watanuki, App.l Phys. Lett. **48**, 1615 (1986); T. Takeda, T. Kurosu, M. Lida and T. Yao, Surf. Sci. **174**, 548 (1986); T. Yao, Jap. J. Appl. Phys. **25**, L942 (1986)
11. M. Pessa, *in Abstracts of The MRS 1986 Fall Meeting,* Boston, USA, p. 632
12. J. Nishizawa, H. Abe and T. Kurabayashi, J. Electrochem. Soc., **132**, 1197 (1985); J. Nishizawa, H. Abe, T. Kurabayashi and N. Sakuri, J. Vac. Sci. Technol. **A4**, 706 (1986); J. Nishizawa and T. Kurabayashi, J. Cryst. Soc. Japan **28**, 133 (1986)
13. S.M. Bedair, M.A. Tischler, T. Katsuyama and N.A. El-Masry, Appl. Phys. Lett. **47**, 51 (1985)
14. M.A. Tischler and S.M. Bedair, Appl. Phys. Lett **48**, 1681 (1986)
15. A. Doi, Y. Aoyagi and S. Namba, Appl. Phys. Lett. **48**, 1787 (1986)
16. A. Usui and S. Sunakawa, Jap. J. App.l Phys. **25**, L212 (1986)
17. M.A. Tischler, N.G. Anderson and S.M. Bedair, Appl. Phys. Lett. **49**, 1199 (1986)
18. M.A. Tischler and S.M. Bedair, J. Crystal Growth **77**, 89 (1986)
19. A. Usui and H. Sunakawa, *in Proceedings of The 13th Int. Symp. on GaAs and Related Compounds,* Las Vegas, 1986
20. J.A. Gaj, R.R. Galazka and M. Nawrochi, Solid State Commun. **25**, 193 (1978)
21. J.K. Furdyna, J. Appl. Phys. **53**, 7637 (1982)
22. N.B. Brandt and V.V. Moshchalkov, Adv. Phys. **33**, 193 (1984)
23. M. Dobrowolska, A.M. Witowski, J.K. Furdyna, T. Ichiguchi, H.D. Drew and P.A. Wolff, Phys. Rev. **B29**, 6652 (1984)
24. A.V. Nurmikko, R.L. Gunshor, and L.A. Kolodziejski, IEEE J. Quantum Eletron. **QE-22**, 1785 (1986)
25. L.A. Kolodziejski, R.L. Gunshor, N. Otsuka, S. Datta, W.M. Becker, and A.V. Nurmikko, IEEE J. Quantum Electron. **QE-22**, 1666 (1986)
26. Y. Hefetz, W.C. Goltsos, D. Lee, A.V. Nurmikko, L.A. Kolodziejski and R.L. Gunshor, Superlatt. and Microstr. **2**, 455 (1986).

* Permanent address: NEC, 1-1, Miyazaki 4-chome, Miyamae-ku, Kawasaki, Japan

AUTHOR INDEX

AUTHOR INDEX